HARMONIC, WAVELET AND p-ADIC ANALYSIS

HARMONIC, WAVELET AND p-ADIC ANALYSIS

editors

N M Chuong
Institute of Mathematics, Vietnamese Acad. of Sci. & Tech., Vietnam

Yu V Egorov
University of Toulouse, France

A Khrennikov
Vaxjo University, Sweden

Y Meyer
ENS-Cachan, France

D Mumford
Brown University, USA

World Scientific

NEW JERSEY • LONDON • SINGAPORE • BEIJING • SHANGHAI • HONG KONG • TAIPEI • CHENNAI

Published by

World Scientific Publishing Co. Pte. Ltd.

5 Toh Tuck Link, Singapore 596224

USA office: 27 Warren Street, Suite 401-402, Hackensack, NJ 07601

UK office: 57 Shelton Street, Covent Garden, London WC2H 9HE

British Library Cataloguing-in-Publication Data
A catalogue record for this book is available from the British Library.

HARMONIC, WAVELET AND p-ADIC ANALYSIS

ISBN-13 978-981-270-549-5
ISBN-10 981-270-549-X

Printed in Singapore by B & JO Enterprise

PREFACE

The mutual influence between mathematics, sciences and technology is more and more widespread. It is both important and interesting to discover more and more profound connections among different areas of Mathematics, Sciences and Technology. Particularly exciting has been the discover in recent years of many relations between harmonic analysis, wavelet analysis and p-adic analysis.

So in 2005, from June 10 to 15, at the Quy Nhon University of Vietnam, an International Summer School on "Harmonic, wavelet and p-adic analysis" was organized in order to invite a number of well known specialists on these fields from many countries to give Lectures to teachers, researchers, and graduate students Vietnamese as well as from foreign institutions.

This volume contains the Lectures given by those invited Professors, including some from Professors who could not come to the School. These Lectures are concerned with deterministic as well as stochastic aspects of the subjects.

The contents of the book are divided in two Parts and four Sections.

Part A deals with wavelets and harmonic analysis. In Section I some mathematical methods, especially wavelet theory, one of the most powerful tools for solution of actual problems of mathematical physics and engineering, are introduced. The connection between wavelet theory and time operators of statistical mechanics is established. Wavelets are also connected to the theory of stochastic processes. Multiwavelet and multiscale approximations and localization operator methods are presented.

Section II is devoted to some of the most interesting aspects of harmonic analysis. The nonlinear spectra based on the so called Fiber spectral analysis with applications are discussed. Here the very famous critical Sobolev problem is developed, too. The representation theory of affine Hecke algebras, the quantized algebras of functions on affine Hecke algebras are reviewed and the so called quantized algebras of functions on affine Hecke algebras of type A and the corresponding q-Schur algebras are defined and their irreducible unitarizable representation are classified. A survey is made of the past 40 years of the Andreotti-Grauert legacy as well as its recent developments (cohomologically q-convex, cohomologically q-complete spaces, strong q-pseudoconvexity, pseudoconvexity of order m) with some new results which did not appear elsewhere.

In Part B some recent developments in deterministic and stochastic analysis over archimedean and non-archimedean fields are introduced. In Section III some Cauchy pseudodifferential problems over p-adic fields, some classes of p-adic Hilbert transformations in some classes of p-adic spaces, say BMO, VMO, are investigated. An analogue of probability theory for probabilities taking values in topological groups is developed. A review is presented of non-Kolmogorovian models with negative, complex, and p-adic probabilities with some applications in physics and cognitive sciences.

Section IV is devoted to archimedean stochastic analysis, more precisely to some recent aspects on stochastic integral equations of Fredholm type, on reflecting stochastic differential equations with jumps, on analytic processes and Levy processes. Here an interesting relation between harmonic analysis, group theory and white noise theory is also developed.

The Editors

CONTENTS

Part A

WAVELET AND HARMONIC ANALYSIS

Chapter I

WAVELETS AND EXPECTATIONS

Harmonic, Wavelet and p-Adic Analysis
Eds. N. M. Chuong *et al.* (pp. 5–22)

§1. WAVELETS AND EXPECTATIONS: A DIFFERENT PATH TO WAVELETS

KARL GUSTAFSON*

Department of Mathematics, University of Colorado, Boulder, CO, USA

Independent of the other communities who have developed theories of wavelets over the last twenty years, we developed over the same period a view of wavelets seen as stochastic processes. That context arose naturally from our theory of Time operators in statistical mechanics. Essential ingredients in our theory included Kolmogorov dynamical systems and conditional expectations. The purpose of the present paper is to come up-to-date on the relationship of our theory to the general theory of wavelets.

Keywords: wavelet; multiresolution analysis; stochastic processes; Kolmogorov systems; conditional expectation; positivity preserving

1. Introduction, Background, and Summary

The usual approaches to wavelets have been found through the intimate connections that wavelet theory has to other parts of mathematics, physics, and engineering. Notable among those have been coherent states in quantum mechanics, spline approximation theory, filter banks, windowed Fourier transforms, phase-space analysis of signal processing, reproducing Hilbert spaces. Essentially independent of those communities, we have developed a theory of wavelets based upon our theory of Time operators in statistical mechanics. The essential ingredients include Kolmogorov dynamical systems and conditional expectations and we viewed wavelets as embedded within the theory of stochastic processes. In fact, we exploited stochastic multiresolution structures 30 years ago when we established the unitary equivalence between continuous parameter regular stationary stochastic processes and Schrödinger quantum mechanical momentum-position couples. Such multiresolution structures also played a key role in our work 20

*This paper is an elaboration of the lecture 'Wavelets and Expectations' by the author at the International Summer School on Harmonic, Wavelet and p-adic Analysis 2005-Quy Nhon, Vietnam, June 10–15, 2005.

years ago on the relations between discrete parameter Kolmogorov systems and Haar systems and a Time operator for both. The purpose of this paper is to quickly review and then come up-to-date on the relationship of our theory[1–14] to the general theory of wavelets. For the latter, see for example,[15–23] among many other books and papers. For stochastic processes and related, see for example.[24–32]

Here is a quick historical background and outline of this paper. This author first saw "wavelets" when working one summer as a college student on a geophysics seismic prospecting boat in the Gulf of Mexico in summer of 1956. Some discussion of the seismic origins of wavelet theory will be given below (Sec. 2). The point is that one sets off an explosion of dynamite and catches the reflections of the various underlying earth strata on recording tapes, from which one tries to devine whether or not oil or natural gas lies below. Then in 1974 B. Misra and this author were trying to formulate models for the decay of quantum mechanical particles and ended up connecting that question to the theory of regular stationary stochastic processes (Sec. 3). The point is that that link best describes this author's (different) path to wavelets. In 1980–85 R. Goodrich and this author tried to extend the notion of such stochastic processes to two and three dimensional parameter space (Sec. 4). This brought us into contact with issues of spectral estimation and spectral factorization as practiced by the electrical engineering signal processing communuity. The point is that we ended up formulating some aspects of higher dimensional wavelets before the wavelet community turned to such. Then at the 1985 Alfred Haar conference in Budapest, this author established the connection of dynamically unstable coarse-grained irreversible processes from statistical mechanics to Kolmogorov systems and to Haar systems. The point is that the mechanism of connection between these fields was the developing theory of internal Time operators (Sec. 5). However, due to other pressing research work (for this author, computational fluid dynamics, optical computing, and neural network projects), this connection was not further pursued at that time. However, our formulations of Time operators in statistical mechanics, and in particular our use of the Foias–Nagy–Halmos dilation theory in our studies of irreversibility, made it clear to us in 1991 that wavelet subspaces were just wandering subspaces, and that we could obtain a Time operator theory of wavelets (Sec. 6). However, we did not publish these results until many years later. Perhaps it should be mentioned that this author had been informally aware of the wavelet theory since a lecture by Ingrid Daubechies at a mathematical physics conference in Birmingham, Alabama, in March 1986. We had a wavelet-based neural network hardware project in our optical computing center here at the University of Colorado during 1988–1992. However, this author's co-worker I. Antoniou in the statistical physics work became overwhelmingly busy managing the institute of a Nobel Prize chemist (I. Prigogine) in Brussels. The point is that it was not until 1998-2000 that we published our theory of wavelets, in several papers. Among those results is a 'Sturm-Liouville' view of wavelets, albeit only with respect to a first order differential equation (Sec. 7). The most comprehensive of those papers is Ref. 12, to which we will often refer in this paper. One issue discussed there is that traditional wavelet theory naturally developed in spaces of infinite measure (e.g., $\mathbb{R}^1$) whereas Kolmogorov theory naturally developed in spaces of finite measure (e.g., probability one). Another issue concerns the fact that wavelet structures and Kolmogorov structures carry different positivity properties. We will come up-to-date on the latter issue here (Sec. 8). In particular we will answer in the negative a speculation this author made in Ref. 13. The point is that it is misleading to think in terms of a positive scaling function which generates in its integer translates a complete orthonormal basis. Section 9 mentions some recent related work by others. Sec. 10 lists some conclusions.

As mentioned above, we summarized our view of wavelets in the predecessor paper Ref. 12 and it is suggested that the reader may wish to consult that paper for more details about several matters to be discussed here. For the most part, we don't want to repeat the discussions and details in that paper here. The goal of the present paper is to come up-to-date from Ref. 12, i.e., to both supplement and complement the results presented earlier in Ref. 12. Therefore, let us just quickly recall the 8 sections of Ref. 12: (2) Wavelets and Kolmogorov automorphisms, (3) Wavelets and regular stochastic processes, (4) Wavelets and continuous parameter processes, (5) Wavelets and martingales, (6) Wavelets and ergodic theory, (7) Wavelets and statistical physics, (8) Historical remarks and comparisons. Next, to expeditiously join[12] to this present paper, let us summarize some multiresolution structures discussed in Ref. 12. The abbreviated descriptions here are this author's and do not do justice to the richness of the mathematical theories mentioned, but is is hoped tht the following summary will be a useful short-hand for the reader who may not wish to read.[12]

Shorthand Summary of Multiresolution Structures in Context

MRA (wavelets)	Meanings/ Interpretations
1) $\mathcal{H}_n \subset \mathcal{H}_{n+1}$	1) Nested Subspaces
2) $\bigcap \mathcal{H}_n = \{0\}$	2) Admissibility
3) $\bigcup \mathcal{H}_n$ dense	3) Cyclic dilation V
4) $f(x) \in \mathcal{H}_n \leftrightarrow f(2x) \in \mathcal{H}_{n+1}$	4) Refinement, Scaling
5) $\exists \phi \in \mathcal{H}_o \ni \phi_n(x) = \phi(x-n)$ is c.o.s., $n \in \mathbb{Z}$	5) Cyclic translation T
K-system $(\Gamma, \mathcal{B}, \mu, S^n)$	**Meanings/Interpretations**
1′) $S^n\mathcal{B}_o = \mathcal{B}_n \subseteq \mathcal{B}_m = S^m\mathcal{B}_0$	1′) Increasing Event Fields
2′) $\bigcap \mathcal{B}_n = \mathcal{B}_{-\infty} =$ trivial σ-algebra	2′) Trivial Starting Field
3′) $\bigcup \mathcal{B}_n = \ =$ full σ-algebra	3′) Full (exact) System
4′) $Vf(x) = f(Sx)$, S measure preserving	4′) Underlying Dynamics
5′) S is K-mixing	5′) K-automorphisms are ergodic
Regular Stochastic Process X_n	**Meanings/Interpretations**
1″) $\mathcal{H}_n \subset \mathcal{H}_{n+1}$	1″) Independent Innovations
2″) $\bigcap \mathcal{H}_n = \{0\}$	2″) Empty Remote Past
3″) $\bigcup \mathcal{H}_n$ dense	3″) Complete Future
4″) $\mathcal{H}_n = V^n\mathcal{H}_o$	4″) Regularity
5″) V has countable multiplicity on $\bigoplus_{n=-\infty}^{\infty}(\mathcal{H}_n \ominus \mathcal{H}_{n-1})$	5″) Bilateral Shift of ∞ multiplicity
Continuous Parameter Process	**Meanings/Interpretations**
1‴) $\mathcal{H}_t \subseteq \mathcal{H}_s$, $t < s$	1‴) Spectral Subspaces
2‴) $\bigcap \mathcal{H}_t = \{0\}$	2‴) Stationarity
3‴) $\bigcup \mathcal{H}_t$ dense	3‴) $\lim_{t\to\infty} P_t = I$
4‴) $\mathcal{H}_t = V_t\mathcal{H}_o$	4‴) Underlying Implementation Group
5‴) V_t is irreducible	5‴) Square Integrable, cyclic

This present paper will be written in the chronological order of the time steps outlined above of our own development of the viewpoints of our wavelet theory. General wavelet theory and techniques and applications are now so developed and the associated literature so extensive that we leave all such to the expositions of others. The value of this paper will be in its different viewpoint, caused by its different historical progression, leading to a different perspective on wavelets. A corollary value will be a broader view of wavelets within both stochastic and multiscale contexts.

2. 1956: Wavelets and Seismic Oil Prospecting

As mentioned in the Introduction, this author first saw wavelets working on an oil prospecting boat in summer 1956. Later, as is well-known to some, the general mathematical theory of wavelets received its key impetus from interest by mathematicians and physicists working with geologists from oil companies. We will briefly discuss both of these historical traces in this section.

In particular, a very important early wavelet paper was that of Grossmann and Morlet Ref. 33. Grossmann was a theoretical physicist and also mentor to the important later work of Daubechies Ref. 19. Morlet was a geologist working for an oil company. Morlet had suggested (with others, see Refs. 34–36) that seismic traces could be analyzed in terms of wavelets of fixed shape. The main idea was[33] "to analyze functions in terms of wavelets obtained by shifts (only in direct space, not in Fourier transformed space) and dilations from a suitable basic wavelet." We will return to the important content within the parentheses, soon, and also later.

In order to make some money in order to continue university education, this author worked in gold mines and on land-survey crews in Alaska the summers of 1954 and 1955 and then on an oil exploration boat in the Gulf of Mexico summer of 1956. There were 4 tasks for us "roughnecks" (a term in the oil business) on the boat and we alternated between them throughout the day. One was to release the winch so that the long cable with seismic recorders would flow out behind the moving boat. The second was to drop the dynamite off another boat, roughly at the midpoint of the seismic cable. The third was to take the recorded seismic traces to the photographic dark room and develop them. The fourth was to hang them to dry on long racks in the boat's hold. Then a company geologist on the boat would interpret them.

During tasks three and four above, this author saw many seismic traces. It was apparent to all that after the initial explosion, the seismic waves did not interfere with each other. Thus as they came back from their reflections off various earth strata under the ocean floor, you could linearly superimpose them one on the other. Moreover, the same explosion "wavelet" profile would appear on all of them, just separated by the time delays off the various geological layers.

That is what is meant by the "direct space" in the parentheses above.

One wants to remain in the natural context, that is, the time domain. It appears that this point has been lost in much of the recent wavelet research. One sees phrases such as "for convenience we will work in the frequency domain." But now you have destroyed the original motivation. Also by taking Fourier transforms, you have insisted on representing everything in an eigenbasis of the Laplacian. Certainly the Fourier transform and the Fourier theory and the Fourier methods are powerful tools that can greatly advance the mathematics in many domains. But one could always take the Fourier transforms of the seismic data to do frequency analyses, before wavelet theory appeared as an alternative.

Much later, after we had established the connection between our work on Kolmogorov dynamical systems and the theory of wavelets, this author went back into the geosciences seismic prospecting literature. In particular, we identify in Ref. 12, Sec. 8.2 how the Predictive Decomposition Theorem (E. Robinson, 1954) uses minimum delay wavelets as a generalization of minimum phase wavelets to separate a stationary stochastic seismic data recording into its deterministic and nondeterministic parts (Wold decomposition). There the response function b_n, "which is the shape of these wavelets," reflects the dynamics of the time series. Let us elaborate a bit here.

First, to Robinson and the others in the mathematical geosciences community then, a wavelet was defined rather generally to be a one-sided function $w(t)$, i.e., $w(t) = 0$ for $t < 0$, and of finite energes, i.e. $\int_0^\infty |w(t)|^2 dt < \infty$. In other words, these could be described in more modern terms as just the $L^2(0,\infty)$ identification of Hardy functions on the upper half plane. On the other hand, the wavelets they were most interested in were the minimum-delay wavelets. As we will explain later (Sec. 4), these correspond to what are now called outer functions. Moreover, the geoscience researchers in the 1950's were really thinking in terms of oscillating specific waveforms that they saw returned from underearth or undersea seismic reflections.

That is the second point. If possible, the reader should access the classic book Deconvolution Ref. 37. There on p. 1 we find "Even in the days of galvanometer cameras and paper records, 'wavelet contractor' electronic input filters were designed to enhance resolution." On p. 43 you can see an actual seismic trace with a single basic waveform repeating itself. When discussing his predictive decomposition theorem, on p. 55 Robinson states "All these wavelets have the same minimum-delay shape." Most important, however, and somewhat against the later speculation of Morley, we find on p. 115 "Nonetheless, a seismic trace is not made up of wavelets which have exactly the same form and which differ only in amplitudes and arrival times." On p. 184 we see the Ricker wavelet, evidently more natural to seismic trace observations, is allowed to be symmetric about 0. These look roughly like upside-down Mexican hat functions. They are looked at there at 75 cps and then at 37.5 cps, i.e., at what we would now call two wavelet scalings.

3. 1976: Quantum Mechanics and Stochastic Processes

In 1974 the theoretical physicist B. Misra and this author were looking at models for the decay of quantum mechanical particles. We ended up Ref. 1 proving that every regular stationary stochastic process ζ_t is unitarily equivalent to the quantum mechanical momentum process $\psi_t = e^{-ipt}\psi$ where p is the one-dimensional selfadjoint momentum operator $p = -i\partial/\partial x$ in $\mathcal{L}^2(-\infty,\infty)$ and where $\psi \in \mathcal{L}^2(-\infty,0]$. Because our proofs used the Stone-Von Neumann theorem about Schrödinger couples, this paper Ref. 1 can in retrospect be seen as the latent beginning of our later theory of Time operators (Sect. 6 here). Moreover, it was through this work that we learned (e.g., see Rosanov Ref. 38) that regular stationary stochastic processes possess the multiresolution properties 1″)–5″), granted that those properties are not always expressed that way in the stochastic process literature. For more about Stochastic processes, see e.g. Refs. 23,25–32.

This work on particle quantum mechanics and quantum measurement theory led to the interesting Zeno paradox Ref. 39 which has current implications Ref. 40 for quantum computing. For this author's contributions, see also Ref. 41.

Thus, for us, quantum mechanics played a key role in our path to wavelets. It did also for Daubechies, see Ref. 43, although in a different way, from the theory of coherent states Ref. 43. However, in the Grossmann, Daubechies, Klauder, et al. evolving theory of wavelets from the coherent state point of view of physics, there was no connection to multiresolution structures, and in particular, no connection to stationary stochastic process theory. Instead, the main underlying motivations were phase space analysis, and in particular, the Weyl-Heisenberg group and the affine group.

4. 1980: Higher Dimensional Spectral Estimation

Because of our interest via[1] in one parameter regular stationary stochastic processes, during 1980–1985 this author tried with R. K. Goodrich to develop a comparable theory of two and three parameter regular stationary processes.[2,3] It should be noted here that it is quite easy to generalize real valued stochastic processes to vector-valued processes, even infinite dimensional, Banach space valued versions, e.g., see work by Masani and others in the literature. But we found the two- and three-dimensional parameter case more difficult. For example, one may want the process labelled by two independent times, or labelled on a spatial sphere. The underlying difficulties can be seen in two collateral ways. First, two parameter processes would need to use the theory of analytic functions in two complex variables much as the one parameter prediction theory has rich connections to the theory of analytic functions. Seen another way, the natural ordering of one parameter processors, e.g., according to $\mathbb{R}^1$ seen as time, is less clear in higher dimensions.

The point of mentioning these investigations here is that, in retrospect, they may be seen as an early formulation of part (the translation part) of what later became higher dimensional wavelet theory. This author has never discussed this connection elsewhere so we would like to briefly mention it here. Also, we were not satisfied when we left these investigations, so perhaps all the recent research on higher dimensional wavelet sets could be helpful to come back to address some of our concerns.

In particular, we were quickly led to the question: what characteristic functions $\chi(S)$ could serve by their translations and rotations to generate a complete orthonormal basis for $\mathcal{L}^2(\mathbb{R}^2)$? Comparable one dimensional considerations can be found in the book Ref. 44. However, inner-outer factorizations in higher dimensions were not generally available. Therefore we went the route of defining two parameter regular stationary stochastic processes and related those to unitary regular representation theory. In retrospect, this is much like later work in wavelet research using the left-regular unitary representations. Among our results were the following. Let $E_{(s,t)}$, E_s, and F_t be the projections onto, respectively, $\overline{\text{sp}}\{U_{(x,y)}\phi \mid x \leqq s, y \leqq t\}$, $\overline{\text{sp}}\{U_{(x,y)}\phi \mid x \leqq s\}$, $\overline{\text{sp}}\{U_{(x,y)}\phi \mid y \leqq t\}$. Here $U_{(x,y)}$ is any continuous unitary representation of $\mathbb{R}^2$ which has a cyclic vector ϕ. So we are trying to generalize the one dimensional situation where ϕ is $\chi[0,1]$, i.e., where ϕ is the Haar function. We showed that such $U_{(x,y)}$ is unitarily equivalent to the regular representation of $\mathbb{R}^2$ iff ϕ has the property that $\bigcap_s \mathcal{R}(E_s) = \{0\} = \bigcap_t \mathcal{R}(F_t)$ where $\mathcal{R}$ denotes the range of the projections. This is a generalization of the emptiness of the infinite remote past 2″) for one dimensional multiresolutions. If ϕ also had the additional property that $E_sF_t = E_{(s,t)}$ for all (s,t) in $\mathbb{R}^2$, we called $U_{(x,y)}$ a regular process. A weakly regular process was defined by the weaker requirement that $E_sF_t = F_tE_s$. We gave some examples [2] but we did not get far in identifying what would now be called wavelet support sets. We did obtain[2,3] some interesting new inner-outer factorization theorems for $\mathbb{R}^2$ and $\mathbb{R}^3$. See also our other papers Refs. 45–48. We were thinking always in terms of what would now be called higher dimensional Haar wavelets, i.e., in terms of characteristic functions $\phi(S)$ where S is a compact set in $\mathbb{R}^2$ or $\mathbb{R}^3$. It did not enter our thoughts to consider other, e.g., oscillatory functions, $\psi(S)$.

Another innovative feature from[3] that may now be seen as predecessor-related to wavelet theory is our use of "look-alike" functions f and $\hat{f}$ to serve as sharp limits on spectral decay theorems. In that connection we suggested a study of functional dependencies $\hat{f}(\lambda) = af(b\lambda + c)$ of transforms on important groups such as the affine and scaling of groups.

5. 1985: Kolmogorov Automorphisms as Wavelet Refinements

This author decided Ref. 4 at the 1985 Alfred Haar Conference to try to connect our result on intertwined reversible-irreversible (think: group U_t versus semigroup W_t) descriptions, which due to the intertwining conditions asserted that the underlying dynamics were necessarily those of Kolmogorov automorphism systems, to what this author regarded then as more general Haar systems. See Sec. 5 of Ref. 4. The idea was to develop new Haar systems for a wide class of Bernoulli systems and other dynamical systems of Kolmogorov type. The mechanism used to connect these two fields was the iterated baker transformations T^n where the baker transformation T iteratively refines the unit square in its well-known kneading operation. This was an early Time operator which satisfied $T\chi_n = n\chi_n$, $U^{-n}TU^n = T+nI$, where $\chi_n = U^n\chi_0$, where χ_0 was -1 on $0 \leqq x \leqq 1/2$, $0 \leqq y \leqq 1$ and χ_n was $+1$ on $1/2 \leqq x \leqq 1$, $0 \leqq y \leqq 1$, and U^n was induced from the baker transformation $b(x,y) = (2x, y/2)$ for $0 \leqq x \leqq 1/2$, $b(x,y) = (2x - 1, (y+1)/2)$ for $1/2 \leqq x \leqq 1$. The point, in retrospect, is that the baker transformation performs the same iterative refinement as do the wavelet approximation spaces $\cdots V_{-2} \subset V_{-1} \subset V_0 \subset V_1 \subset V_2 \cdots$. But within the Kolmogorov system context of chaotic maps, the dynamics is the thing, whereas in wavelet multiresolution theory, approximation is the thing.

Our work on probabilistic versus deterministic descriptions is rather technical and we refer the reader to Refs. 5–8 for more details. In particular, Part II of the book Ref. 8 does a pretty good job relating the irreversibility theory of Prigogine to the Kolmogorov dynamics systems and to multiresolution analyses. We needed to dilate our semigroup evolutions to unitary evolutions and we used a lot of mathematical dilation machinery. This author recently surveyed that work in [49] and we will not go into further detail here. However, one operator dilation theory we used was that of Foias–Nagy Ref. 50, see also Halmos Ref. 51. Thus it was apparent to us rather early (1991) that wavelet subspaces were just wandering subspaces. We will return to this point in the next section.

Thus by 1985 we were very familiar with the multiresolution structures $1'')$–$5'')$ and $1')$–$5')$. Let me also mention the multiresolution structures $1''')$–$5''')$. These are also mentioned briefly in the book Ref. 8, Part II, Remark 2.2.2 and discussed in Ref. 12. From $1''')$–$5''')$ we may define a continuous multiresolution analysis. The motivation is that regular stochastic processes X_n in the structure $1'')$–$5'')$ have as well continuous parameter versions X_t, e.g., see Ref. 25 and we would like the same concept extended to wavelets. We later found a few other papers with this idea, see Ref. 12. One can think of the axioms $1''')$–$3''')$ just as a resolution of the identity for some selfadjoint operator. Then, if there was an underlying measure preserving dynamic transformation S_t as in the Kolmogorov theory, then $4''')$ could be restated as: $f \in \mathcal{H}_t$ iff $f(S_{-s}) \in \mathcal{H}_{t-s}$. This is vague but more general than $4''')$ and the idea is to generalize all the other multiresolution structures fourth property to allow unitary V's to be generated as unitary representations of groups other than the dilation groups. Condition $5''')$ also presumes that V_t has Lebesgue spectrum of coutable multiplicity.

One example of this situation is Lax–Phillips scattering theory.[52] That theory may be seen to be a multiresolution analysis. The incoming and outgoing subspaces form a continuous multiresolution analysis with projections upon them giving the subspaces $1''')$–$3''')$. The scaling property $4''')$ is replaced by a geometric domain-of-dependence condition. By use of the canonical commutation relations one has a countable Lebesgue multiplicity and irreducibility $5''')$. However, we have not investigated how general the system $1''')$–$5''')$ or its modifications may be. In some sense, this theory remains unfinished, in somewhat the same and related way as our theory discussed in Sec. 4 is also unfinished.

6. 1991: Wavelets as Wandering Subspaces for Time Operators

Due to all our related work with earlier multiresolution structures as discusssed in the sections above, we rather immediately saw in 1991 that we could obtain a Time operator for wavelets and that the wavelet subspaces are the age eigenstates of the Time operator. This early understanding is documented in the 1992 NATO grant proposal[5] which is cited here just to affirm that we saw the wandering subspace view of wavelets independently and as early as others[53] who also saw it. As stated in Ref. 5, "A four-way connection between Kolmogorov Systems, Wavelet Multiresolution Analyses, Bilateral Shifts with wandering cyclic subspaces, and regular stochastic processes has been established. A general Time operator from the theory of irreversible statistical mechanics has been constructed for any discrete wavelet structure."

It should be mentioned that we are much more interested in the connection to Kolmogorov systems and in seeing wavelet multiresolutions as embedded within stochastic processes theory, than about any priority about the wandering subspace connection. However, the Foias–Nagy–Halmos theory Refs. 50,51 of unitary dilations of contractions on Hilbert space, with which we were so familiar because of our work with the Kolmogorov–Prigogine theories interconnections, just immediately reveals that the wavelet subspaces are a special case of wandering subspaces. Indeed, on p. 1 of Foias–Sz. Nagy Ref. 50, we find wandering subspaces $\mathcal{W}$ defined for any isometry V if $V^n(\mathcal{W}) \perp V^m(\mathcal{W})$ for any integers $n \neq m$. We only learned much later (1998) of the wavelet wandering subspace observation of Goodman, et al. Ref. 53. To the credit of the latter authors, they were interested in using the wandering subspace concept to generalize from translation-unitary and dilation-unitary wavelets to more general unitary operator wavelets. In particular, the object of[53] was to extend Mallat's Ref. 54 multiresolution formulation of wavelets to wavelets generated by a finite number of functions. Our motivation was completely different: the wavelet subspaces W_n are age eigenspaces of our Time operator.

Because our Time operator theory of wavelets is set out adequately in the book Ref. 8 and our papers Refs. 9–14, we do not want to repeat those expositions here. The (oversimplified) point is that by use of the Stone Von Neumann theorem for unitary equivalence to Schrödinger couples and (related) use of systems of imprimitivity from group representation theory, one arrives at $V^{-n}TV^n = T + nI$ where V is the scaling or other operator of property 4 of a multiresolution structure. With P_n the projections onto the 'approximation subspaces' $\mathcal{H}_n$ of properties 1 through 3 of the multiresolution structure, T may be defined as $T = \Sigma_n n(P_{n+1} - P_n)$. One may proceed similarly in the continuous parameter setting. To determine cyclic generating vectors and hence the 5th property of countable irreducibility is a little more technical. In that connection one may direct the reader to the recent paper Ref. 14, which in its writing, led this author to Marshall Stone's great book Refs. 55, where one finds Stone's proof of a cyclic vector to be quite involved, even mysterious, until you can finally chase it down.

7. 1998: Haar and Wavelet Bases and Differential Equations

Our Time operator theory of wavelets reveals that T is in the role (up to unitary equivalence) of the 'position' operator q within the Schrödinger couple theory. The point is, for any wavelet multiresolution analysis, the Time operator is not defined in some *ad hoc* manner, but rather the wavelet time operator T is determined naturally from all 5 multiresolution properties in exactly the same way as the Time operator is determined in statistical physics. Increasing position is increasing age is increasing detail refinement.

For any wavelet multiresolution, but especially for the Haar basis, we see that the Time operator is the natural operator for which the wavelet basis is its eigenbasis. If we permit one more unitary equivalence (Fourier Transform), we see that the Haar basis is the eigenbasis of a first order (momentum) differential operator. More correctly, the Haar basis is the eigenbasis of a pseudo-differential operator. In the discrete parameter case, T is canonically conjugate (under Fourier-Mellin transform) to the first order differential generator of the dilations. See Ref. 10 for more details.

A point to be mentioned here is that Haar's original tasking was to find a complete orthonormal set that was *not* the eigenbasis of any differential operator. Now we have found a differential operator for the Haar basis. But it is *not* a second order Sturm–Liouville operator, from which most of the important eigenbases of physics come. And by its definition, our wavelet time operator T is defined in terms of the 'mother' wavelet ψ, and not directly in terms of some given scaling function ϕ. We will return to this point in the next section. Remember that eigenbases from Sturm–Liouville second order differential operators have a general property that the first eigenfunction has no interior nodes, e.g., it is everywhere positive in the interior of its domain of definition. Thus, although one certainly has many oscillatory properties in wavelet bases as in Sturm–Liouville bases, they are not the same creatures.

8. 2005: Positivity in Kolmogorov and Wavelet Structures

Partially related to the discussion in the previous section, let us observe here both some positivity and nonpositivity properties of wavelet structures. In particular, we want to negate a speculation this author put forth in Ref. 12.

Let us get to that issue right away.

First let us recall that Kolmogorov systems possess many nice positivity properties. Positivity here means positivity preserving: $f(x) \geq 0 \Rightarrow Pf \geq 0$, where P is the operator of interest. This lattice property is essential if probability densities are to be retained as probability densities under the transformation P. In particular, conditional expectation operators have this *p.p.* property. Another way to say this is that when one does functional analysis on the K-system 1′)–5′), one superimposes over the phase space dynamics S probabilistic function spaces which enjoy not only Banach space properties but also Banach lattice properties. See for example the discussion in the book Ref. 8, Part II, Chapter 1.

When we turn to wavelet multiresolution structures, there is no inherent (e.g., probabilistic) reason to expect lots of *p.p.* properties. Nonetheless, the two principal unitary operations in standard wavelet multiresolutions are indeed *p.p.* We may make this precise here as follows.

Lemma 8.1. *The wavelet MRA translations $T_n f(x) = f(x+n)$ of MRA property 5), and the dilations $Df(x) = \sqrt{2}f(2x)$ of MRA property 4), are positivity preserving unitary operators on $\mathcal{L}^2(\mathbb{R})$. The MRA property 1) embeddings $\mathcal{H}_n \subset \mathcal{H}_{n+1}$ are positivity preserving embeddings. The 'empty remote past' MRA property 2) does not disturb any positivity preservation. The denseness MRA property 3) ensures that positivity preservation common to all $\mathcal{H}_n$ carries into the completion $\mathcal{L}^2(\mathbb{R})$.*

Lemma 8.2. *The scaling projections $P_n : \mathcal{L}^2(\mathbb{R}) \to \mathcal{H}_n$ are all positivity preserving iff P_0 is positivity preserving.*

Verification of Lemma 8.1 is straightforward. Lemma 8.2 follows from the positivity preservation inherent to the relation $P_{n+1} = DP_nD^{-1}$. This author had somewhat casually observed these properties and when writing Ref. 12, p. 94 there was made the speculation "Most wavelet projectors are positivity preserving." This speculation is not correct, as will be clear from the following discussion. In returning to this question in 2001 and the Lemmas 8.1 and 8.2 above, it was easy to also observe the following positivity preserving property.

Lemma 8.3. *When the scaling function $\phi(x)$ of wavelet MRA property 5) is everywhere nonnegative, the P_0, hence all P_n, are positivity preserving.*

The proof of Lemma 8.3 resides in the fact that the $\phi(x+n)$ form a complete orthonormal set. Let $f(x) \geq 0$, $f \in \mathcal{L}^2(\mathbb{R})$, and $\phi(x) \geqq 0$, $\phi \in \mathcal{L}^2(\mathbb{R})$.

Then $P_0 f(x) = \Sigma_n \langle f(x), \phi(x+n)\rangle \phi(x+n) \geqq 0$ because the inner products are all nonnegative real $\mathcal{L}^1$ integrals. It may also be observed that when ϕ is not everywhere nonnegative, P_0 need not be *p.p.* Not wishing to employ the Fourier transform, one may look at the Daubechies scaling function $\phi^2(x)$ Ref. 17, p. 198; or Ref. 19, p. 197. Let $f(x)$ be the characteristic function $\chi[0,1]$. Then $P_0 f(x) = \langle f, \phi(x)\rangle \phi(x) + \langle f, \phi(x+1)\rangle \phi(x+1) + \langle f, \phi(x+2)\rangle \phi(x+2)$ and $P_0 f(2) = \int_0^1 \phi(s) ds \cdot (1-\sqrt{3})/2 < 0$ since the integral is positive.

Still wanting as much positivity preservation in wavelet structures as possible, the goal being to see the wavelet projectors as coarse-graining conditional expectations, this author circulated his rough draft to a few friends. Costas Karanikas responded in 2001 via his results [56–58] on Gibbs effects in wavelet expansions, indicating that the speculation was ill-founded.

Returning to this issue now at this writing and in this section, let us note that the question of which P_o are positivity preserving via the sufficient condition of Lemma 8.3 is misleading on two counts. First, how many scaling functions $\phi(x)$ are nonnegative? Second, related and more subtle, the usual way that MRA property 5) is stated in much of the literature is misleading. The point is that there are not many scaling functions ϕ in wavelet practice whose translates form a complete orthonormal basis. This contrasts with the fact that many wavelets ψ with their translates do form orthogonal bases. What one does get from the ϕ translates are usually Riesz bases. They then can be orthogonalized but then one loses positivity in the translates.

Thus the speculation centers on the issue of whether there are any nonnegative scaling functions $\phi(x)$ satisfying the MRA property 5), other than the Haar function. The important result that there are not, this author has found recently in the paper by A. J. E. M. Janssen Ref. 59.

9. Recent Related Work by Others

Kubrusly and Levan Refs. 60–63 have followed our approach to show, among others, that with respect to an orthonormal wavelet $\psi \in \mathcal{L}^2(\mathbb{R})$, any $f \in \mathcal{L}^2(\mathbb{R})$ is the sum of its "layers of detail" over all time shifts. This contrasts with the conventional wavelet view of f as the sum of its layers of details over all scales. Thus one may write $f(x) = \sum_m \sum_n \langle f, \psi_{m,n}\rangle \psi_{m,n}(x) = \sum_n \sum_m \langle f, \psi_{m,n}\rangle \psi_{m,n}$. One could describe this as a kind of 'Fubini' theorem, which, as is well known, one does not usually expect in the absence of absolute summability. Their approach also includes reducing subspaces of the dilation operator.

P. Jorgensen Ref. 64 has independently investigated connections between wavelet multiresolutions and Lax-Phillips scattering theory.

This author recently found the important Janssen result[59]from a web-search which led first to the papers of Walter and Shen Refs. 65,66. As this author was, these authors are concerned with the loss of positivity preservation when trying to approximate $f(x) \geq 0$ with wavelet expansions. Their approach is to use an Abel summability method or that in combination with sampling techniques to construct a positive scaling function ρ_r from a given orthonormal scaling function ϕ for the subspace V_0. In this way they also are able to remove the Gibbs effect.

10. Conclusions

We may summarize this short survey with the following list of conclusions/opinions, which roughly are chronological by section of this paper.

1. Multiresolution structures are common to regular stationary Stochastic processes, Kolmogorov automorphism dynamical systems, Lax-Phillips scattering theory, and standard wavelet formulations. Their occurence in each of these mathematical theories enriches our understanding of the others.

2. Ideally and insofar as possible, wavelets should be seen, used, and studied in terms of themselves as the expansion and transform basis. This was the original motivation from seismic data analysis. Fourier transforming them into the frequency domain, with all its advantages and available heavy machinery, amounts to subsuming wavelet theory into standard harmonic analysis.

3. The general theory of Time Operators, and in particular our Time operator of wavelets, orginated in our use of the canonical commutation relations of quantum mechanics and our connection of those to regular stationary stochastic processes.

4. Our early work on higher dimensional (parameter) regular stochastic processes may be seen as an early predecessor to what is now called higher dimensional wavelet support sets. We left some interesting open questions in that work.

5. Kolmogorov dynamical systems possess rich ergodic properties due to their underlying measure preserving and chaotic phase space maps. Wavelet structures, and indeed much of signal processing, tend to ignore the physical dynamics which generates the signals.

6. Our approach to wavelets can be described as that of "representation-free theory of shifts" in which we deal with shift V (dilation) first via its wandering subspaces W_n. This allows us to get a Time operator. The

conventional wavelet theory does not do that.

7. Haar's basis, originally designed to not come from any selfadjoint second order Sturm-Liouville differential operator, is now seen naturally as an eigenbasis of a first order selfadjoint differential operator.

8. Most wavelet projectors may not be seen as coarse-graining conditional expectations in the Kolmogorov dynamical systems sense.

Acknowledgements

Useful discussions about conditional expectations with I. Antoniou led to my discussions in 2001 with Costas Karanikas who showed me his Gibb's effects results and consequent loss of the positivity preserving property in most wavelet projections. Nhan Levan in 1998 at the Ralph Phillips conference in California pointed out to me the independent formulations[53] of wavelet subspaces as wandering subspaces. Hans Primas in 2004 pointed out to me that many practitioners tacitly assume that regular second order stochastic processes always correspond to K-flows with positive entropy, although we do not take that point of view. Primas also pointed out a typo in Ref. 12, p. 89, line 6: the Property $5'$) should be Property $5''$) there. Anatoly Vershik corresponded with me in 2004 about mixing versus nonmixing assumptions in polymorphic Kolmogorov systems. Palle Jorgensen sent me some of his papers and interest in 2003 following my publication of Ref. 14. I also had useful communications from Adhemar Butheel in 2001 concerning the extent to which wavelets had made it into the JPEG standards. Hans van den Berg communicated his interest in our approach to us in 2001 and we appreciated his kind comments. Similarly we appreciate the interest of Daniel Alpay, 2005. I had interesting discussions with Zuowei Shen and Peter Massopust about spline wavelets at the 2005 Quy Nhon meetings.

References

1. K. Gustafson, B. Misra, *Lett. Math. Phys.* **1**, 275 (1976).
2. R. Goodrich, K. Gustafson, *J. Approximation Theory* ***31,*** 268(1981).
3. R. Goodrich, K. Gustafson, *J. Approximation Theory* ***48***, 272 (1986).
4. K. Gustafson, R. Goodrich, *Colloquia Mathematica Societatis Janos Bolyai* ***49***, 401 (1985).
5. K. Gustafson, I. Antoniou, *Wavelets and Kolmogorov Systems* (NATO Grant) (2004) http://www.auth.gr/chi/PROJECTSWaveletsKolmog.html.
6. I. Antoniou, K. Gustafson, *Physica* A **197**, 153 (1993).
7. I. Antoniou, K. Gustafson, Z. Suchanecki, *Physica* A **252**, 345 (1998).

8. K. Gustafson, *Lectures on Computational Fluid Dynamics, Mathematical Physics, and Linear Algebra* (World-Scientific, Singapore, 1997).
9. K. Gustafson, *IMACS Series in Computational and Applied Mathematics*, J. Wang, M. Allen, B. Chen, T. Mathew (Eds.), **4** (IMACS, New Brunswick, NJ, 1998).
10. I. Antoniou, K. Gustafson, *Differential Equations 34*, 829 (1998).
11. K. Gustafson, *Proc. Workshop on Wavelets and Wavelet-based Technologies*, M. Kobayashi, S. Sakakibara, M. Yamada (Eds.)(Tokyo, 29–30 October 1998, IBM Japan/University of Tokyo, 1998).
12. I. Antoniou, K. Gustafson, *Mathematics and Computers in Simulation* **49**, 81 (1999).
13. I. Antoniou, K. Gustafson, *Solitons and Fractals* **11**, 443 (2000).
14. K. Gustafson, *Rocky Mt. J. Math.* **33**, 661 (2003).
15. G. Strang, T. Nguyen, *Wavelets and Filter Banks* (Wellesley–Cambridge Press, Boston, MA, 1996).
16. M. Holschneider, *Wavelets: An Analysis Tool* (Oxford Univ. Press, Oxford, 1995).
17. G. Kaiser, *A Friendly Guide to Wavelets* (Birkhauser, Boston, 1994).
18. C. Chui, *An Introduction to Wavelets* (Academic Press, Boston, 1992).
19. I. Daubechies, *Ten Lectures on Wavelets* (SIAM Publications, Philadelphia, 1992).
20. A. Louis, P. Maas, A. Rieder, *Wavelets: Theory and Applications* (Wiley, New York, 1997).
21. P. Wojtaszczyk, *A Mathematical Introduction to Wavelets* (Cambridge Univ. Press, Cambridge, 1997).
22. M. Ruskai, et. al. (Eds.). *Wavelets and Their Applications* (Jones and Bartlett, Boston, 1992).
23. J. C. Van den Berg (Ed.), *Wavelets in Physics* (Cambridge Univ. Press, Cambridge, 1999).
24. P. Kopp, *Martingales and Stochastic Integrals* (Cambridge Univ. Press, Cambridge, 1984).
25. J. Doob, *Stochastic Processes* (Wiley, New York, 1953).
26. M. Loève, *Probability Theory*, 3rd Edition, D. Van Nostrand (Princeton, NJ, 1963; 4th Edition (2 volumes), 1977).
27. R. Bahadur, *Proc. Amer. Math. Soc.* **6**, 565 (1955).
28. S. T. C. Moy, *Pacific J. Math.* **4**, 47 (1954).
29. P. Dodds, C. Huijsmans, B. de Pagter, *Pacific J. Math.* **141**, 55 (1990).
30. Y. Abramovich, C. Aliprantis, O. Burkinshaw, *J. Math. Anal. Applic.* **177**, 641 (1993).
31. P. Halmos, *Measure Theory*, D. Van Nostrand (Princeton, NJ, 1950).
32. R. Dudley, *Real Analysis and Probability* (Chapman and Hall, New York, 1989).
33. A. Grossmann, J. Morlet, *SIAM J. Math. Anal.* **15**, 723 (1984).
34. J. Morlet, *Proc. 51st Annual Meeting of the Society of Exploration Geophysicists* (Los Angeles, 1981).
35. J. Morlet, G. Arens, E. Fourgeau, D. Giard, *Geophys.* **47**, 222 (1982).

36. P. Groupillaud, A. Grossmann, J. Morlet, *Geoexploration* **23**, 85 (1984).
37. G. Webster (Ed.), *Society of Exploration Geophysics* (Tulsa, OK, 1978).
38. Y. Rozanov, *Innovation Processes* (Wiley, New York, 1967).
39. B. Misra, E. C. G. Sudarshan, *J. Math. Phys.* **18**, 756 (1977).
40. I. Antoniou, V. Sadovnichy, H. Walther (Eds.), *The Physics of Communication* (World Scientific, Singapore, 2003).
41. K. Gustafson, *International J. of Theoretical Physics* (2005) (to appear).
42. I. Daubechies, A. Grossmann, Y. Meyer, *J. Math. Phys.* **27**, 1271 (1986).
43. J. Klauder, B. S. Skagerstam, *Coherent States* (World Scientific, Singapore, 1985).
44. H. Dym, H. McKean, *Gaussian Processes, Function Theory, and the Inverse Spectral Problem* (Academic Press, New York, 1976).
45. R. Goodrich, K. Gustafson, *Quantum Mechanics in Mathematics, Chemistry, and Physics, Plenum Press* (Eds.)(New York, 1981).
46. R. Goodrich, K. Gustafson, *Spectral Theory of Differential Operators*, I. Knowles, R. Lewis (Eds.) (North Holland, Amsterdam, 1981).
47. R. Goodrich, K. Gustafson, *Colloquia Mathematica Societatis Janos Bolyai* **35**, 539 (1983).
48. R. Goodrich, K. Gustafson, *Colloquia Mathematica Societatis Janos Bolyai* **49**, 383 (1985).
49. K. Gustafson, *Between Choice and Chance*, H. Atmanspacher, R. Bishop (Eds.) (Imprint Academic, Thorverton, UK, 2002).
50. B. Sz. Nagy, C. Foias, *Harmonic Analysis of Operators in Hilbert Space* (North Holland, Amsterdam, 1970).
51. P. Halmos, *A Hilbert Space Problem Book* (Springer, New York, 1982).
52. P. Lax, R. Phillips, *Academic Press* (New York, 1967).
53. T. Goodman, S. Lee, W. S. Tang, *Trans. Amer. Math. Soc.* **338**, 639 (1993).
54. S. Mallat, *Trans. Amer. Math Soc.* **315**, 69 (1989).
55. M. H. Stone, *Amer. Math. Soc.* (Providence, RI, 1932).
56. C. Karanikas, *Results of Math.* **34**, 330 (1998).
57. N. Atreas, C. Karanikas, *J. Fourier Anal. Appl.* **5**, 575 (1999).
58. N. Atreas, C. Karanikas, *J. Comp. Anal. Appl.* **2**, 89 (2000).
59. A. J. E. M. Janssen, *IEEE Trans. Information Theory* **38**, 884 (1992).
60. N. Levan, C. Kubrusly, *Math. Comput. Simulation* **63**, 73 (2003).
61. N. Levan, C. Kubrusly, *International J. Wavelets, Multiresolution and Information Processing* **2**, 237 (2004).
62. C. Kubrusly, N. Levan, *Math. Comput. Simulation* **65**, 607 (2004).
63. C. Kubrusly, N. Levan, *Abstract wavelets generated by Hilbert Space shift operators*, (2005) (to appear).
64. P. E. T. Jorgensen, *Clifford Algebras in Analysis and Related Topics, CRC Press* J. Ryan (Ed.) (Boca Raton, FL, 1996).
65. G. G. Walter, X. Shen, *Wavelets, Multiwavelets and Their Applications, Contemporary Math*, A. Aldroubi, E. B. Lin (Eds.), *Amer. Math. Soc.* **216**, 63 (1998).
66. G. G. Walter, X. Shen, *J. Appl. Comp. Harmonic Analysis* **12**, 150 (2002).

Harmonic, Wavelet and p-Adic Analysis
Eds. N. M. Chuong et al. (pp. 23–36)

§2. CONSTRUCTION OF UNIVARIATE AND BIVARIATE EXPONENTIAL SPLINES

XIAOYAN LIU
University of La Verne
email: liux@ulv.edu

In this paper, univariate and bivariate exponential spline functions with compact supports are constructed by the integral iteration formulas . The properties of exponential splines are explored. In addition, hyperbolic splines are formed as linear combinations of exponential splines and properties are surveyed. Furthermore, an orthonormal exponential spline on small compact support is constructed. The integral formulas can also be used to build high dimensional exponential splines, which are not cross products of univariate exponential splines and have their own advantages.

Keywords: Spline functions, Exponential functions, Exponential Splines

1. Introduction

Univariate exponential splines are piecewise exponential polynomials of forms

$$\sum_{k=0}^{n} a_k e^{k\alpha x}$$

in each interval (where $\alpha \neq 0$ is a real number) and they are nature extensions of polynomial splines. Needless to say, exponential splines have their own advantages. For example, since the most prominent functions in continuous-time signals-and-systems theory are exponential functions, the exponential splines would have more impact on continuous-time signal processing than the polynomial splines. A number of paper have appeared to study the properties of the exponential splines and exponential B-splines (cf. Refs. 2–5,8–12). However, as far as I am aware of, only a few paper have appeared to explore the bivariate or higher dimensional exponential splines (cf. Refs. 2,4,8,11). Furthermore, there is an elegant integral iteration formula for constructing polynomial splines (cf. Refs.6,7). I have not seen the use of it in the construction of exponential splines. Therefore, in this paper, I am going to investigate the exponential splines built by integral iteration formula and the properties. In the Sec. 2, the integral iteration formula for constructing univariate exponential splines with compact support will be introduced. In the Sec. 3, the properties of the exponential splines will be established. In addition, hyperbolic splines were formed as linear combinations of exponential splines and properties are surveyed in the Sec. 4. In the Sec. 5, an example of univariate orthonormal exponential splines with compact support will be given. It shows one more advantage of the exponential splines. In the Sec. 6, the integral iteration formula will be extended to the two-dimensional case.

2. Integral Iteration Formulas for Constructing Univariate Exponential Splines

Let us first consider the one dimensional case and splines with small compact support. We consider the equal distance (=h) partition of the real number line.

Let

$$\mu_h(x) = \begin{cases} 1, & -\frac{h}{2} < x \leq \frac{h}{2} \\ 0, & \text{elsewhere} \end{cases} .$$

This is the zero degree polynomial B-spline, as well as the zero degree exponential spline. Let $B_{0,h}(x) = \mu_h(x)$, to get the higher degree polynomial B-splines, the following integral iteration formula was used (cf. Ref.[1]):

$$B_{n,h}(x) = \int_{-\frac{h}{2}}^{\frac{h}{2}} B_{n-1,h}(x+t)dt, \qquad n = 1, 2, 3, ...$$

Similarly, to get higher degree exponential splines, let $E_{0,h}(x) = \mu_h(x)$, and

$$I_{n,h}\beta(x) = \int_{-\frac{h}{2}}^{\frac{h}{2}} \beta(x+t)e^{-n\alpha t}\, dt, \quad n = 1, 2, ...;$$

$$E_{n,h}(x) = \frac{n\alpha}{2\sinh(n\alpha h/2)} I_{n,h}E_{n-1,h}(x), n = 1, 2, ...; \tag{1}$$

$E_{n,h}(x)$ is indeed an exponential spline of degree n with a compact support $\left(\frac{-(n+1)h}{2}, \frac{(n+1)h}{2}\right)$.

Proposition 2.1. *Let $E_{n,h}(x)$ be defined as above. Then $E_{n,h}(x) \in$*

$C^{n-1}(-\infty,\infty)$ *and they are piece-wise exponential splines of degree n, i. e.*

$$E_{n,h}(x) = \begin{cases} \sum_{m=0}^{n}\left(t_{n,m,l}e^{m\alpha x}\right), & x \in (lh,(l+1)h) \text{ for } n \text{ odd}, \\ & l = -\frac{n+1}{2}, -\frac{n-1}{2}, ..., 0, ..., \frac{n-1}{2}; \\ \sum_{m=0}^{n}\left(t_{n,m,l}e^{m\alpha x}\right), & x \in \left(\left(l-\frac{1}{2}\right)h, \left(l+\frac{1}{2}\right)h\right) \text{ for } n \text{ even}, \\ & l = -\frac{n}{2}, -\frac{n-2}{2}, ..., 0, ..., \frac{n}{2}; \\ 0 & \text{for other } .x \end{cases}$$

$n = 1, 2, 3, ...$

Proof. Apparently, $E_{n,h}(x) \in C^{n-1}(-\infty,\infty)$ because it is constructed by integrating a continuous function $n-1$ times. Besides, $E_{n,h}(x) = 0$ for $x > \frac{n+1}{2}h$ or $x < -\frac{n+1}{2}h$ from its integration formula.

We can easily prove that $E_{n,h}(x)$ is a piecewise exponential function by the induction. It is obviously true for $E_{1,h}(x)$ (the explicit form is given later in this section). Assume it is true for $n = k$. When $n = k+1$, for any given x, if $x \le -\frac{k+2}{2}h$, or $x \ge \frac{k+2}{2}h$, and $-\frac{h}{2} < v < \frac{h}{2}$, then $E_{k,h}(x+v) = 0$. Suppose that $-\frac{k+2}{2}h < x < \frac{k+2}{2}h$, if k is odd, there must be an integer j such that $-\frac{h}{2} \le x - jh < \frac{h}{2}$. Then

$$\begin{aligned} E_{k+1,h}(x) &= \frac{(k+1)\alpha}{2\sinh((k+1)\alpha h/2)}\int_{-\frac{h}{2}}^{\frac{h}{2}} E_{k,h}(x+v)e^{-(k+1)\alpha v}dv \\ &= \frac{(k+1)\alpha e^{(k+1)\alpha x}}{2\sinh((k+1)\alpha h/2)}\int_{x-\frac{h}{2}}^{x+\frac{h}{2}} E_{k,h}(u)e^{-(k+1)\alpha u}du \\ &= \frac{(k+1)\alpha e^{(k+1)\alpha x}}{2\sinh((k+1)\alpha h/2)}\left(\int_{x-\frac{h}{2}}^{jh}\sum_{m=0}^{k}(a_{k,m,j-1}e^{m\alpha u})e^{-(k+1)\alpha u}du\right. \\ &\quad \left.+\int_{jh}^{x+\frac{h}{2}}\sum_{m=0}^{k}(a_{k,m,j}e^{m\alpha u})e^{-(k+1)\alpha u}du\right) \\ &= \sum_{m=0}^{k+1}(a_{k+1,m,j}e^{m\alpha x}); \end{aligned}$$

In the case that k is even, the proof is similar when we substitute jh by $jh - \frac{h}{2}$ in the process above. Therefore, the resulting functions are exponential functions of degree $k+1$ in each interval. □

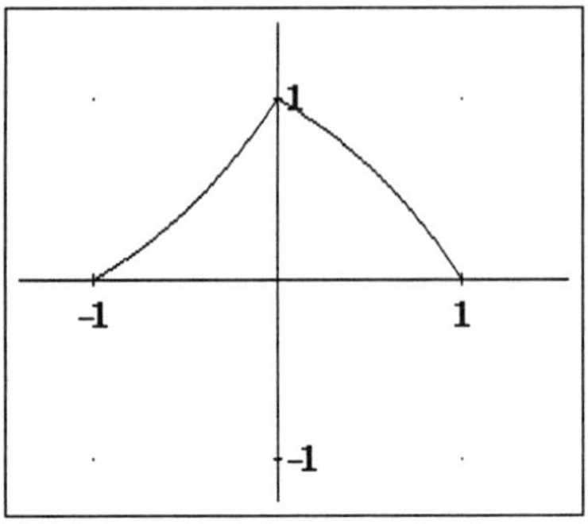

Fig. 1. The graph of $E_{1,h}(x)$.

2.1. *An Example*

$$E_{1,h}(x) = \frac{I_1 E_{0,h}(x)}{I_1 E_{0,h}(0)} = \frac{\alpha}{2\sinh(\alpha h/2)} \int_{-\frac{h}{2}}^{\frac{h}{2}} \mu_h(x+t) e^{-\alpha t} dt$$
$$= \begin{cases} \frac{1}{2\sinh\frac{\alpha h}{2}} \left(e^{\frac{\alpha h}{2}} - e^{\alpha x - \frac{\alpha h}{2}} \right), & \text{for } 0 \le x \le h; \\ \frac{1}{2\sinh\frac{\alpha h}{2}} \left(e^{\alpha x + \frac{\alpha h}{2}} - e^{\frac{-\alpha h}{2}} \right), & \text{for } -h \le x \le 0; \\ 0 & \text{for } x > h \text{ or } x < -h. \end{cases}$$

3. Properties of Exponential Splines

The exponential splines constructed here have very nice approximation properties. I will demonstrate that the proper linear combinations of exponential splines $E_{n,h}(x)$ preserve the exponential functions of degrees up to n, i. e. $\{1, e^{\alpha x}, e^{2\alpha x}, ..., e^{n\alpha x}\}$.

Proposition 3.1. *Let $E_{n,h}(x)$ be defined as in* (1). *Then*

$$\sum_{j=-\infty}^{\infty} E_{n,h}(x - jh) = 1 \ , \ n = 1, 2, 3, \cdots \ , \qquad (2)$$

for any real x.

Proof. Obviously,

$$\sum_{j=-\infty}^{\infty} \mu_h(x - jh) = 1 \ .$$

Consequently, for any given x (notice that the infinite sum is actually finite because $E_{n,h}(x)$ has compact support),

$$\begin{aligned}\sum_{j=-\infty}^{\infty} E_{1,h}(x-jh) &= \sum_{j=-\infty}^{\infty} \frac{\alpha}{2\sinh(\alpha h/2)} \int_{-\frac{h}{2}}^{\frac{h}{2}} \mu_h(x-jh+t)e^{-\alpha t}\, dt \\ &= \frac{\alpha}{2\sinh(\alpha h/2)} \int_{-\frac{h}{2}}^{\frac{h}{2}} e^{-\alpha t} \sum_{j=-\infty}^{\infty} \mu_h(x-jh+t)\ dt \\ &= \frac{\alpha}{2\sinh(\alpha h/2)} \int_{-\frac{h}{2}}^{\frac{h}{2}} e^{-\alpha t}\ dt = 1.\end{aligned}$$

So (2) is true for $n = 1$. Assume that $\sum_{j=-\infty}^{\infty} E_{n-1,h}(x-jh) = 1$, we deduct that

$$\begin{aligned}\sum_{j=-\infty}^{\infty} E_{n,h}(x-jh) &= \frac{n\alpha}{2\sinh(n\alpha h/2)} \sum_{j=-\infty}^{\infty} \int_{-\frac{h}{2}}^{\frac{h}{2}} E_{n-1,h}(x-jh+t)e^{-n\alpha t}\, dt \\ &= \frac{n\alpha}{2\sinh(n\alpha h/2)} \int_{-\frac{h}{2}}^{\frac{h}{2}} \left(\sum_{j=-\infty}^{\infty} E_{n-1,h}(x-jh+t)\right) e^{-n\alpha t}\, dt \\ &= \frac{n\alpha}{2\sinh(n\alpha h/2)} \int_{-\frac{h}{2}}^{\frac{h}{2}} e^{-n\alpha t}\ dt = 1.\end{aligned}$$

The identity (2) is confirmed by induction. □

Proposition 3.2.

$$\int_{-\infty}^{\infty} E_{n,h}(x) = 1\ , n = 0, 1, 2, 3, \cdots . \tag{3}$$

Proof. The proof is very similar to the proof of Pro. 3.2, we only need to change the infinite sum to the infinite integral in each expression. □

Proposition 3.3. *Let $E_{1,h}(x)$ be defined as above. Then*

$$\sum_{j=-\infty}^{\infty} e^{j\alpha h} E_{1,h}(x-jh) = e^{\alpha x}. \tag{4}$$

Furthermore, for $n = 1, 2, 3, 4, \cdots$,

$$\sum_{j=-\infty}^{\infty} \frac{\sinh(n\alpha h/2)}{n\sinh\left(\frac{\alpha h}{2}\right)} e^{j\alpha h} E_{n,h}(x-jh) = e^{\alpha x}\ . \tag{5}$$

Proof. The identity (4) can be proved by direct computation as follows. For any given x, if k is an integer such that $0 \le x - kh < h$, then for $-\frac{h}{2} \le v \le \frac{h}{2}, -\frac{h}{2} \le x - kh + v < \frac{3h}{2}$. By the definition of $E_{0,h}(x)$ we recognize that $E_{0,h}(x - jh + v) = 0$, for $j < k$ or $j > (k+1)$. Thus

$$
\begin{aligned}
&\sum_{j=-\infty}^{\infty} e^{\alpha hj} E_{1,h}(x - jh) \\
&= \sum_{j=-\infty}^{\infty} \frac{\alpha}{2\sinh(\alpha h/2)} \int_{-\frac{h}{2}+x-kh}^{\frac{h}{2}+x-kh} e^{\alpha hj} E_{0,h}(u) e^{-\alpha(u-x+kh)} du \\
&= \frac{\alpha}{2\sinh(\alpha h/2)} \int_{-\frac{h}{2}+x-kh}^{\frac{h}{2}+x-kh} \left(e^{\alpha hk} E_0(u) + e^{\alpha kh+\alpha h} E_0(u-h)\right) e^{-\alpha u+\alpha x-k\alpha h} du \\
&= \frac{\alpha}{2\sinh(\alpha h/2)} \left(\int_{-\frac{h}{2}+x-kh}^{\frac{1}{2}h} e^{-\alpha u+\alpha x} du + \int_{\frac{1}{2}h}^{\frac{h}{2}+x-kh} e^{\alpha h} e^{-\alpha u+\alpha x} du \right) = e^{\alpha x}.
\end{aligned}
$$

To verify (5), the mathematical induction can be employed. When $n = 1$, it is the same as (4). Assume (5) is true for $n = k$. Then for $n = k+1$,

$$
\begin{aligned}
&\sum_{j=-\infty}^{\infty} \frac{\sinh((k+1)\alpha h/2)}{(k+1)\sinh\left(\frac{\alpha h}{2}\right)} e^{j\alpha h} E_{k+1,h}(x - jh) \\
&= \sum_{j=-\infty}^{\infty} \frac{\alpha}{2\sinh\left(\frac{\alpha h}{2}\right)} \int_{-\frac{h}{2}}^{\frac{h}{2}} e^{j\alpha h} E_{k,h}(x + v - jh) e^{-(k+1)\alpha v} dv \\
&= \frac{k\alpha}{2\sinh(k\alpha h/2)} \int_{-\frac{h}{2}}^{\frac{h}{2}} \sum_{j=-\infty}^{\infty} \frac{\sinh(k\alpha h/2)}{k\sinh\left(\frac{\alpha h}{2}\right)} e^{j\alpha h} E_{k,h}(x + v - jh) e^{-(k+1)\alpha v} dv \\
&= \frac{k\alpha}{2\sinh(k\alpha h/2)} \int_{-\frac{h}{2}}^{\frac{h}{2}} e^{\alpha x+\alpha v} e^{-(k+1)\alpha v} dv = e^{\alpha x}.
\end{aligned}
$$

Therefore, the identity holds for all integer $n \ge 2$. □

Proposition 3.4.

$$
\sum_{j=-\infty}^{\infty} e^{n\alpha jh} E_{n,h}(x - jh) = \beta_n e^{n\alpha x} \tag{6}
$$

for any real x, $n = 2, 3, 4, \ldots$, where β_n is a constant not depending on x and $\beta_0 = \beta_1 = \beta_2 = \beta_3 = \beta_4 = 1$.

Proof. Let

$$
\begin{aligned}
e_n(x) &= \sum_{j=-\infty}^{\infty} e^{n\alpha jh} E_{n,h}(x-jh) \\
&= \frac{n\alpha}{2\sinh(n\alpha h/2)} \sum_{j=-\infty}^{\infty} e^{n\alpha jh} \int_{-\frac{h}{2}+x-jh}^{\frac{h}{2}+x-jh} E_{n-1,h}(u) e^{-n\alpha(u-x+jh)} du.
\end{aligned}
$$

Then

$$
\begin{aligned}
\frac{d}{dx} e_n(x) &= \frac{n\alpha}{2\sinh(n\alpha h/2)} \sum_{j=-\infty}^{\infty} e^{n\alpha jh} E_{n-1,h}(\frac{h}{2}+x-jh) e^{-n\alpha(\frac{h}{2})} \\
&\quad - \frac{n\alpha}{2\sinh(n\alpha h/2)} \sum_{j=-\infty}^{\infty} e^{n\alpha jh} E_{n-1,h}(-\frac{h}{2}+x-jh) e^{n\alpha(\frac{h}{2})} \\
&\quad + n\alpha \frac{n\alpha}{2\sinh(n\alpha h/2)} \sum_{j=-\infty}^{\infty} \int_{-\frac{h}{2}+x-jh}^{\frac{h}{2}+x-jh} E_{n-1,h}(u) e^{-n\alpha(u-x+jh)} du.
\end{aligned}
$$

The first two sums are actually the same so they cancel each other out and the following equation is implied

$$
\frac{d}{dx} e_n(x) = n\alpha e_n(x).
$$

Solving this differential equation, we reach the conclusion:

$$
e_n(x) = \sum_{j=-\infty}^{\infty} e^{n\alpha jh} E_{n,h}(x-jh) = \beta_n e^{n\alpha x},
$$

where β_n is a constant not depending on x.
Note. By direct calculations we obtain $\beta_0 = \beta_1 = \beta_2 = \beta_3 = \beta_4 = 1$. □

Proposition 3.5.

$$
\sum_{j=-\infty}^{\infty} e^{k\alpha jh} E_{k+m,h}(x-jh) = \prod_{l=1}^{m} \frac{(k+l)}{\sinh(\frac{k+l}{2}\alpha h)} \frac{\sinh\left(\frac{l\alpha h}{2}\right)}{l} \beta_k e^{k\alpha x}, \tag{7}
$$

for $m = 1, 2, 3, ..., k = 1, 2, ...$

Proof. The identities can be proved by induction again. It is similar to the proof of Pro. 3.4, so we omit it. □

4. Hyperbolic Splines

It is easy to see that if we let $C_{n,h}(x) = \frac{1}{2}\left(E_{n,h}(x) + E_{n,h}(-x)\right)$ and $S_{n,h}(x) = \frac{1}{2}\left(E_{n,h}(x) - E_{n,h}(-x)\right)$, then $C_{n,h}(x)$ and $S_{n,h}(x)$ are hyperbolic function $(\cosh(\alpha x), \sinh(\alpha x))$ splines. In addition, we find that

Proposition 4.1.

$$\sum_{j=-\infty}^{\infty} C_{n,h}(x-jh) = 1 \qquad \textit{and} \qquad \sum_{j=-\infty}^{\infty} S_{n,h}(x-jh) = 0\ , \tag{8}$$

$$\sum_{j=-\infty}^{\infty} \left(\cosh(n\alpha jh)\, C_{n,h}(x-jh) + \sinh(n\alpha jh)\, S_{n,h}(x-jh)\right) = \beta_n \cosh(n\alpha x)\,, \tag{9}$$

$$\sum_{j=-\infty}^{\infty} \left(\sinh(n\alpha jh)\, C_{n,h}(x-jh) + \cosh(n\alpha jh)\, S_{n,h}(x-jh)\right) = \beta_n \sinh(n\alpha x)\,, \tag{10}$$

where $n = 0, 1, 2, \cdots$; $(\beta_0 = \beta_1 = \beta_2 = \beta_3 = \beta_4 = 1)$.

Proof. Identities (8) follows straight from Pro. 3.2.

Next, applying identities from Pro. 3.4:

$$\sum_{j=-\infty}^{\infty} e^{k\alpha jh} E_{k,h}(x-jh) = \beta_k e^{k\alpha x};$$

$$\sum_{j=-\infty}^{\infty} e^{-k\alpha jh} E_{k,h}(-x+jh) = \beta_k e^{-k\alpha x},$$

we confirm that for $n = 0, 1, 2, \cdots$,

$$\begin{aligned}
&\sum_{j=-\infty}^{\infty} \left(\cosh(n\alpha jh)\, C_{n,h}(x-jh) + \sinh(n\alpha jh)\, S_{n,h}(x-jh)\right) \\
&= \frac{1}{4}\sum_{j=-\infty}^{\infty} \Big(\left(e^{n\alpha jh} + e^{-n\alpha jh}\right)\left(E_{n,h}(x-jh) + E_{n,h}(-x+jh)\right) \\
&\quad + \left(e^{n\alpha jh} - e^{-n\alpha jh}\right)\left(E_{n,h}(x-jh) - E_{n,h}(-x+jh)\right)\Big) \\
&= \frac{1}{4}\sum_{j=-\infty}^{\infty} \left(2e^{n\alpha jh} E_{n,h}(x-jh) + 2e^{-n\alpha jh} E_{n,h}(-x+jh)\right) \\
&= \frac{1}{2}\beta_n\left(e^{n\alpha x} + e^{-n\alpha x}\right) = \beta_n \cosh(n\alpha x)\,.
\end{aligned}$$

Identity (10) could be deduced by the similar analysis. □

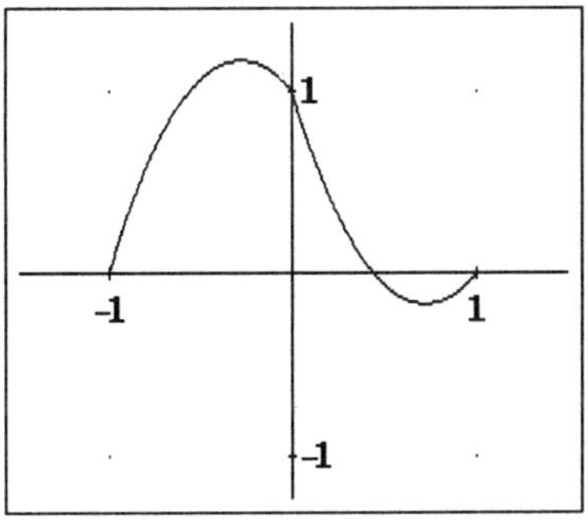

Fig. 2. The graph of $G(x)$.

5. A Univariate Orthonormal Exponential Spline with Minimum Support

Example 5.1. The univariate continuous orthonormal exponential spline of degree 1 with minimum support

Let $\alpha = 1, E_1(x) = E_{1,1}(x)$ be defined as above. Let $G(x) = cE_1(x) + dE_1(-x)$, where, c and d are given by:

$$c = -\frac{\left(e^2 - 2e - 1\right)}{2\sqrt{2\,(e+1) - e^2}\sqrt{2\sqrt{2\,(e+1) - e^2} - e^2 + 2e + 3}}, \tag{11}$$

$$d = \frac{2\sqrt{2\,(e+1) - e^2} - e^2 + 2e + 3}{2\sqrt{2\,(e+1) - e^2}\sqrt{2\sqrt{2\,(e+1) - e^2} - e^2 + 2e + 3}}.$$

Then the orthonormal conditions are satisfied:

$$\int_{-\infty}^{\infty} G(x)\, G(x)\, dx = 1 \quad \text{and}$$
$$\int_{-\infty}^{\infty} G(x)\, G(x-k)\, dx = 0 \quad , k = \pm 1, \pm 2, \pm 3, \ldots.$$

Furthermore,

$$\sum_{j=-\infty}^{\infty} G(x-j) = 1 \quad \text{and} \quad \int_{-\infty}^{\infty} G(x)\, dx = 1.$$

6. Integral Iteration Formulas for Constructing Bivariate Exponential Splines

For partitions on the plane, we consider the type I triangulation (which means one diagonal is added to each cell of the rectangular partition) and assume equal distance ($= h$) on the x direction and the y direction ($= l$). For simplicity, let $h = l$ here. Then, we will let

$$A(x,y) = \begin{cases} 1, & -\frac{h}{2} < x \leq \frac{h}{2}, -\frac{h}{2} < y \leq \frac{h}{2} \\ 0 & \text{elsewhere} \end{cases},$$

$$B_0(x,y) = \int_{-\frac{h}{2}}^{\frac{h}{2}} A(x+t, y+t)dt.$$

Define

$$I_1 f(x,y) = \int_{x-\frac{h}{2}}^{x+\frac{h}{2}} f(u,y)du,$$

$$I_2 f(x,y) = \int_{y-\frac{h}{2}}^{y+\frac{h}{2}} f(x,u)du,$$

$$I_3 f(x,y) = \int_{-\frac{h}{2}}^{\frac{h}{2}} f(x+t, y+t)dt.$$

Then $B_{3n}(x,y) = I_1^n I_2^n I_3^n B_0(x,y) \in S_{3n+1}^{2n}(\Delta_1)$ are polynomial B-splines (cf Ref. 1).

To build exponential splines, we let

$$E_{1,h}(x,y) = \frac{\alpha}{2\sinh\left(\frac{\alpha h}{2}\right)} \int_{-\frac{h}{2}}^{\frac{h}{2}} A(x+t, y+t)e^{-\alpha t}dt \tag{12}$$

then $E_{1,h}(x,y)$ is a continuous bivariate exponential spline function with the following explicit expression

$$E_{1,h}(x,y) = \frac{e^{\frac{\alpha h}{2}}}{2\sinh\left(\frac{\alpha h}{2}\right)} \begin{cases} 1 - e^{\alpha x - \alpha h}, & 0 \leq x \leq h, 0 \leq y \leq h, y \leq x; \\ e^{\alpha y} - e^{\alpha x - \alpha h}, & 0 \leq x \leq h, -h \leq y \leq 0, y \leq x; \\ e^{\alpha y} - e^{-\alpha h}, & -h \leq x \leq 0, -h \leq y \leq 0, y \leq x; \\ e^{\alpha x} - e^{-\alpha h}, & -h \leq x \leq 0, -h \leq y \leq 0, x \leq y; \\ e^{\alpha x} - e^{\alpha y - \alpha h}, & -h \leq x \leq 0, 0 \leq y \leq h, x \leq y; \\ 1 - e^{\alpha y - \alpha h}, & 0 \leq x \leq h, 0 \leq y \leq h, x \leq y. \end{cases}$$

If we define

$$J_{1j}f(x,y) = \int_{-\frac{h}{2}}^{\frac{h}{2}} f(x+u,y)e^{-(3j-1)\alpha u}du,$$

$$J_{2j}f(x,y) = \int_{-\frac{h}{2}}^{\frac{h}{2}} f(x,u+y)e^{-3j\alpha u}du,$$

$$J_{3j}f(x,y) = \int_{-\frac{h}{2}}^{\frac{h}{2}} f(x+u,y+u)e^{-(3j+1)\alpha u}du;$$

$$\eta_j = \int_{-\frac{h}{2}}^{\frac{h}{2}}\int_{-\frac{h}{2}}^{\frac{h}{2}}\int_{-\frac{h}{2}}^{\frac{h}{2}} e^{-(3j+1)\alpha t}dte^{-(3j)\alpha v}dve^{-(3j-1)\alpha u}du,$$

$$I_j = \frac{1}{\eta_j}J_{11j}J_{12j}J_{13j}, \qquad j = 1,2,....$$

then we get

Proposition 6.1. *Let* $T_{0,h}(x,y) = E_{1,h}(x,y)$, $T_{n,h}(x,y) = I_n...I_1E_{1,h}(x,y)$, $n = 1,2,...$, *then they are exponential spline functions* $\in C^{2n}(R^2)$ *on the type I triangulation* Δ_1.

Proof. The proof of piece-wise spline functions can be written similarly as the proof of Pro. 2.1.

The proof the smoothness can be done by the mathematical induction.

When n=0, $T_{0,h}(x,y) = E_{1,h}(x,y)$ is continuous. Assume $T_{n,h}(x,y) \in C^{2n}(R^2)$. For $k = n+1$, we can write expressions explicitly for all second partial derivatives. However, to save the space, we omit the unnecessary details. (Please contact the author if you have any question.) It is easy to see that $\frac{\partial^2}{\partial x^2}T_{n+1,h}(x,y)$ is simply a linear combination of integrals:

$$\frac{3n-1}{\eta_n}\int_{-\frac{h}{2}}^{\frac{h}{2}}\int_{-\frac{h}{2}}^{\frac{h}{2}} T_{n,h}(x\pm\frac{h}{2}+t,y+v+t)e^{-(3n+1)\alpha t}dte^{-3n\alpha v}dve^{-(3n-1)\alpha(\pm\frac{H}{2}))}$$

and

$$\frac{1}{\eta_n}\int_{-\frac{h}{2}}^{\frac{h}{2}} T_{n,h}(x+\frac{h}{2}\pm\frac{h}{2},y+v\pm\frac{h}{2})e^{-3n\alpha v}dve^{-(3n+1)\alpha(\pm\frac{h}{2})}e^{-(3n-1)\alpha(\pm\frac{h}{2}))}$$

and

$$\frac{1}{\eta_n}\int_{-\frac{h}{2}}^{\frac{h}{2}} T_{n,h}(x+\frac{h}{2}+t,y\pm\frac{h}{2}+t)e^{-(3n+1)\alpha t}dte^{-3n\alpha(\pm\frac{h}{2})}e^{-(3n-1)\alpha(\pm\frac{h}{2}))}.$$

It also is straight forward to show that $\frac{\partial^2}{\partial x\partial y}T_{n+1,h}(x,y)$ is a linear combination of some of integrals above. Therefore $\frac{\partial^2}{\partial x^2}T_{n+1,h}(x,y)$,

$\frac{\partial^2}{\partial x \partial y} T_{n+1,h}(x,y) \in C^{2n}\left(R^2\right)$ by induction assumption. We can prove that $\frac{\partial^2}{\partial y^2} T_{n+1,h}(x,y) \in C^{2n}\left(R^2\right)$ by the same analysis.

In conclusion, $T_{n+1,h}(x,y) \in C^{2n+2}\left(R^2\right)$. The proof is completed. □

The bivariate exponential splines have some basic properties as the univariate exponential splines.

Proposition 6.2.

$$\sum_{j=-\infty}^{\infty} \sum_{m=-\infty}^{\infty} T_{n,h}(x - jh, y - mh) = 1. \tag{13}$$

Proof. The proof can be done by the induction again. For any given (x, y), the sum on the left side is a finite sum. So we can exchange the sum and the integration freely. When $n = 0$,

$$\begin{aligned}
&\sum_{j=-\infty}^{\infty} \sum_{m=-\infty}^{\infty} T_{0,h}(x - jh, y - mh) \\
&= \sum_{j=-\infty}^{\infty} \sum_{m=-\infty}^{\infty} \frac{\alpha}{2\sinh\left(\frac{\alpha h}{2}\right)} \int_{-\frac{h}{2}}^{\frac{h}{2}} A(x - jh + t, y - mh + t) e^{-\alpha t} dt \\
&= \frac{\alpha}{2\sinh\left(\frac{\alpha h}{2}\right)} \int_{-\frac{h}{2}}^{\frac{h}{2}} \left(\sum_{j=-\infty}^{\infty} \sum_{m=-\infty}^{\infty} A(x - jh + t, y - mh + t) \right) e^{-\alpha t} dt \\
&= \frac{\alpha}{2\sinh\left(\frac{\alpha h}{2}\right)} \int_{-\frac{h}{2}}^{\frac{h}{2}} 1 e^{-\alpha t} dt = 1.
\end{aligned}$$

Assume that for $n = k$, (13) is true. Then for $n = k + 1$, we employ the definition of $T_{k+1,h}(x, y)$, exchange the sum to arrive

$$\begin{aligned}
&\sum_{j=-\infty}^{\infty} \sum_{m=-\infty}^{\infty} T_{k+1,h}(x - jh, y - mh) \\
&= \frac{1}{\eta_k} \int_{-\frac{h}{2}}^{\frac{h}{2}} \int_{-\frac{h}{2}}^{\frac{h}{2}} \int_{-\frac{h}{2}}^{\frac{h}{2}} \left(\sum_{j=-\infty}^{\infty} \sum_{m=-\infty}^{\infty} T_{k,h}(x - jh + u + t, y - mh + v + t) \right) \\
&\quad e^{-(3k+1)\alpha t} dt \, e^{-(3k)\alpha v} dv \, e^{-(3k-1)\alpha u} du \\
&= \frac{1}{\eta_k} \int_{-\frac{h}{2}}^{\frac{h}{2}} \int_{-\frac{h}{2}}^{\frac{h}{2}} \int_{-\frac{h}{2}}^{\frac{h}{2}} e^{-(3k+1)\alpha t} dt e^{-(3k)\alpha v} dv e^{-(3k-1)\alpha u} du = \frac{\eta_k}{\eta_k} = 1.
\end{aligned}$$

Consequently, the proposition holds for any positive integer n. □

Proposition 6.3.

$$\sum_{j=-\infty}^{\infty}\sum_{m=-\infty}^{\infty} e^{j\alpha h}E_{1,h}(x-jh,y-mh)=e^{\alpha x}; \tag{14}$$

$$\sum_{j=-\infty}^{\infty}\sum_{m=-\infty}^{\infty} e^{m\alpha h}E_{1,h}(x-jh,y-mh)=e^{\alpha y}. \tag{15}$$

Proof. Let (p,q) be a pair of integers such that $ph \leq x < (p+1)\,h, qh \leq y < (q+1)\,h$. Then $E_{1,h}(x-jh,y-mh) \neq 0$ when $(j,m) = (p,q)$ or $(j,m) = (p+1,q)$ or $(j,m) = (p+1,q+1)$; $E_{1,h}(x-jh,y-mh) = 0$ for all other integer pairs (j,m). Furthermore, if $y-qh \leq x-ph$, then

$$\begin{aligned}
&\sum_{j=-\infty}^{\infty}\sum_{m=-\infty}^{\infty} e^{j\alpha h}E_{1,h}(x-jh,y-mh)\\
&= \frac{1}{2\sinh\left(\frac{\alpha h}{2}\right)}(e^{p\alpha h}E_{1,h}(x-ph,y-qh)\\
&\quad +e^{(p+1)\alpha h}E_{1,h}(x-(p+1)\,h,y-(q+1)\,h)\\
&\quad +e^{(p+1)\alpha h}E_{1,h}(x-(p+1)\,h,y-qh))\\
&= \frac{1}{2\sinh\left(\frac{\alpha h}{2}\right)}(e^{p\alpha h}\left(e^{\frac{\alpha h}{2}}-e\alpha^{\alpha x-p\alpha h-\frac{\alpha h}{2}}\right)\\
&\quad +e^{(p+1)\alpha h}\left(e^{\alpha y-(q+1)\alpha h+\frac{\alpha h}{2}}-e^{-\frac{\alpha h}{2}}\right)\\
&\quad +e^{(p+1)\alpha h}\left(e^{\alpha x-(p+1)\alpha h+\frac{\alpha h}{2}}-e^{\alpha y-q\alpha h-\frac{\alpha h}{2}}\right))\\
&= \frac{1}{2\sinh\left(\frac{\alpha h}{2}\right)}e^{\alpha x}\left(e^{\frac{\alpha h}{2}}-e^{-\frac{\alpha h}{2}}\right)=e^{\alpha x};
\end{aligned}$$

if $y-qh > x-ph$, then

$$\begin{aligned}
&\sum_{j=-\infty}^{\infty}\sum_{m=-\infty}^{\infty} e^{j\alpha h}E_{1,h}(x-jh,y-mh)\\
&= \frac{1}{2\sinh\left(\frac{\alpha h}{2}\right)}(e^{p\alpha h}E_{1,h}(x-ph,y-qh)\\
&\quad +e^{(p+1)\alpha h}E_{1,h}(x-(p+1)\,h,y-(q+1)\,h)\\
&\quad +e^{p\alpha h}E_{1,h}(x-ph,y-(q+1)\,h))
\end{aligned}$$

$$
\begin{aligned}
&= \frac{1}{2\sinh\left(\frac{\alpha h}{2}\right)}(e^{p\alpha h}\left(e^{\frac{\alpha h}{2}} - e^{\alpha y - q\alpha h - \frac{\alpha h}{2}}\right) \\
&\quad + e^{(p+1)\alpha h}\left(e^{\alpha x - (p+1)\alpha h + \frac{\alpha h}{2}} - e^{-\frac{\alpha h}{2}}\right) \\
&\quad + e^{p\alpha h}\left(e^{\alpha y - (q+1)\alpha h + \frac{\alpha h}{2}} - e^{\alpha x - p\alpha h - \frac{\alpha h}{2}}\right)) \\
&= \frac{1}{2\sinh\left(\frac{\alpha h}{2}\right)} e^{\alpha x}\left(e^{\frac{\alpha h}{2}} - e^{-\frac{\alpha h}{2}}\right) = e^{\alpha x}.
\end{aligned}
$$

Hence the identity (14) is true. The identity (15) can be proved in the same way. □

References

1. C. K. Chui, *Multivariate Splines* (SIAM, Philadelphia, 1988).
2. W. Dahman and C. A. Micchelli, *On theory and application of exponential splines,* in *Topics in Multivariate Approximation,* Eds., C. K. Chui, L. L. Schumaker, and F. I. Utreras (Academic Press, New York, 1987) pp. 37–46.
3. J. W. Jerome, *J. of Approximation Theory,* **7**, 143 (1973).
4. B. J. McCartin, *J. of Approximation Theory,* **66**, 1 (1991).
5. S. Karlin and Z. Ziegler, *SIAM J. Numerical Analysis,* **3**,514 (1966).
6. X. Liu, *Bivariate Cardinal Spline Functions for Digital Signal Processing, Trends* in *Approximation Theory,* Eds., K. Kopotum, T. Lyche and M. Neamtu (Vanderbilt University Press, Vanderbilt, 2001).
7. X. Liu, *Journal of Computational and Applied Mathematics* (to appear).
8. A. Ron, *Constructive Approximation,* **4**, 357 (1988).
9. A. Ron, *Rocky Mountain Journal of Mathematics,* **22**, 331 (1992).
10. L. L. Schumaker, *J. Math. Mech.,* **18**, 369 (1968).
11. A. Sharma, J. Tzimbalario, *SIAM J. Math. Anal.,* **7** (1976).
12. J. D. Young, *The Logistic Review,* **4**, 17 (1968).

Harmonic, Wavelet and p-Adic Analysis
Eds. N. M. Chuong *et al.* (pp. 37–57)

§3. MULTIWAVELETS: SOME APPROXIMATION-THEORETIC PROPERTIES, SAMPLING ON THE INTERVAL, AND TRANSLATION INVARIANCE

PETER R. MASSOPUST

GSF - Institute for Biomathematics and Biometry
Neuherberg, Germany
and
Centre of Mathematics, M6
Technical University of Munich
Garching, Germany
E-mail: massopust@ma.tum.de

In this survey paper, some of the basic properties of multiwavelets are reviewed. Particular emphasis is given to approximation-theoretic issues and sampling on compact intervals. In addition, a translation invariant multiwavelet transform is discussed and the regularity and approximation order of the associated correlation matrices, which satisfy a particular matrix-valued refinement equation, are presented.

Keywords: Refinable function vectors, multiwavelet transform, translation invariant wavelets, correlation functions.

1. Introduction

During the last decade, wavelet analysis has become a powerful analyzing and synthesizing tool in pure and applied mathematics. The ability of wavelets to resolve different scales and to transfer information back and forth between these scales has been successfully applied to signal processing, data and image compression.[7,34] The behavior of the continuous or discrete wavelet transform at different levels of resolution is one of the key features of the theory. The continuous wavelet transform gives a highly redundant two-dimensional representation of a function whereas the discrete (orthogonal) transform yields a more efficient representation in an appropriate sequence space.

More recently, multiwavelets have improved the performance of wavelets for several applications by providing added flexibility.[35] Multiwavelets are bases of $L^2(\mathbb{R}^n)$ consisting of more than one base function or generator.[14,15,17,28,34] One of the advantages of multiwavelets is that unlike in the case of a single wavelet, the regularity and approximation order can be improved by increasing the number of generators instead of lengthening the support. These additional generators then provide more flexibility in approximating a given function.

In this article presents an introduction to and an overview of the theory of multiwavelets stating some of their approximation-theoretic properties. The emphasis will be on regularity, approximation order, and vanishing moments. In addition, it is shown how sampling with multiwavelets on compact intervals is done and how multiwavelets may be employed to construct bases on $L^2[0,1]$ without adding additional boundary functions or modifying existing ones. A translation invariant multiwavelet transform is introduced and it is shown how the existence of more than one generator adds a new feature to the representation of a function in terms of so-called redundant projectors. Whereas in the case of a single wavelet this redundant representation depends explicitly only on the autocorrelation functions, the cross-correlation functions enter implicitly into the representation if more than one wavelet is used. It will be seen that this is a direct consequence of the matrix-valued refinement equation satisfied by the correlation functions associated with a multiwavelet. Finally, some results related to the regularity and vanishing moments of a translation invariant multiwavelet system are stated.

The structure of this article is as follows. In Sec. 2 a brief review of multiwavelet theory is provided, the relevant terminology and notation is introduced. Shift-invariant and refinable spaces are defined as they are the natural setting for wavelets, and some approximation-theoretic results are presented. Sec. 3 deals with the issue of sampling data with multiwavelets. Multiwavelets on the interval are briefly introduced by consider one particular example, namely the GHM scaling vector and the DGHM multiwavelet. In Sec. 5, a translation invariant multiwavelet transform is introduced and its properties presented and discussed. The results are then applied to the particular example from Sec. 4, namely the DGHM multiwavelet system.

2. Notation and Preliminaries

In this section we give a brief review of the theory of multiwavelets. For a more detailed presentation of multiwavelets, the reader is referred to the references given in the bibliography.[15,17,18,28,34]

2.1. *Shift-invariant spaces*

Let $n \in \mathbb{N}$ and let $\Lambda \subset \mathbb{R}^n$ be a lattice of full rank, i.e., $\Lambda = M\,\mathbb{Z}^n$ for an invertible real $n \times n$ matrix. For $\lambda \in \Lambda$, the mapping $T_\lambda : L^2(\mathbb{R}^n) \to L^2(\mathbb{R}^n)$ defined by

$$T_\lambda f(x) := f(x - \lambda)$$

is called a *translation along the lattice* Λ*)*. A closed subspace $V \subseteq L^2(\mathbb{R})$ is called *shift-invariant with respect to* Λ if

$$\forall \lambda \in \Lambda : T_\lambda V \subset V$$

Now suppose $\Phi := \{\varphi_1, \ldots, \varphi_r\}$ is a finite collection of $L^2(\mathbb{R}^n)$ functions. The space

$$S[\Phi] := \mathrm{cl}_{L^2}\,\mathrm{span}\,\{T_\lambda\,\varphi_i \;:\; 1 \le i \le r, \lambda \in \Lambda\}$$

is called a *finitely generated shift-invariant space.* If $r = 1$, $V[\varphi]$ is called a *principal shift-invariant space.* The elements of Φ are called the *generators* of $S[\Phi]$.

As an example, consider $n = 1$, $\Lambda := \mathbb{Z}$ and let $\varphi(x) := (1 - |x|)_+$. Then $S[\varphi]$ constitutes the shift-invariant space of all piecewise linear functions in $L^2(\mathbb{R})$ supported on integer knots, i.e., the spline space $\mathcal{S}^1(\mathbb{Z})$.

2.2. *Refinable spaces*

Let $A \in \mathrm{GL}\,(n, \mathbb{R})$, the linear group of invertible $n \times n$ matrices with real entries and let D_A be the unitary operator on $L^2(\mathbb{R}^n)$ defined by

$$D_A f(x) := |\det A|^{1/2}\, f(Ax).$$

A closed subspace $V \subseteq L^2(\mathbb{R})$ is called *refinable* if

$$|\det A| > 1 \quad \text{and} \quad V \subset D_A V$$

As a simple example, consider $n = 1$ and let $Ax := 2\,x$. Then the space $S[\varphi]$ defined above is refinable: For $\varphi(x) = \varphi(2x) + (1/2)[\varphi(2x+1) + \varphi(2x-1)]$. This type of equation is referred to as a *refinement* or *two-scale dilation equation.*

Assume that $\Phi = \{\varphi_1, \ldots, \varphi_r\} \subset L^p(\mathbb{R}^n)$, $p \in [1, \infty]$, and that $V[\Phi]$ is a refinable space for the unitary operator D_A. Then as $V \subset D_A V$, there exists a sequence $\{P(\lambda)\colon \lambda \in \Lambda\} \in \ell^p(\mathbb{R}^{r\times r})$ of $r \times r$ matrices with the property that

$$\Phi(x) = \sum_{\lambda\in\Lambda} P(\lambda)\, \Phi(D_A\, T_\lambda\, x).$$

It should be noted that shift-invariant space $\not\Leftrightarrow$ refinable space.

2.3. *Multiwavelets*

Let $A \in \mathrm{GL}(n, \mathbb{Z})$ and assume all the eigenvalues have modulus greater than one. A finite collection of real-valued L^2-functions $\Psi := (\psi_1, \ldots, \psi_s)^T$ is called a *multiwavelet* if the two-parameter family $\{\Psi_{jk} := |\det M|^{j/2}\, \Psi(A^j \cdot -k) : j \in \mathbb{Z},\, k \in \mathbb{Z}^n\}$ forms a Riesz basis of $L^2(\mathbb{R}^n)$.

One way to construct a multiwavelet is through multiresolution analysis, which consists of a nested sequence $V_j \subset V_{j+1}$, $j \in \mathbb{Z}$, of closed subspaces of $L^2(\mathbb{R}^n)$ with the property that the closure of their union is $L^2(\mathbb{R}^n)$ and their intersection is the trivial subspace $\{0\}$. Furthermore, each subspace V_j is spanned by the A-dilates and integer translates of a finite set of *scaling functions* $\Phi := \{\phi_i : i = 1, \ldots, r\}$, sometimes also called the *generators* of the multiresolution analysis. In other word, $V_j = S[\Phi \circ D_A^j]$, where D_A is the unitary operator corresponding to A and $\Lambda = \mathbb{Z}^n$. Typically, the *scaling vector* or *refinable function vector* $\Phi = (\phi_1, \ldots, \phi_r)^T$ has compact support or decays rapidly enough at infinity. (Here, the support of Φ is defined as the union of the supports of its individual components.) The number s is related to r via the equation $s = (|\det A| - 1)r$. For $r = 1$ we obtain the classical wavelet systems as defined and discussed in, for instance,[8,12,30]

The condition that the spaces V_j be nested implies that the scaling vector Φ satisfies a two-scale matrix dilation equation or matrix refinement equation

$$\Phi(x) = \sum_{k\in\mathbb{Z}^n} P(k)\, \Phi(Ax - k), \tag{1}$$

where the sequence $\{P(k)\}_{k\in\mathbb{Z}}$ of $r \times r$ matrices is sometimes called the *mask* or the *filter coefficient matrices* corresponding to Φ. As seen in the previous subsection, these matrices satisfy $\sum_{k\in\mathbb{Z}} \|P(k)\|_{\ell^2(\mathbb{R}^{r\times r})} < \infty$.

Define $W_j := V_{j+1} \ominus V_j$ and $W_j \perp_{L^2} V_j$, then it can be shown that there exists a set of generators Ψ, called a *multiwavelet*, such that $W_j = S[\Psi \circ D_A^j]$. Moreover, the multiwavelet satisfies a two-scale matrix dilation equation of the form

$$\Psi(x) = \sum_{k\in\mathbb{Z}^n} Q(k)\Phi(Ax - k), \tag{2}$$

where the $r \times r$ matrices $\{Q(k)\}_{k\in\mathbb{Z}^n}$ are again in $\ell^2(\mathbb{R}^{r\times r})$.

The pair (Φ, Ψ) will be called a *multiwavelet system.* As the emphasis in this paper is entirely on compactly supported and orthogonal multiwavelets with dyadic refinement, i.e., $A = 2Ix$, we assume that scaling vectors and multiwavelets satisfy the following conditions.

Compact Support: Both Φ and Ψ have compact support. This implies that the sums in (1) and (2) are finite.

L^2-Orthogonality: The scaling vectors and multiwavelets are L^2-orthonormal in the following sense:*

$$\begin{aligned} \langle \Phi_{jk}, \Phi_{j\ell}\rangle &= \delta_{k\ell} I_r, \\ \langle \Psi_{jk}, \Psi_{m\ell}\rangle &= \delta_{jm,k\ell} I_r, \\ \langle \Phi_{jk}, \Psi_{j\ell}\rangle &= O_r, \end{aligned} \tag{3}$$

where I and O denotes the identity and zero matrix, respectively. Here we defined the inner product of two vector-valued functions F and G by $\langle F, G\rangle := \int_{\mathbb{R}^n} F(x)\, G^T(x)\, dx$. For complex-valued L^2 functions, the transpose operator T has to be replaced by the hermitian conjugate operator $*$.

In terms of the filter coefficient matrices the above orthogonality conditions read

$$\begin{aligned} &\sum_{k\in\mathbb{Z}^n} P(k)P^T(k+2\ell) = \delta_{0\ell} I_r \\ &\sum_{k\in\mathbb{Z}^n} Q(k)Q^T(k+2\ell) = \delta_{0\ell} I_r \\ &\sum_{k\in\mathbb{Z}^n} P(k)Q^T(k+2\ell) = O_r, \qquad \text{for all } \ell\in\mathbb{Z}^n. \end{aligned}$$

For $n = 1$ there exists a relationship between the number r of scaling functions or wavelets, the number $N+1$ of nonzero terms in (1), and the degree of regularity s of Φ and Ψ, namely, $r(N-1) \geq s$. Unlike in the case $r = 1$, the regularity s may be increased not only by increasing N or, equivalently, the length of support of Φ and Ψ, but also by increasing the number r of generators.

*For a multivariate vector-valued function Θ, $\Theta_{jk} := 2^{nj/2}\Theta(2^j \cdot -k)$.

2.4. Reconstruction and decomposition algorithm

Since $V_{j+1} = V_j \oplus W_j$, every function $f_{j+1} \in V_{j+1}$ can be *decomposed* into an "averaged" component $f_j \in V_j$ and a "difference" or "fine-structure" component $g_j \in W_j$: $f_{j+1} = f_j + g_j$. (Note that (1) describes a weighted average of Φ in terms of $\Phi \circ D_A$.) This decomposition can be continued until f_{j+1} is decomposed into a coarsest component f_0 and j difference components g_m, $m = 1, \ldots, j$:

$$f_{j+1} = f_0 + g_1 + \ldots + g_j. \tag{4}$$

This *decomposition algorithm* can be reversed to give a *reconstruction algorithm*: Given the coarse components together with the fine structure components one reconstructs any $f_j \in V_j$ via reversal of (4). Note that both algorithms are usually applied to the expansion coefficients (in terms of the underlying basis) of f and g and that they involve the matrices $P(k)$ and $Q(k)$. More precisely, the decomposition algorithm applied to $f \in V_j$ gives

$$\begin{aligned} f_{j+1} &= \sum_k \langle f, \Phi_{j+1,k}\rangle\, \Phi_{j+1,k} \\ &= \sum_k \langle f, \Phi_{jk}\rangle\, \Phi_{jk} + \sum_k \langle f, \Psi_{jk} \langle\, \Psi_{jk}, \end{aligned} \tag{5}$$

where the inner products $\langle f\Phi_{j+1,k}\rangle$, $\langle f, \Phi_{jk}\rangle$, and $\langle f, \Psi_{jk}\rangle$ are related via

$$\langle f, \Phi_{jk}\rangle = \sum_m \langle f, \Phi_{j+1,k}\rangle\, P^T(m-2k) \tag{6}$$

and

$$\langle f, \Psi_{jk}\rangle = \sum_m \langle f, \Phi_{j+1,k}\rangle\, Q^T(m-2k). \tag{7}$$

Conversely, the reconstruction algorithm applied to a function $f_j \in V_j$, $f_j = \sum_k \langle f_j, \Phi_{jk}\rangle\, \Phi_{jk}$ and $g_j \in W_j$, $g_j = \sum_k \langle f_j, \Phi_{jk}\rangle\, \Psi_{jk}$ yields

$$\langle f_{j+1}, \Phi_{j+1,k}\rangle = \sum_m \langle f_j, \Phi_{jm}\rangle P(k-2m) + \langle f_j, \Psi_{jm}\rangle Q(k-2m). \tag{8}$$

Introducing the column vectors $\boldsymbol{c}_{jk} := \langle f, \Phi_{jk}\rangle^T$ and $\boldsymbol{d}_{jk} := \langle f, \Psi_{jk}\rangle^T$, one can write (5) in the form

$$f_{j+1} = \sum_k \boldsymbol{c}_{jk}^T\, \Phi_{jk} + \boldsymbol{d}_{jk}^T\, \Psi_{jk},$$

and (6) and (7) as

$$\boldsymbol{c}_{jk} = \sum_m P(m-2k)\, \boldsymbol{c}_{j+1,m} \quad \text{and} \quad \boldsymbol{d}_{jk} = \sum_m Q(m-2k)\, \boldsymbol{c}_{j+1,m},$$

while (8) is given by

$$\boldsymbol{c}_{j+1,k} = \sum_m P^T(k-2m)\,\boldsymbol{c}_{jm} + Q^T(k-2m)\,\boldsymbol{d}_{jm}.$$

Introducing the column vectors $\boldsymbol{C}_j := (\boldsymbol{c}_{jk})$ and $\boldsymbol{D}_j := (\boldsymbol{d}_{jk})$, the decomposition and reconstruction algorithm may be schematically presented as follows.

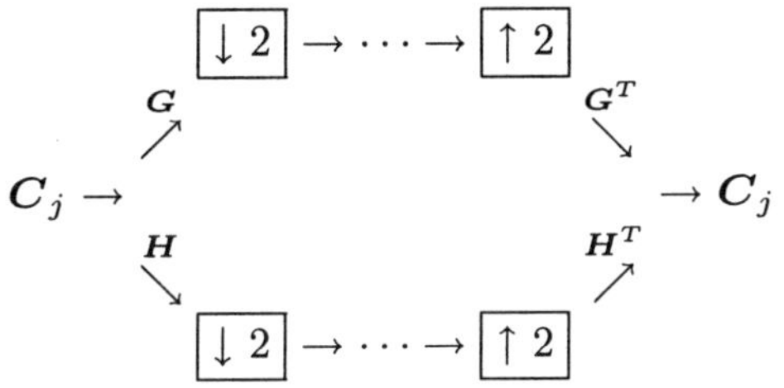

where $\boldsymbol{G}$ and $\boldsymbol{H}$ are sparse TOEPLITZ matrices with matrix entries (P and Q, respectively). One commonly refers to the matrices $\boldsymbol{G}$ and $\boldsymbol{H}$ as a *low pass* and *high pass filter*, respectively. The *downsampling operator* $\downarrow$ uses only the even indices $(2m)$ at level $j+1$ to obtain the coefficients at level j. The *upsampling operator* $\uparrow$ inserts zero between consecutive indices at level j before $\boldsymbol{G}$ and $\boldsymbol{H}$ are applied to obtain the coefficients at level $j+1$. As a consequence of the decomposition algorithm, any function $f \in L^2(\mathbb{R}^n)$ may be represented as a *multiwavelet series* of the form

$$\begin{aligned} f &= \sum_{k\in\mathbb{Z}^n}\sum_{i=1}^{r} \langle f, \phi_{i,jk}\rangle\, \phi_{i,jk} + \sum_{m\geq j}\sum_{k\in\mathbb{Z}^n}\sum_{i=1}^{r} \langle f, \psi_{i,mk}\rangle\, \psi_{i,mk} \\ &= \mathcal{P}^j[f] + \sum_{m\geq j} \mathcal{Q}^m[f], \end{aligned} \tag{9}$$

where $\mathcal{P}^j$ and $\mathcal{Q}^j$ denote the orthogonal projectors of $L^2(\mathbb{R})$ onto V_j and W_j, $j \in \mathbb{Z}$, respectively.

2.5. *Stability of projectors*

It is known that the projectors $\mathcal{P}^j$ and $\mathcal{Q}^j$ are uniformly bounded and uniformly ℓ^p-stable in the following sense. (Cf. for instance Ref. 24.)

Proposition 2.1. *Assume that $\Phi, \Psi \in L^p(\mathbb{R}^n)^r$, $r \in \mathbb{N}$, $p \in [1,\infty]$, are compactly supported. Then, for any $f \in L^p(\mathbb{R}^n)$,*

$$\|\mathcal{P}^j f\|_{L^p(\mathbb{R}^n)} \lesssim \|f\|_{L^p(\mathbb{R}^n)} \qquad \textit{and} \qquad \|\mathcal{Q}^j f\|_{L^p(\mathbb{R}^n)} \lesssim \|f\|_{L^p(\mathbb{R}^n)}.$$

In addition,

$$\left\| \sum_{k\in\mathbb{Z}^n} c(k)\,\Phi_{jk} \right\|_{L^p(\mathbb{R}^n)} \sim 2^{nj(\frac{1}{2}-\frac{1}{p})} \|\{c(k)\}\|_{\ell^p(\mathbb{R}^{r\times r})}.$$

Here $A \lesssim B$ and $A \gtrsim B$ means $A \leq C_1\, B$ and $A \geq C_2\, B$, respectively, for constants C_1 and C_2 not depending on any of the variables or parameters appearing in the expressions for A and B. Note that the value of the constants may change from context to context. $A \sim B$ stands for $A \lesssim B$ and $A \gtrsim B$.

Remark 2.1.

- The above results holds for any $A \in \mathrm{GL}(n, \mathbb{Z})$ whose eigenvalues have modulus greater than one.
- ℓ^p stability implies that the mapping

$$L^p(\mathbb{R}^n) \ni \sum_k c(k)\,\Phi_{jk} \mapsto \{c(k)\}_{k\in\mathbb{Z}^n} \in \ell^p(\mathbb{Z}^n)$$

  is an isomorphism.

2.6. *Approximation order and smoothness*

For approximation-theoretic purposes, the spaces V_j are usually required to reproduce polynomials up to a certain degree $D-1$, i.e., $\Pi_d \subset V_0 = S[\Phi]$, where Π_d denotes the space of real-valued polynomials of degree $d-1$ or order d. As the multiwavelet space W_0 is orthogonal to V_0, $\Pi_d \perp W_0$:

$$\langle (\,\cdot\,)^p, \Psi \rangle = \int_{\mathbb{R}^n} x^p \Psi(x)\, dx = 0, \qquad p = 0, \ldots, D-1.$$

Such a multiwavelet system will be called a *multiwavelet system of order D*. For the remainder of this paper, we assume that we always deal with a multiwavelet system of order $D > 0$. Note that if f is a polynomial of degree at most $D-1$, then its representation (9) reduces to $f = \mathcal{P}^j[f]$. In the case $r = 1$ this in particular implies that the span of ϕ contains all polynomials of degree $< D$. For $r > 1$, the span of each individual scaling function ϕ_i may in general not contain all such polynomials. (See Ref. 26,32 for examples and details.)

In general, the projection $\mathcal{P}^j[f]$ is at least as smooth as the most irregular component of the scaling vector Φ. In particular, if ϕ_i is in the Sobolev space $H^\alpha(\mathbb{R})$ for $i = 1, \ldots, r$ then $\mathcal{P}^j[f] \in H^\alpha(\mathbb{R})$ for each $j \in \mathbb{Z}$.

It is well-known that a multiwavelet system of order D satisfies the following JACKSON-type inequality.

Proposition 2.2. *Suppose that $f \in C^n(\mathbb{R})$, $1 \le n \le D$, is compactly supported. Then*

$$\| f - \mathcal{P}^j[f] \|_{L^2} \le C\, 2^{-jn},$$

for a positive constant C independent of j and n.

The exact relationship between the reproduction of polynomials by the integer shifts of Φ and the L^p-approximation order of $\mathcal{P}^j$ is discussed in Ref. 29.

In addition, multiwavelet systems provide a nice characterization of BESOV spaces. To this end, recall that the M-th order difference operator Δ_h^M of step size $h \in \mathbb{R}^n$ is defined by

$$(\Delta_h^M f)(x) := \sum_{m=0}^{M} (-1)^{M-m} \binom{M}{m} f(x + m\,h).$$

Definition 2.1. Let $0 < p, q \le \infty$, let $\sigma_p := \left(\frac{1}{\min\{p,1\}} - 1\right)$, and suppose that $s > \sigma_p$. Suppose $M \in \mathbb{N}$ is such that $M > s \ge M - 1$. Then a function $f \in L^p$ belongs to the BESOV space $B_q^s(L^p(\mathbb{R}^n))$ iff

$$\|f\|_{B_q^s(L^p(\mathbb{R}^n))} := \begin{cases} \|f\|_{L^p} + \left(\int_{\mathbb{R}^n} |h|^{-sq} \|\Delta_h^M f\|_{L^p}^q \frac{dh}{|h|^n} \right)^{\frac{1}{q}} < \infty, & 0 < q < \infty \\ \\ \|f\|_{L^p} + \sup_{0 \ne h \in \mathbb{R}^n} |h|^{-s} \, \|\Delta_h^M f\|_{L^p} < \infty, & q = \infty. \end{cases}$$

Note that $B_q^s(L^p(\mathbb{R}^n))$ is a BANACH space for $1 \le p, q \le \infty$; otherwise a quasi-BANACH space.

Theorem 2.1. *Assume that $A \in GL(n, \mathbb{Z})$ is similar to a diagonal matrix $diag(\rho_1, \ldots, \rho_n)$ with $|\rho_1| = \ldots = |\rho_n| =: \varrho$. Furthermore, assume that the multiwavelet system (Φ, Ψ) is compactly supported and in $C^{M-1}(\mathbb{R}^n) \times C^{M-1}(\mathbb{R}^n)$. Then,*

$$f \in B_q^s(L^p(\mathbb{R}^n)) \iff \left(\sum_{k \in \mathbb{Z}^n} |\langle f, \Phi(\cdot - k) \rangle|^p \right)^{\frac{1}{p}}$$

$$+ \left(\sum_{j \in \mathbb{Z}_0^+} \left(\varrho^{js} \,|\det A|^{j(1/2 - 1/p)} \,\|\langle f, \Psi_{j,(\cdot)} \rangle\|_{\ell^p} \right)^q \right)^{1/q} < \infty.$$

(Usual modifications when $p = \infty$ and $q = \infty$.)

An application that makes use of the scaling behavior of wavelet coefficients in BESOV spaces is discussed in Ref. 4.

3. Sampling with Multiwavelets

Representing discretely sampled data in terms of multiwavelets requires special care since there is more than one generator for the spaces V_j. Here we consider the case $n = 1$ and $A = 2Ix$. Suppose that $\boldsymbol{f} \in \ell^2(\mathbb{Z})$ is a discrete scalar signal representing the samples of a function $f \in L^2(\mathbb{R})$, and that the resolution of the samples is such that one has a representation of the form $f = \sum_k \boldsymbol{c}_k^T \Phi_{jk}$. Next, we discuss how the samples in $\boldsymbol{f}$ are assigned to the coefficients $\boldsymbol{c}$. For this purpose, we consider the *polyphase form* $\boldsymbol{F} \in (\ell^2(\mathbb{Z}))^r$ of $\boldsymbol{f}$ defined by

$$\boldsymbol{F}(k) := \begin{pmatrix} \boldsymbol{f}(rk) \\ \vdots \\ \boldsymbol{f}(rk + r - 1) \end{pmatrix},$$

where $\boldsymbol{f}(i)$ denotes the ith component of $\boldsymbol{f} \in \ell^2(\mathbb{Z})$. Now define a mapping $\mathfrak{Q} : (\ell^2(\mathbb{Z}))^r \to (\ell^2(\mathbb{Z}))^r$ by $\boldsymbol{c}^T = \mathfrak{Q}(\boldsymbol{F})$. To proceed, the following result is needed. (For a proof see, for instance,[16])

Theorem 3.1. *Suppose $\mathfrak{L} : \ell^2(\mathbb{Z}) \to \ell^2(\mathbb{Z})$ is a bounded, shift-invariant linear transformation. Then there exists a $Q \in \ell^2(\mathbb{Z})$ such that*

$$\mathfrak{L}(\zeta) = Q * \zeta, \qquad \forall \zeta \in \ell^2(\mathbb{Z}).$$

Here $* : \ell^2(\mathbb{Z}) \times \ell^2(\mathbb{Z}) \to \ell^2(\mathbb{Z})$ denotes the *convolution operator* defined by:

$$\{\zeta(\nu)\} * \{\mathfrak{Q}(\nu)\} := \left\{ \sum_\mu \zeta(\mu)\mathfrak{Q}(\nu - \mu) \right\}_{\nu=-\infty}^{+\infty}.$$

Thus, if $\mathfrak{Q}$ is a bounded linear shift-invariant transformation with an inverse $\mathfrak{Q}^{-1}$ satisfying the same conditions, then both can be represented as a convolution:

$$\mathfrak{Q}(\zeta) = Q * \zeta, \qquad \text{and} \qquad \mathfrak{Q}^{-1}(\zeta) = \widetilde{q} * \zeta$$

where the sequences of $r \times r$-matrices Q and $\widetilde{Q}$ are called a *prefilter for* Φ and *postfilter for* Φ, respectively.

In order to exploit the full power of filter banks, the filters Q and $\widetilde{Q}$ should be orthogonal (preserving the L^2-norm or energy of the signal) and preserve the approximation order D of the multiwavelet system. In Ref. 22 such pre- and postfilters are constructed and applied to image compression. The construction of multiwavelet filters and the design for optimal orthogonal prefilters can be found in Refs. 1,23, respectively.

3.1. *The GHM scaling vector and DGHM multiwavelet*

Next we consider a special scaling vector and associated multiwavelet that is being used later in this paper. This so-called *GHM scaling vector* and *DGHM multiwavelet* were first introduced in Refs. 15,17 and later in Ref. 28. This particular multiwavelet system was the first example exhibiting wavelets that are compactly supported, continuous, orthogonal, and possess symmetry. Both the scaling vector and the multiwavelet are two-component vector functions $\Phi = (\phi_1, \phi_2)^T$ and $\Psi = (\psi_1, \psi_2)^T$ with the following properties.

- $\operatorname{supp} \phi_1 = [0, 1]$ and $\operatorname{supp} \phi_2 = \operatorname{supp} \psi_1 = \operatorname{supp} \psi_2 = [-1, 1]$.
- The scaling vector Φ and the associated multiwavelet Ψ satisfy (3).
- The wavelets ψ_1 and ψ_2 are antisymmetric and symmetric, respectively.
- The multiwavelet system (Φ, Ψ) is of order $D = 2$, i.e., has approximation order two: $\Pi^2 \subset S[\Phi]$ and $\langle (\,\cdot\,)^p, \Psi \rangle = 0$, $p = 0, 1$.
- $\Phi, \Psi \in C^{0,1}(\mathbb{R}) \times C^{0,1}(\mathbb{R})$. Hence all four component functions possess a weak first derivative.
- The GHM scaling vector is interpolatory: Given a set of interpolation points $Z := \{Z_i\}$ supported on $\frac{1}{2}\mathbb{Z}$, there exists a set of vector coefficients $\{\boldsymbol{\alpha}_k\}$ such that $\sum_k \boldsymbol{\alpha}_k^T \Phi(x-k)$ interpolates Z. (Note that $\phi_1(1/2) = 1 = \phi_2(0)$.)
- The DGHM multiwavelet system can be easily modified to obtain a multiresolution analysis on $L^2[0, 1]$ *without* the addition of boundary functions.
- The length of support of Φ and Ψ, $\operatorname{supp} \Phi = 3$, and the approximation order are the same as that of the Daubechies ${}_2\phi$ scaling function and ${}_2\psi$ wavelet, but the GHM scaling vector and DGHM multiwavelet have slightly higher regularity. It turns out that the Daubechies wavelet system $({}_2\phi, {}_2\psi)$ and the (GHM,DGHM) wavelet system are the only two having with approximation order two and local dimension 3.[19]

Figure 1 shows the graphs of the GHM scaling vector and the DGHM multiwavelet.

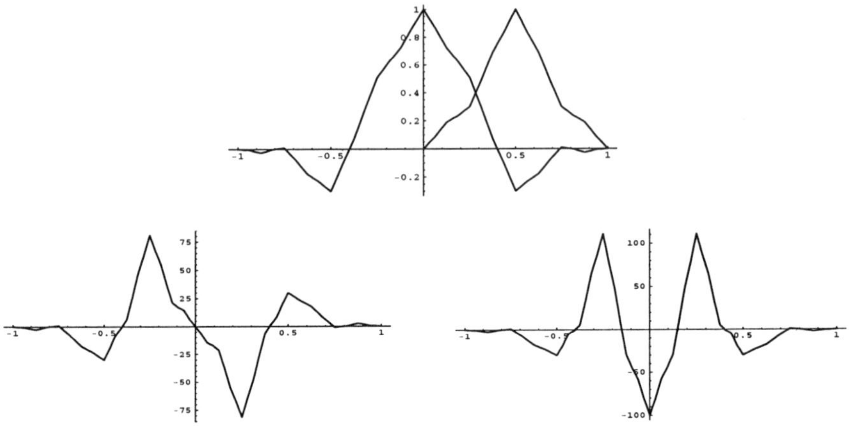

Fig. 1. The orthogonal GHM scaling vector (top) and the orthogonal DGHM multiwavelet: ψ_1 (bottom left) and ψ_2 (bottom right).

4. Multiwavelets on the Interval

It is possible to obtain a multiresolution analysis on an interval by modifying the DGHM multiwavelet system. The process involved in obtaining bases on say $[0,1]$ without introducing additional boundary functions, as is the case for other wavelet constructions, only has to make use of the fact that the GHM scaling vector and the associated DGHM multiwavelet are *piecewise fractal functions.*[15,17,28] The main idea is as follows. At any given level of approximation $j \geq 0$, take as a basis the restrictions to $[0,1]$ of all the translates of ϕ_1 and ϕ_2, respectively, ψ_1 and ψ_2 at level j whose support has nonempty intersection with the open interval $(0,1)$. More precisely, if

$$\phi^*_{i,jk} := \phi_{i,jk}\big|_{[0,1]} \qquad \text{and} \qquad \psi^*_{i,jk} := \psi_{i,jk}\big|_{[0,1]},$$

then the following, easily verified, theorem holds.[15,28]

Theorem 4.1. *For all $j \in \mathbb{Z}_0^+$, the set $\mathcal{B}^*_{\Phi,j} := \{\phi^*_{i,jk} : i = 1,2;\, k = 0,1,\ldots,2^j - 2 + i\}$ is an orthonormal basis for $V_j^* := V_j \cap L^2[0,1]$ and $\mathcal{B}^*_{\Psi,j} := \{\psi^*_{i,jk} : i = 1,2;\, k = i-1,\ldots,2^j+1-i\}$ constitutes an orthonormal bases for $W_j^* := W_j \cap L^2[0,1]$. Moreover, card$\mathcal{B}^*_{\Phi,j} = 2^{j+1}+1$, card$\mathcal{B}^*_{\Psi,j} = 2^{j+1}$ and $L^2[0,1] = V_0^* \bigcup_{j\geq 0} W_j^*$.*

We remark that the elements in V_j^* provide interpolation on the lattice $2^{-(j+1)}\mathbb{Z}$: The scaling function $\phi_{2,jk}$ interpolates at $2^{-j}\mathbb{Z}$, whereas the function $\phi_{1,jk}$ interpolates in-between, i.e., on $2^{-(j+1)}\mathbb{Z}$.

The construction on the interval was generalized to triangulations in $\mathbb{R}^n$ in Refs. 14,20. The interested reader is referred to these publications and the references given therein.

4.1. *Function sampling on* $[0, 1]$

In many applications one deals with a finite amount of data that needs to be analyzed or stored in a buffer for later retrieval. In order to employ a multiscale decomposition of the type introduced above, one chooses a finest level of approximation, say $J > 0$, and takes $2^{J+1}+1$ data points or samples. (This is the number of GHM scaling functions on $[0, 1]$ at level $J > 0$ with data supported on $2^{-(J+1)}\mathbb{Z}$.) In this case, we denote the collection of samples by $\boldsymbol{f}_J = (f_i)_{i=0}^{2^{J+1}}$. Using the elements in $\mathcal{B}^*_{\Phi,j}$, which we express in the form

$$\boldsymbol{\phi}^* := (\phi^*_{2,J,0}, \phi^*_{1,J,0}, \phi^*_{2,J,1}, \ldots, \phi^*_{1,J,2^J}, \phi^*_{2,J,2^J+1})^T,$$

we need to assign a data vector $\boldsymbol{c}_J$ to this collection of samples. This is done via the polyphase representation applied now to the case $r = 2$. In Ref. 22, orthogonal pre– and post filters that preserve the approximation order $D = 2$ of the DGHM multiwavelet system were constructed. Employing these filters yields the required assignment $\boldsymbol{f}_J \mapsto \boldsymbol{c}_J$.

Applying the decomposition and reconstruction algorithm to a finite set of data such as $\boldsymbol{c}_J$ is now straightforward. The length of the data vector $\boldsymbol{c}_J$ equals $2^{J+1} + 1$ and application of the matrices $\boldsymbol{G}$ and $\boldsymbol{H}$, followed by downsampling $\downarrow 2$, produces two data vectors $\boldsymbol{c}_{J-1}$ and $\boldsymbol{d}_{J-1}$ of length $2^J + 1$ and 2^J, respectively. The data vector $\boldsymbol{c}_{J-1}$ may be regarded as a weighted average with respect to the filter coefficients in $\boldsymbol{G}$ of the original data vector $\boldsymbol{c}_J$, and the vector $\boldsymbol{d}_{J-1}$ carries the information that was lost in the averaging procedure $\boldsymbol{c}_J \mapsto \boldsymbol{c}_{J-1}$. Thus, the data vector $\boldsymbol{d}_{J-1}$ contains the detail or fine structure of the original data $\boldsymbol{f}$. The data vector $\boldsymbol{c}_{J-1}$ may further be decomposed according to the scheme

$$\begin{array}{ccccccc} \boldsymbol{c}_J & \rightarrow & \boldsymbol{c}_{J-1} & \rightarrow & \cdots & \rightarrow & \boldsymbol{c}_L \\ & \searrow & & \searrow & \cdots & \searrow & \\ & & \boldsymbol{d}_{J-1} & & & & \boldsymbol{d}_L \end{array}$$

The mapping $\mathcal{W} : \ell^2(\mathbb{Z}) \to \ell^2(\mathbb{Z})$, $\mathcal{W}(\boldsymbol{c}_J) := (\boldsymbol{d}_{J-1}, \boldsymbol{c}_{J-1})$, is called the discrete *(multi)wavelet transform*. Repetitively applying $\mathcal{W}$ until a coarsest level $L < J$ is reached, yields a multiscale representation of the original data vector $\boldsymbol{c}_J$ in the form

$$\boldsymbol{c}_J = (\boldsymbol{d}_{J-1}, \boldsymbol{d}_{J-2}, \ldots, \boldsymbol{d}_L, \boldsymbol{c}_L)$$

where the lengths of the multiscale components are $(2^J, 2^{J-1}, \ldots, 2^{L+1}, 2^{L+1}+1)$.

Reconstruction proceeds according to the scheme

$$\begin{array}{ccccccc} \boldsymbol{c}_L & \rightarrow & \boldsymbol{c}_{L+1} & \rightarrow \cdots \rightarrow & \boldsymbol{c}_{J-1} & \rightarrow & \boldsymbol{c}_J \\ & \nearrow & & \nearrow \cdots & & \nearrow & \\ \boldsymbol{d}_L & & \boldsymbol{d}_{L+1} & \cdots & \boldsymbol{d}_{J-1} & & \end{array}$$

Note that for the reconstruction, the data vectors $\boldsymbol{c}_j$ and $\boldsymbol{d}_j$ need to be upsampled, $\uparrow 2$, in order to generate $\boldsymbol{c}_{j+1}$, $L \le j < J$.

5. Translation Invariance

Orthogonal wavelet and multiwavelet transform lack translation invariance. In the case $r = 1$, this lack is overcome by considering all continuous shifts of the orthogonal wavelet transform. This naturally leads to so-called redundant representations of $L^2(\mathbb{R})$ functions.[6,9] Here we extend the approach presented in Ref. 6 to the case $r > 1$.

For $h \in \mathbb{R}$, denote the continuous shift operator (by h) on $L^2(\mathbb{R})$ by $\mathcal{S}_h$. Associated with $\mathcal{S}_h$, introduce *redundant projectors* for functions $f \in L^2(\mathbb{R})$ by

$$\mathcal{P}^j_R[f] := 2^j \int_0^{2^{-j}} \mathcal{S}_{-h}\mathcal{P}^j\mathcal{S}_h[f]\,dh \quad \text{and} \quad \mathcal{Q}^j_R[f] := 2^j \int_0^{2^{-j}} \mathcal{S}_{-h}\mathcal{Q}^j\mathcal{S}_h[f]\,dh,$$

where $\mathcal{P}^j$ and $\mathcal{Q}^j$ are the orthogonal projectors defined in (9). The following result is shown in Ref. 5.

Proposition 5.1. *The redundant projectors $\mathcal{P}^j_R$ and $\mathcal{Q}^j_R$, $j \in \mathbb{Z}$, are translation invariant, i.e.,*

$$\mathcal{P}^j_R[f(\cdot+\delta)](y) = \mathcal{P}^j_R[f](y+\delta) \quad \textit{and} \quad \mathcal{Q}^j_R[f(\cdot+\delta)](y) = \mathcal{Q}^j_R[f](y+\delta),$$

for all $\delta \in \mathbb{R}$, and yield the following representation of a function $f \in L^2(\mathbb{R})$:

$$\begin{aligned} f &= 2^j \sum_{i=1}^r \int_{\mathbb{R}} \varrho_{ii}(2^j(x-\cdot))f(x)dx + \sum_{m\ge j}\sum_{i=1}^r 2^m \int_{\mathbb{R}} \tau_{ii}(2^m(x-\cdot))f(x)dx \\ &= \mathcal{P}^j_R[f] + \sum_{m\ge j} \mathcal{Q}^m_R[f], \end{aligned} \tag{10}$$

where ϱ_{ii} and τ_{ii} are the autocorrelation functions of the components of the scaling vector Φ and the multiwavelet Ψ:

$$\varrho_{ii}(x) := \int_{\mathbb{R}} \phi^i(y)\phi^i(y+x)\,dx \quad \textit{and} \quad \tau_{ii}(x) := \int_{\mathbb{R}} \psi^i(y)\psi^i(y+x)\,dx.$$

It was observed in Ref. 5 that in order to obtain a refinement equation for the autocorrelation functions ϱ_{ii} and τ_{ii}, the cross-correlation functions ϱ_{ij} and τ_{ij} are needed, although they do not explicitly appear in (10). Following the terminology introduced in Ref. 5, the above representation (10) is termed a *autocorrelation transformation* or a *hidden basis multiwavelet representation.*

5.1. *Matrix-valued refinement*

In the case $r = 1$, the autocorrelation functions satisfy a refinement equation where the filter coefficients are the so-called *à-trous filters.*[3,21,31,33] If $r > 1$, these refinement equations become matrix-valued as shown below. (Cf. Ref. 5 for proofs.)

Theorem 5.1. *Let*

$$\Theta(x) := \int_{\mathbb{R}} \Phi(y)\Phi^T(y+x)\,dy$$

be the $r \times r$ correlation matrix of the scaling vector Φ. The elements of Θ are the correlation functions of the component functions of Φ, i.e.,

$$\varrho_{ij}(x) := \int_{\mathbb{R}} \phi^i(y)\phi^j(y+x)\,dy, \quad i,j = 1,\ldots,r.$$

Then Θ satisfies a matrix-valued refinement equation *of the form*

$$\Theta(x) = \frac{1}{2}\sum_{k\in\mathbb{Z}}\sum_{\ell\in\mathbb{Z}} P(k+\ell)\,\Theta(2x-\ell)\,P^T(k). \tag{11}$$

The lack of commutativity in the algebra of matrices requires the following approach to express (11) in the usual form (1). Regard the $r \times r$ matrix Θ as a column vector Γ of length r^2:

$$\Gamma = \left(\varrho_{11}\ \cdots\ \varrho_{1r}\ \cdots\ \varrho_{r1}\ \cdots\ \varrho_{rr}\right)^T$$

and define an operator $\mathcal{T} : \ell^2(\mathbb{Z}^{r\times r}) \to \ell^2(\mathbb{Z}^{r\times r})$ by

$$(\mathcal{T}\Upsilon)(x) = \frac{1}{2}\sum_{k\in\mathbb{Z}}\sum_{\ell\in\mathbb{Z}} P(k+\ell)\,\Upsilon(2x-\ell)\,P^T(k).$$

Then there exists a finite sequence of $r \times r$ matrices $\{A(k)\}$ such that

$$\Gamma(x) = \sum_k A(k)\,\Gamma(2x-k). \tag{12}$$

Analog to the definition of Θ, one defines a correlation matrix associated with the multiwavelet Ψ by

$$\Xi(x) := \int_{\mathbb{R}} \Psi(y)\Psi^T(y+x)\, dy$$

satisfying a refinement equation of the form

$$\Xi(x) = \frac{1}{2} \sum_{k\in\mathbb{Z}} \sum_{\ell\in\mathbb{Z}} Q(k+\ell)\, \Xi(2x-\ell)\, Q^T(k),$$

which can also be rewritten in the form (12). The pair (Θ, Ξ) is called a *translation invariant multiwavelet system.*

An important feature of autocorrelation functions $\alpha_\phi(s) = \langle \phi, \phi(\cdot + s)\rangle$ in classical wavelet theory is their interpolation property as exhibited in Refs. 12,13. This property is equivalent to $\alpha_\phi(n) = \delta_{0n}$, $n \in \mathbb{Z}$, which implies that the function values of α_ϕ can be computed exactly at the dyadic rationals using the refinement equation for α_ϕ. This interpolation property also holds for $r > 1$, as was shown in Ref. 5.

Proposition 5.2. *The correlation matrix $\Theta(x)$ is skew-symmetric in x and interpolatory, i.e.,*

$$\begin{aligned} \Theta_{ij}(x) &= \Theta_{ji}(-x), \\ \Theta(n) &= \delta_{0n} I_r, \qquad n \in \mathbb{Z}. \end{aligned}$$

It should be pointed out that the elements ϱ_{ij} of Θ are in general not interpolatory.

5.2. *Regularity and moments*

The regularity properties of wavelet systems and their ability to reproduce polynomials are fundamental to many applications such as compression and denoising. The regularity of Φ depends on the decay rate of the infinite product

$$\widehat{\mathfrak{P}}(u) = \lim_{n\to\infty} \left[\boldsymbol{P}\left(\frac{u}{2}\right) \cdots \boldsymbol{P}\left(\frac{u}{2^n}\right) \widehat{\Phi}(0)\right],$$

where $\boldsymbol{P}(u) := \frac{1}{2}\sum_k P(k)\, e^{-iuk}$ denotes the symbol of $\{P(k)\}$. In Ref. 10 it is shown that the above limit exists and that the finite product $\Pi_n(u)\widehat{\Phi}(0) = \boldsymbol{P}\left(\frac{u}{2}\right) \cdots \boldsymbol{P}\left(\frac{u}{2^n}\right) \widehat{\Phi}(0)$ converges pointwise for all $u \in \mathbb{R}$ and uniformly on compact sets. In particular, the following theorem holds.

Theorem 5.2. *Let $\boldsymbol{P}$ be an $r \times r$ matrix of the form*

$$\boldsymbol{P}(u) = \frac{1}{2^m} C_0(2u) \cdots C_{m-1}(2u)\, \boldsymbol{P}^{(m)}(u)\, C_{m-1}(u)^{-1} \cdots C_0(u)^{-1},$$

where the C_i are certain $r \times r$ matrices and $\boldsymbol{P}^{(m)}$ is an $r \times r$ matrix with trigonometric polynomials as entries. Suppose that the spectral radius of $\boldsymbol{P}^{(m)}(0)$ is strictly less than two. For $k \geq 1$, let

$$\gamma_k = \frac{1}{k} \log_2 \sup_u \left\| \boldsymbol{P}^{(m)}\left(\frac{u}{2}\right) \cdots \boldsymbol{P}^{(m)}\left(\frac{u}{2^k}\right) \right\|.$$

Then there exists a positive constant C such that for all $u \in \mathbb{R}$

$$\|\widehat{\mathfrak{P}}(u)\| \leq C\,(1 + |u|)^{-m+\gamma_k}.$$

As in the case $r = 1$ the rate of decay m determines the smoothness of the components of Φ in terms of SOBOLEV norms. Details and the precise matrix factorization are found in Refs. 10,29.

The above results can be generalized to obtain estimates on the regularity of translation invariant multiwavelet systems.[5] To this end, note that the Fourier transform of Eq. (11) is given by

$$\widehat{\Theta}(u) = \boldsymbol{P}\left(\frac{u}{2}\right) \widehat{\Theta}\left(\frac{u}{2}\right) \boldsymbol{P}\left(\frac{u}{2}\right)^T.$$

Lemma 5.1. *Suppose that $\boldsymbol{P}$ satisfies $\|\boldsymbol{P}(u) - \boldsymbol{P}(0)\| \leq C\,|u|^\alpha$ for some $\alpha > 0$ and that $\|\boldsymbol{P}(0)\| < 2^\alpha$. Then the infinite product*

$$\widehat{\Omega}(u) = \lim_{n\to\infty} \left[\boldsymbol{P}\left(\frac{u}{2}\right) \cdots \boldsymbol{P}\left(\frac{u}{2^n}\right)\right] \widehat{\Theta}(0) \left[\boldsymbol{P}\left(\frac{u}{2}\right) \cdots \boldsymbol{P}\left(\frac{u}{2^n}\right)\right]^T$$

converges pointwise for all $u \in \mathbb{R}$ and uniformly on compact subsets.

For the proof of this lemma and the next theorem, we again refer the reader to.[5]

Theorem 5.3. *Let $\boldsymbol{P}$ be an $r \times r$ matrix of the form*

$$\boldsymbol{P}(u) = \frac{1}{2^m} C_0(2u) \cdots C_{m-1}(2u)\, \boldsymbol{P}^{(m)}(u)\, C_{m-1}(u)^{-1} \cdots C_0(u)^{-1},$$

where the C_i are certain $r \times r$ matrices and $\boldsymbol{P}^{(m)}$ is an $r \times r$ matrix with trigonometric polynomials as entries. Suppose that the spectral radius of $\boldsymbol{P}^{(m)}(0)$ is strictly less than two. For $k \geq 1$, let

$$\gamma_k = \frac{1}{k} \log_2 \sup_u \left\| \boldsymbol{P}^{(m)}\left(\frac{u}{2}\right) \cdots \boldsymbol{P}^{(m)}\left(\frac{u}{2^k}\right) \right\|.$$

Then there exists a positive constant C such that for all $u \in \mathbb{R}$

$$\|\widehat{\Omega}(u)\| \leq C\,(1 + |u|)^{2(-m+\gamma_k)}.$$

Note that in complete analogy to the case $r = 1$, the decay rate of $\widehat{\Omega}$ is increased by a factor of 2 and thus also the approximation order. This doubling follows readily from the form of the Fourier transform as a product $\widehat{\varrho}_{ii}(u) = \widehat{\phi}_i(u)\overline{\widehat{\phi}_i(u)}$. As the correlation functions τ_{ii} have an analogous product representation in terms of the wavelet functions ψ_i in the Fourier domain, they have vanishing moments up to order $2D$. This then implies that the autocorrelation transformation (10) has approximation order $2D$. Since the cross-correlation functions τ_{ij}, $i \neq j$, are products of the form $\widehat{\psi}_i\overline{\widehat{\psi}_j}$, their moments up to order $2D$ also vanish. Hence, it is said that the translation invariant multiwavelet system (Θ, Ξ) has approximation order $2D$.

It is worthwhile mentioning that the above results hold for semiorthogonal, biorthogonal, and oblique wavelet systems as long as the approximation spaces V_j contain polynomials and are orthogonal to the wavelet spaces W_j.

5.3. *Coiflet property*

In the case $r = 1$, the translation invariant wavelet system has the *coiflet property of order* $2D$. That is, if $\tau(x) := \int_{\mathbb{R}} \psi(y)\psi(y+x)\,dx$ and $\varrho(x) := \int_{\mathbb{R}} \phi(y)\phi(y+x)\,dx$ then for $p = 0, 1, \ldots, 2D-1$ and $q = 1, \ldots, 2D-1$

$$\langle (\,\cdot\,)^p, \tau\rangle = 0 \qquad \text{and} \qquad \langle (\,\cdot\,)^q, \varrho\rangle = 0.$$

This property also holds in the case $r > 1$ for the functions ϱ_{ii}, but *not necessarily* for the cross-correlation functions ϱ_{ij}. Since $\widehat{\Theta}(u) = \widehat{\Phi}(u)\widehat{\Phi}^T(-u)$, the translation invariant multiwavelet system (Θ, Ξ) has a *multicoiflet property of order* $2D$, only if the off-diagonal terms in the matrix

$$\sum_{\nu=0}^{n} \binom{n}{\nu} \widehat{\Phi}^{(n-\nu)}(0)\widehat{\Phi}^{(\nu)T}(0), \qquad n = 1, \ldots, 2D-1,$$

vanish identically. For example, for $n = 1$ and $r = 2$, the scaling functions ϕ_1 and ϕ_2 must satisfy

$$\widehat{\phi}_1{}'(0)\widehat{\phi}_2(0) - \widehat{\phi}_2{}'(0)\widehat{\phi}_1(0) = 0. \tag{13}$$

Similar to orthogonal multiwavelet systems, the smoothness of the projection $\mathcal{P}_R^j[f]$ is, in general, determined by the smoothness of the functions ϱ_{ii}. For instance, if $\phi_i \in H^\alpha(\mathbb{R})$ for $i = 1, \ldots, r$ then $\varrho_{ii} \in H^{2\alpha}(\mathbb{R})$. As the redundant projection $\mathcal{P}_R^j[f]$ is a sum of convolutions, a result in Ref. 27 shows that $\mathcal{P}_R^j[f]$ is in the Sobolev space $H^{2\alpha+\beta}(\mathbb{R})$ whenever $f \in H^\beta(\mathbb{R})$.

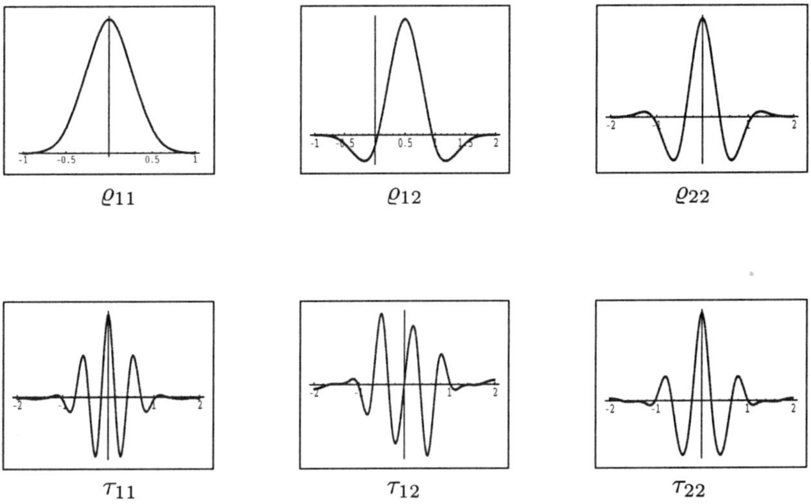

Fig. 2. The correlation functions for the DGHM multiwavelet

5.4. *An example: The DGHM multiwavelet*

The DGHM multiwavelet system has approximation order 2 and $\Phi, \Psi \in C^{0,1}(\mathbb{R})$ which implies that $\mathcal{P}^j[f] \in C^{0,1}(\mathbb{R})$. The calculation of the regularity based on the estimates given in Theorem (5.2) is done in Ref. 10 and yields $m = 2$ and $\gamma_k < 1$, for k large enough. Moreover, the individual wavelet functions have vanishing moments up to order 2.

The translation invariant multiwavelet system (Θ, Ξ) associated with the DGHM multiwavlet contains the eight functions ϱ_{ij} and τ_{ij}, $i, j = 1, 2$, where $\varrho_{12}(x) = \varrho_{21}(-x)$ and $\tau_{12}(x) = \tau_{21}(-x)$. The graphs of these functions are depicted in Fig. 2.

Employing the results stated in the previous section, it follows that Θ has approximation order four and that Ξ has vanishing moments up to order four. Moreover, the first to third moments of Θ vanish. The system (Θ, Ξ) does not have the multicoiflet property of order $2D$, since, e.g., condition (13) is not satisfied. The elements of Θ are in $C^{1,1}(\mathbb{R})$ and as a consequence, the redundant projections $\mathcal{P}^j_R[f]$ are in the Sobolev space $H^{2+\beta}(\mathbb{R})$ whenever $f \in H^\beta(\mathbb{R})$.

References

1. K. Attakitmongcol, D. Hardin and D. Wilkes, *IEEE Trans. Image Proc.* **10**, 1476 (2001).

2. M. F. Barnsley, J. Elton, D. P. Hardin, and P. R. Massopust, *SIAM J. Math. Anal.* **20**, 1218 (1989).
3. G. Beylkin, *SIAM J. Numer. Anal.* **6**, 1716 (1992).
4. K. Berkner, M. Gormish, and E. Schwartz, *Appl. Comp. Harm. Anal.* **11**, 2 (2001).
5. K. Berkner and P. Massopust, *Technical Report CML TR 98-06* (Rice University, 1998).
6. K. Berkner and R. O. Wells, Jr., *Technical Report CML TR 98-01* (Rice University, 1998).
7. C. S. Burrus, R. A. Gopinath, and H. Guo, *Introduction to Wavelets and Wavelet Transforms* (Prentice Hall, Englewood Cliffs, HJ, 1998).
8. C. Chui, *An Introduction to Wavelets* (Academic Press, San Diego, 1992).
9. R. R. Coifman and D. L. Donoho, *Translation invariant denoising,* in *Wavelets and Statistics*, ed., A. Antoniades, (Springer Lecture Notes, Springer Verlag, 1995).
10. A. Cohen, I. Daubechies, G. Plonka, *The Journal of Fourier Analysis and Applications*, **3**, 295 (1997).
11. I. Daubechies, *Commun. Pure and Applied Math.* **41**, 909 (1988).
12. I. Daubechies, *Ten Lectures on Wavelets*, SIAM, Vol. 61 (Philadelphia, 1992).
13. G. Deslauriers and S. Dubuc, *Constr. Appr.* **5**, 49 (1989).
14. G. Donovan, J. Geronimo, and D. Hardin, *Constr. App.* **16**, 201 (2000).
15. G. Donavan, J. S. Geronimo, D. P. Hardin, and P. R. Massopust, *SIAM J. Math. Anal.* **27**, 1158 (1996).
16. M. Frazier, *An Introduction to Wavelets Through Linear Algebra* (Springer Verlag, New York, 1999).
17. J. S. Geronimo, D. P. Hardin, and P. R. Massopust, *J. Approx. Th.* **78**, 737 (1994).
18. T. N. T. Goodman and S. L. Lee, *Trans. Amer. Math. Soc.* **342**, 307 (1994).
19. D. Hardin and T. Hogan, *Constructing orthogonal refinable function vectors with prescribed approximation order and smoothness,* in *Wavelet Analysis and Applications, Guangzhou 1999* (2002), pp. 139–148.
20. D. Hardin and D. Hong, *J. Comput. Appl. Math.* **155**, 91 (2003).
21. M. Holschneider, R. Kronland-Martinet, J. Morlet, and P. Tchamitchian, *A real-time algorithm for signal analysis with the help of the wavelet transform,* in *Wavelets: Time-Frequency Methods and Phase Space* (Springer Verlag, Berlin, 1989), pp. 286–297.
22. D. P. Hardin and D. Roach, *IEEE Trans. Circ. and Sys. II: Anal. and Dig. Sign. Proc.* **45**, 1106 (1998).
23. D. P. Hardin X.-G. Xia, J. Geronimo and B. Suter, *IEEE Trans. on Signal Processsing* **44**, 25 (1996).
24. M. Lindemann, *Approximation Properties of Non-Separable Wavelet Bases with Isotropic Scaling Matrices*, PhD Dissertation (University of Bremen, Germany, 2005).
25. M. Lang, H. Guo, J. E. Odegard, C. S. Burrus, and R. O. Wells, Jr., *IEEE Sig. Proc. Letters*, **3**, 10 (1996).
26. J. Lebrun and M. Vetterli, *Higher order balanced multiwavelets* (IEEE

ICASSP, 1998).

27. A. K. Louis, P. Maaß, and A. Rieder, *Wavelets* (Teubner Verlag, Stuttgart, Germany, 1994).
28. P. R. Massopust, *Fractal Functions, Fractal Surfaces, and Wavelets* (Academic Press, San Diego, 1994).
29. G. Plonka, *Constr. Approx.* **13**, 221 (1997).
30. H. L. Resnikoff and R. O. Wells, Jr, *Wavelet Analysis and Scalable Structure of Information* (Springer-Verlag, New York) (to appear).
31. N, Saito and G. Beylkin, *IEEE Trans. Sig. Proc.*, **14**, 3548 (1993).
32. I. Selesnik, *Multiwavelet bases with extra approximation properties,* (Technical Report, Department of Electrical and Computer Engineering, Rice University, 1997).
33. M. J. Shensa, *IEEE Trans. Sig. Proc.* **40**, 3464 (1992).
34. G. Strang and T. Nguyen, *Wavelets and Filter Banks* (Wellesley-Cambridge Press, 1996).
35. V. Strela, P. Heller, G. Strang, P. Topiwala, and C. Heil, *IEEE Trans. on Image Proc.* (to appear).
36. V. Strela and A. T. Walden, *Signal and Image Denoising via Wavelet Thresholding: Orthogonal and Biorthogonal, Scalar and Multiple Wavelet Transforms,* (Preprint 1998).
37. P. Wojtaszczyk, *A Mathematical Introduction to Wavelets* (Cambridge University Press, London, UK, 1997).

Harmonic, Wavelet and p-Adic Analysis
Eds. N. M. Chuong *et al.* (pp. 59–81)

§4. MULTI-SCALE APPROXIMATION SCHEMES IN ELECTRONIC STRUCTURE CALCULATION

REINHOLD SCHNEIDER and TORALF WEBER†

Fakcultat fur Mathematik
Technische Universitat ChemnitzZwickau
D-09009 Chemnitz, Gremany
†*E-mail: twe@numerik.uni-kiel.de*

1. Introduction

The numerical simulation of molecular structures is of growing importance for modern developments in technology and science, like molecular biology and nano-sciences, semiconductor devices etc. On microscopic scales classical mechanics must be replaced by the laws of quantum mechanics. Therefore reliable computational tools should be based on *First Principles* of quantum mechanics for simulating the quantum mechanical phenomena accurately. In these *ab initio* computations, the model equations are derived on the basis of only very few fundamental laws of quantum mechanics, namely the many particle *Schrödinger equation* as a commonly accepted fundamental basis.

Based on the fundamental work of Dirac, Hartree and Slater and others during 70 years history in quantum chemistry impressive progress has been achieved. The impressive success of recent ab initio computations is the result of systematic developments in quantum chemistry using Gaussian type basis functions and additionally the development of density functional theory by Kohn and co-authors, which simplifies the equations drastically. In particular, the work of Pople and Kohn was awarded in 1998 by the noble prize in chemistry.

Gaussian type basis functions are commonly used in computational quantum chemistry. Already relatively few of these basis functions provide highly accurate results. They have been optimized up to an impressive efficiency. In density functional theory, i.e. for the numerical solution of Kohn Sham equations, systematic basis functions based on Cartesian grids are also used in practice. In fact extremely large systems, in particular metallic systems, are computed with plane wave basis sets, finite differences, splines and wavelets in conjunction with pseudo-potentials. In fact, the use of pseudo potentials reduces the number of those basis functions drastically.

For atomic orbital functions like Gaussian type orbitals or Slater type orbitals rigorous convergence and approximation estimates are not proved yet. And due to its nature, it will be hard to obtain such estimates. Alternatively for methods which are based on Cartesian grids like plane wave basis functions, B-splines or multiresolution spaces, e.g. wavelets, or finite difference methods the approximation property of the basis functions is known. Due to the fact that the supports of the basis functions overlap, the Galerkin method requires matrices representing the potentials which are asymptotically sparse, but practically still contain several thousands of entries in each row. This is in strong contrast to finite difference methods where these matrices are diagonal (for local potentials). This means the complexity of the matrix vector multiplication differs by a factor of 100 to 1000. For interpolating basis functions, an alternative projection method namely the collocation method also yields diagonal matrices for representing local potentials. Even if it is not mentioned explicitly in the literature the collocation method is involved when using plane wave basis sets. Also many finite difference methods can be cast into the framework of collocation methods on shift invariant function spaces, e.g. multi-resolution spaces. The collocation method for a single particle Schrödinger operator or for the Hamilton Fock operator consists in the solution of the following finite dimensional eigenvalue problem

$$(Hu_h)(x_k) - Eu_h(x_k) = 0 \;, \forall k \;. \tag{1}$$

Both, the collocation method as well as the Galerkin method are projection methods. In contrast to the Galerkin method the collocation method is not variational. As a consequence convergence estimates cannot be obtained by min-max principles. The convergence theory of projection methods for eigenvalue problems has been considered by several authors, see e.g. Refs. 1,41. A comprehensive treatment can be found in Ref. 11. However this theory is incomplete, because these papers are mainly dealing with compact operators. For instance, eigenvalue problems for elliptic partial differential operators on compact domains can be cast into this framework. Unfortunately, the Schrödinger operators and the Hamilton Fock operators on $\mathbb{R}^3$ do not fit into this framework. Typically those operators permit beside a discrete spectrum also a continuous spectrum. This fact makes the convergence theory much more complicated. The well established classical convergence theory about eigenvalue computation via projection methods does not apply directly for the computation of molecules.

All these methods are scaling at least cubically w.r.t. the number of electrons N. This scaling is a bottleneck for computing large systems including several thousands of electrons. Recently ideas have been proposed claiming linear scaling. These methods are working quite well for insulating systems and small sets of highly localized Gaussian basis functions.[20] Nevertheless, including more and more diffusive Gaussian basis functions would ruin the efficiency of the linear scaling methods completely. For the computation of extremely large systems within the framework of Density Functional Theory, wavelet basis functions might offer a perspective.

The present article aims beside a very brief introduction into electronic structure calculation and effective one particle models like Hartree Fock or Kohn Sham, to focus on the convergence theory for projection methods, in particular, the collocation method involved in the numerical solution of the Hartree Fock and Kohn Sham equations. Since both equations are nonlinear and must be solved iteratively each iteration step requires the solution of a linear eigenvalue problem for a single particle Schrödinger type operator. We consider the convergence of projection methods for these linear operators. The convergence theory for the full nonlinear problem is still in its infancy, see e.g. Ref. 10 for further comments. Due to the lack of space we will only provide a road map for this theory and sketch the proofs. The detailed proofs we will be published in a separate paper.

2. Electronic Structure Calculation

The description of a wide range of molecular phenomena requires only very few postulates to establish the corresponding quantum mechanical formulation. In what follows we will confine ourselves to stationary and non-relativistic theory. I.e. we do not consider an explicit dynamic behavior and we neglect relativistic quantum phenomena.

The behavior of a system of N identical particles with spin s_i, $i = 1, \ldots, N$, is completely described by a *state-* or *wave-function*

$$(\mathbf{x}_1, s_1; \ldots; \mathbf{x}_N, s_N) \mapsto \Psi(\mathbf{x}_1, s_1; \ldots; \mathbf{x}_N, s_N) \in \mathbb{C}.$$

For each particle i there are the corresponding spatial coordinates $\mathbf{x}_i = (x_{i,1}, x_{i,2}, x_{i,3}) \in \mathbb{R}^3$ and a spin variable s_i.

In quantum mechanics identical particles cannot be distinguished. Therefore, the state functions Ψ must be either

(1) symmetric for bosons ($s_i \in \mathbb{Z}$), or
(2) antisymmetric for fermions ($s_i = \pm\frac{1}{2}$),

with respect to any permutation between identical particles. This is the well known *Pauli principle.* In particular electrons are fermions, and the spin s_i can be either $\frac{1}{2}$ or $-\frac{1}{2}$. Therefore the state-function of an electronic system $\Psi(\mathbf{x}_1, s_1; \ldots; \mathbf{x}_N, s_N)$ is anti-symmetric: For $(\mathbf{x}_1, s_1; \ldots; \mathbf{x}_N, s_N) \mapsto \Psi(\mathbf{x}_1, s_1; \ldots; \mathbf{x}_N, s_N)$ there holds

$$\Psi(\ldots, \mathbf{x}_i, s_i; \ldots; \mathbf{x}_j, s_j \ldots) = -\Psi(\ldots; \mathbf{x}_j, s_j; \ldots; \mathbf{x}_i, s_i; \ldots).$$

The state function Ψ is an eigenfunction of the Hamilton operator $\mathcal{H}$, i.e. $\mathcal{H}\Psi = E\Psi$. The eigenvalues E are the total energy of the system in the state Ψ. For an eigenfunction Ψ one uses the normalization condition

$$\langle \Psi, \Psi \rangle := \int_{\mathbb{R}^{3N}} \sum_{s_i \in \{\pm\frac{1}{2}\}} \Psi(\mathbf{x}_1, s_1; \cdots; \mathbf{x}_N, s_N)\overline{\Psi(\mathbf{x}_1, s_1; \cdots; \mathbf{x}_N, s_N)} d\mathbf{x}_1 \cdots d\mathbf{x}_N = 1 \, .$$

Since the effective mass of a nucleus is much larger than the mass of an electron, the nuclei can be treated as classical particles. For the stationary computations, they are fixed at the centers of the atoms $\mathbf{R}_j \in \mathbb{R}^3$ and have the total charge Z_j, $j = 1, \ldots, M$ for each atom. Consequently, they are modeled by an exterior potential

$$V_{core} = \sum_{i=1}^{N} \sum_{j=1}^{M} \frac{-Z_j}{|\mathbf{x}_i - \mathbf{R}_j|}.$$

This idealization is called the *Born-Oppenheimer-approximation.* If we are using *atomic units*, we obtain a partial differential equation of eigenvalue type, i.e., Ψ satisfies the (stationary) **Schrödinger equation**

$$\mathcal{H}\Psi := \sum_{i=1}^{N} [-\frac{1}{2}\Delta_i - \sum_{j=1}^{M} \frac{Z_j}{|\mathbf{x}_i - \mathbf{R}_j|} + \sum_{j<i}^{N} \frac{1}{|\mathbf{x}_i - \mathbf{x}_j|}]\Psi = E\Psi, \qquad (2)$$

with $\Psi \in L_2((\mathbb{R}^3 \times \{\pm\frac{1}{2}\})^N)$. The input parameters are only the centers of the atoms $\mathbf{R}_j \in \mathbb{R}^3$ together with the total charge Z_j, $j = 1, \ldots, M$ of the atoms and the total number of electrons N. One is mostly interested in the lowest eigenvalue, the so called ground state energy E_0. For example, inner atomic forces can be computed from the gradient of E_0 with respect the variation of the location of the nuclei (Hellman Feynman forces). Moreover a stable geometric configuration of a molecule is found by optimizing the ground state energy with respect to the different geometric positions of the nuclei (geometry optimization).

From a pure mathematical point of view the linear differential operator $\mathcal{H}$ has a relatively simple structure. Therefore important results about existence and regularity are available, many of them since more than 30 years. We refer to the book[36] as well as e.g. several survey articles like.[25,38]

In particular, it is known that the operator $\mathcal{H}$ admits a discrete spectrum below the continuous spectrum. Therefore there exists a lowest eigenvalue E_0 with an eigenfunction in the Sobolev space $H^1((\mathbb{R}^3 \times \{\pm\frac{1}{2}\})^N)$. The subspace of antisymmetric functions in $L_2((\mathbb{R}^3 \times \{\pm\frac{1}{2}\})^N)$ will be denoted by $\bigwedge_{i=1}^N L_2(\mathbb{R}^3 \times \{\pm\frac{1}{2}\})$. Consequently, the state function Ψ is from the space

$$H_a^1 = H^1\Big(\Big(\mathbb{R}^3 \times \Big\{\pm\frac{1}{2}\Big\}\Big)^N\Big) \cap \bigwedge_{i=1}^N L_2\Big(\mathbb{R}^3 \times \Big\{\pm\frac{1}{2}\Big\}\Big). \tag{3}$$

Moreover it is known that the corresponding eigenfunction function is exponentially decaying at infinity.

In contrast to its simplicity this equation seems to be nearly intractable by deterministic numerical methods. Because the electronic Schrödinger equation is posed in extremely high dimensions, numerical approximation is hampered by the curse of dimensions. Actually, the number of dimensions is (neglecting the spin variable) $3N$, where N is the total number of particles, in our case electrons, inside the system. The anti-symmetry constraint, formulated by the Pauli principle, is posing additional difficulties. Last but not least the state function are not completely smooth. They admit singularities in the derivatives, so called cusps. In fact, existing deterministic methods like *full CI* usually are scaling exponentially with the number of electrons N. There are some recent approximation theoretic concepts, namely *sparse grids* or *hyperbolic cross approximation*,[8,44] which can partially circumvent the curse of dimensions.

Despite these difficulties, after more than 50 years of development in quantum chemistry, and quantum physics nowadays there are tools available to compute the ground state energy of relatively large systems up to a considerable accuracy. This progress has been awarded by the noble price in chemistry given to Noble for the development incorporated in the software package GAUSSIAN and to R.V. Kohn for the development of density functional theory. Perhaps a historical survey even has to consider more than 20 outstanding scientist who made milestone contributions to this successful development of numerical methods. Due to the success and also the limitations of the Hartree Fock approximation, one branch is trying to compute the ground state energy from the solution of a nonlinear and coupled system in only one particle variable, i.e. in $\mathbb{R}^3 \times \{\pm\frac{1}{2}\}$ or even $\mathbb{R}^3$. These used methods allow the treatment of rather large system because one has excluded the problem of high dimensional approximation. Nevertheless there remains an intrinsic modeling error since no existing model is completely equivalent to the original electronic Schrödinger equation. Due to their efficiency the methods are widely and successfully used for large systems, in particular for the computation of bulk crystals in solid state physics. In the present paper we will focus only on those *effective particle methods.* Very recent methods are scaling, in a very rough sense, linearly with respect to the number of particles N.

3. Effective One-Particle Models

3.1. *Hartree-Fock equations*

Since $L_2((\mathbb{R}^3 \times \{\pm\frac{1}{2}\})^N) = \bigotimes_{i=1}^N L_2(\mathbb{R}^3 \times \{\pm\frac{1}{2}\})$ is a tensor-product space, the subspace of antisymmetric functions $\bigwedge_{i=1}^N L_2(\mathbb{R}^3 \times \{\pm\frac{1}{2}\})$ is spanned by *Slater determinants* of the form

$$\Psi_{SL}(\mathbf{x}_1, s_1; \ldots; \mathbf{x}_N, s_N) = \frac{1}{\sqrt{N!}} \det(\varphi_i(\mathbf{x}_j, s_j)) \quad , \quad \langle \varphi_i, \varphi_j \rangle = \delta_{i,j},$$

with $\langle \varphi_i, \varphi_j \rangle = \int_{\mathbb{R}^3} \sum_{s=\pm\frac{1}{2}} \varphi_i(\mathbf{x}, s) \overline{\varphi_j(\mathbf{x}, s)} d\mathbf{x}$. A Slater determinant Ψ_{SL} is an (anti-symmetric) product of N orthonormal functions, $\varphi_i : \mathbb{R}^3 \times \{\pm\frac{1}{2}\} \to \mathbb{C}$ resp. $\mathbb{R}$, $i = 1, \ldots, N$, called *orbitals*, where N is the number of particles. A fairly simple approximation is found by approximating Ψ by a single Slater determinant. This approximation leads to the well known *Hartree-Fock model.* Due to the Ritz principle the lowest eigenvalue E_0 is the minimum of the Rayleigh quotient

$$E_0 = \min\{\langle \mathcal{H}\Psi, \Psi \rangle : \Psi \in H_a^1 \ , \ \langle \Psi, \Psi \rangle = 1\}.$$

The minimization of the above quadratic functional using only one Slater determinant has to incorporate the orthogonality constraint condition. The Lagrange formalism then leads to the Hartree-Fock equations as a necessary condition, Ref. 39. For the sake of simplicity, we consider in the sequel solely *closed shell systems*, which have an even number of electrons. In the *Restricted Closed Shell Hartree Fock Model* (RHF) one considers pairs of electrons with spin $\pm\frac{1}{2}$. Thus the number of orbitals N will be the number of electron pairs.[24,39] Moreover, each orbital depends only on the spatial variable $\mathbf{x} \in \mathbb{R}^3$. For notational convenience, let us define the so called *density matrix* $\rho(\mathbf{x},\mathbf{y}) := \sum_{i=1}^N \varphi_i(\mathbf{x})\overline{\varphi_i(\mathbf{y})}$ together with the electron density $n(\mathbf{x}) := 2\rho(\mathbf{x},\mathbf{x})$.

With the ansatz $\Psi = \Psi_{SL} = \frac{1}{\sqrt{N!}}\det(\varphi_i(\mathbf{x}_j))$ as a single Slater-determinant one can compute the energy $E(\Psi_{SL}) = \langle \mathcal{H}\Psi_{SL}, \Psi_{SL}\rangle$ explicitly[39]

$$E(\Psi_{SL}) = \sum_{i=1}^{N}[\langle\nabla\varphi_i,\nabla\varphi_i\rangle + 2\langle\mathcal{V}_{core}(\mathbf{x})\varphi_i,\varphi_i\rangle + \langle\mathcal{V}_H(\mathbf{x})\varphi_i,\varphi_i\rangle - \frac{1}{2}\langle\mathcal{W}\varphi_i,\varphi_i\rangle]$$

with core potentials $\mathcal{V}_{core}(\mathbf{x}) = \sum_{j=1}^{M}\frac{-Z_j}{|\mathbf{x}-\mathbf{R}_j|}$. Here the *Hartree potential* $\mathcal{V}_H$ is given by $\mathcal{V}_H(\mathbf{x}) = \int_{\mathbb{R}^3}\frac{n(\mathbf{y})}{|\mathbf{x}-\mathbf{y}|}d\mathbf{y}$, and the *exchange energy* term is

$$\mathcal{W}u(\mathbf{x}) = \int_{\mathbb{R}^3}\frac{2\rho(\mathbf{x},\mathbf{y})}{|\mathbf{x}-\mathbf{y}|}u(\mathbf{y})d\mathbf{y} \ .$$

For $Z := \sum_{j=1}^{M} Z_j \geq N$ the existence of a minimizer of the Hartree-Fock energy

$$\Phi = (\varphi_1,\ldots,\varphi_N) = \operatorname{argmin}\{\mathcal{J}_{HF}(\Phi) = E(\Psi_{SL}) : \varphi_i \in H^1(\mathbb{R}^3), \langle\varphi_i,\varphi_j\rangle = \delta_{ij}\} \tag{4}$$

with the corresponding (approximate) ground state energy

$$E_{HF} = \min\{\mathcal{J}_{HF}(\Phi) = E(\Psi_{SL}) : \varphi_i \in H^1(\mathbb{R}^3)\ ,\ \langle\varphi_i,\varphi_j\rangle = \delta_{ij}\}$$

was shown by Lieb and Simon Refs. 30,31.

A necessary condition for a minimizer are the following *Hartree-Fock equations*: There exists a unitary matrix $\mathbf{U}$, such that the functions $(\breve{\varphi}_i)_{i=1,\ldots,N} = \breve{\Phi} = \mathbf{U}\Phi$ satisfy the eigenvalue problem

$$\mathcal{H}_\Phi\breve{\varphi}_i = \lambda_i\breve{\varphi}_i,$$

with the Hamiltonian

$$\mathcal{H}_\Phi = -\frac{1}{2}\Delta + \mathcal{V}_{core} + \mathcal{V}_{H,\Phi} - \frac{1}{2}\mathcal{W}_\Phi.$$

The orbitals are the eigenfunctions corresponding to the N lowest eigenvalues of $\mathcal{H}_\Phi$, $\lambda_1 \leq \lambda_2 \leq \ldots \leq \lambda_N < 0$. This is called the *aufbau principle.*

It is also known that the orbitals are smooth, $\varphi_i \in C^\infty(\mathbb{R}^3\backslash\{\mathbf{R}_1,\ldots,\mathbf{R}_M\})$, and have exponential decay at infinity $\varphi_i(\mathbf{x}) = \mathcal{O}(e^{-a|\mathbf{x}|})$ if $|\mathbf{x}| \to \infty$, $a > 0$, Ref. 30.

3.2. Density functional theory

Although the approximation by a single Slater-determinant seems to be rather poor, experiences have shown that this approximation is surprisingly good in many situations. For this reason the Hartree Fock model is the basic and representative model equation for ab initio methods, which has to be considered in any kind of analysis of the numerical methods used in electronic structure calculations. A Hartree Fock computation is the basis for all post Hartree-Fock methods in quantum chemistry, like CI methods, perturbation methods and Coupled Cluster.[24] Furthermore the Hartree-Fock model provides a prototype for a whole bunch of equations arising from density functional theory, i.e. Kohn-Sham equations.

Density functional theory is based on the observation that the *ground state energy* E_0 of the electronic Schrödinger equation depends solely on the electron density $n(\mathbf{x}) = 2\rho(\mathbf{x},\mathbf{x})$. This result was first discovered by Kohn and Hohenberg and is known as the *Kohn-Hohenberg theorem.*[15]

This observation has led Kohn and Sham to the following modification of the Hartree-Fock model,

$$E_{KS}(\Phi) = \sum_{i=1}^{N}[\langle\nabla\varphi_i,\nabla\varphi_i\rangle + 2\langle\mathcal{V}_{core}\varphi_i,\varphi_i\rangle + \langle\mathcal{V}_H\varphi_i,\varphi_i\rangle + \langle\mathcal{V}_{XC}\varphi_i,\varphi_i\rangle].$$

Since the exchange term $\mathcal{W}$ in the Hartree-Fock model depends on the full density matrix, and not only on the electron density, this term is replaced by an exchange correlation potential term $\mathcal{V}_{XC}(\mathbf{x})$.

Unfortunately this term is not known explicitly. However many properties of this expression are known, and since this term must be universal for all electronic systems, there exist several successful clues how to realize this term $n \mapsto \mathcal{V}_{XC}$. There is a long list of correlation exchange functionals satisfying known properties more or less. The simplest approximations have the form $\mathcal{V}_{XC}(\mathbf{x}) = -C_{TF}\,\rho(\mathbf{x},\mathbf{x})^{1/3} +$ correction terms. These functionals have been proved to be successful in many situations and they are widely accepted. In benchmark computations merging between Hartree-Fock and Kohn Sham equations, so called hybrid models (e.g. Refs.5,7) of the form

$$E_{HF/KS}(\Phi) = \sum_{i=1}^{N}(\langle\nabla\varphi_i,\nabla\varphi_i\rangle + 2\langle\mathcal{V}_{core}\varphi_i,\varphi_i\rangle + \langle\mathcal{V}_H\varphi_i,\varphi_i\rangle + \alpha\langle\mathcal{V}_{XC}\varphi_i,\varphi_i\rangle - \frac{1}{2}\beta\langle\mathcal{W}\varphi_i,\varphi_i\rangle)$$

where $\alpha = 0$ leads to the Hartree-Fock equations, and $\beta = 0$ to the Kohn-Sham equations, have been shown to perform best. Nevertheless the exact form is not known, and even with best numerical approximation there remains a modeling error. In contrast to Hartree Fock, where the approximate state function is given by a Slater-determinant built by the orbitals, the orbitals from Kohn Sham equations are not related to the wave function Ψ. The relevant output quantity is only the ground state energy E_0.

Some existence results are known also for the Kohn Sham equations based on the local density approximation.[27] There several nonlinear terms have been slightly modified to guarantee sufficient regularity for an analytical treatment. Since the correct exchange correlation term is not known, such modifications may be accepted. Since the Kohn Sham equations are very similar to the Hartree Fock equations, it is usual practice to assume that the Kohn Sham system has similar properties like the Hartree Fock system. In particular it is assumed that the aufbau principle holds. It is also common practise to consider systems at a finite temperature $T > 0$. In this case the electron density is defined by $n(\mathbf{x}) = 2\sum_{k\geq 1} \epsilon(k)|\varphi_k(\mathbf{x})|^2$ where the occupation numbers $\epsilon(k)$ are given by the Fermi statistic.

The solutions of these effective one particle models can be assumed to be $C^\infty(\mathbb{R}^3 \backslash \{\mathbf{R}_j : j = 1, \ldots, M\})$. The singularities of the solutions degrade the convergence rate of the discretization methods. It is common practise in physics to replace the core potential $\mathcal{V}_{core}(\mathbf{x}) = \sum_{j=1}^{M} \frac{-Z_j}{|\mathbf{x}-\mathbf{R}_j|}$ for N electrons by an effective potential (operator) $\mathcal{V}_{eff}$, the so-called *pseudo-potential* for the valence electrons only. These pseudo-potentials reduce the particle number N and smooth the core singularity and oscillations in the core region. Nevertheless there remains a substantial modeling error. Relativistic phenomena have to be treated by the *Dirac equation.* These effects become relevant for heavy atoms and for certain chemical systems, in most cases they are neglected. Pseudo potentials offer a relatively simple way to incorporate relativistic corrections without using the Dirac equations explicitly.

3.3. *Self-consistent field approximation*

An N-tuple $\Phi = (\varphi_1, \ldots, \varphi_N)$ of H^1-functions is called self consistent or aufbau solution, if it satisfies the equations

$$\mathcal{H}_\Phi \varphi_i = \lambda_i \varphi_i \ , \ \lambda_1 \leq \ldots \leq \lambda_N < \lambda_{N+1}.$$

The effective one particle equations, namely Hartree-Fock or Kohn Sham, can be viewed as a fixed point problem for the set of N orbitals.

This suggests the following iteration scheme, the *self consistent field approximation* $\Phi^{(n)} \to \Phi = (\varphi_1, \ldots, \varphi_N)$:

$$\mathcal{H}_{\Phi^{(n)}}\varphi_i^{(n+1)} = \lambda_i^{(n+1)}\varphi_i^{(n+1)}, \quad \lambda_1^{(n+1)} \le \lambda_2^{(n+1)} \le \ldots \le \lambda_N^{(n+1)} < \lambda_{N+1}^{(n+1)}. \tag{5}$$

It is important to observe that in this linearization of the full nonlinear scheme of N unknown functions, for all $\varphi_i^{(n+1)}$ the resulting linear operator is the same. The operator $\mathcal{H}_{\Phi^{(n)}}$ is called *Fock operator* for the Hartree Fock equations or generally *Hamilton Fock operator*. In particular, for the Kohn-Sham equations, the Hamilton Fock operator has the form of a single particle Schrödinger operator with a potential

$$\mathcal{V} := \mathcal{V}_{core}(\mathbf{x}) + \mathcal{V}_H(\mathbf{x}) + \mathcal{V}_{XC}(\mathbf{x}) \ .$$

In this respect, the Kohn-Sham equations are much simpler than the Hartree-Fock equations, because they do not contain a nonlocal operator.

The aufbau solution is in the self consistent limit the solution of the following linear system of partial differential equations of eigenvalue type for N orthonormal functions $\Phi = (\varphi_1, \ldots, \varphi_N)$

$$\mathcal{H}_\Phi\varphi_i(\mathbf{x}) = -\frac{1}{2}\Delta\varphi_i(\mathbf{x}) + \mathcal{V}_{core}\varphi_i(\mathbf{x}) + \mathcal{G}(\Phi)\varphi_i(\mathbf{x}) = \lambda_i\varphi_i(\mathbf{x}) \ , \ i = 1, \ldots, N \, .$$

In the present paper we consider the convergence of the numerical solution of this linear problem obtained by projection methods, in particular by the collocation method, which is mainly used when dealing with Cartesian grids.

The self consistent field approximation is only the simplest prototype of similar iteration schemes. There are many cases where this simple Roothaan scheme fails to converge. Cances and Le Bris[9] have introduced an improved scheme for which they proved convergence. In all these schemes, each step requires the solution of the linearized Kohn Sham or Hartree Fock equations according to the aufbau principle.

It is also worthwhile to notice that instead of the eigenfunctions φ_i we only need a basis of the corresponding invariant space $E = \text{span}\{\varphi_1, \ldots, \varphi_N\}$, or more precisely we only need the orthogonal projection P_E onto the corresponding eigenspace E. In particular this projection is defined by the density matrix: $P_E u = \int \rho(\mathbf{x}, \mathbf{y}) u(\mathbf{y}) d\mathbf{y}$.

3.4. *Projection methods*

For the Hamilton Fock operator acting from $X := H^s(\mathbb{R}^3)$ to $Y := H^{s-2}(\mathbb{R}^3)$, let us consider families of finite dimensional subspaces $V_h \subset X$

and $Y_h \subset Y$, $h > 0$, spanned by basis functions: $V_h = \text{span}\{\phi_k : k \in \mathcal{I}_h\}$ and $Y_h = \text{span}\{\theta_k : k \in \mathcal{I}_h\}$. Projections onto these subspaces can be defined by biorthogonal basis functions $\widetilde{\phi}_k \in X'$, $\widetilde{\theta}_k \in Y'$, i.e.

$$\widetilde{\phi}_l(\phi_k) = \widetilde{\theta}_l(\theta_k) = \delta_{l,k} \quad , \quad l, k \in \mathcal{I}_h \ .$$

Then by

$$P_h f := \sum_{k \in \mathcal{I}_h} \widetilde{\theta}_k(f)\theta_k$$

we define a projector from Y onto Y_h. The corresponding projection method for the numerical solution of an operator equation for $u \in X$,

$$Hu = f \quad , \quad f \in Y$$

is defined by solving the finite dimensional operator equation

$$P_h H u_h = P_h f \ , \tag{6}$$

with unknown function $u_h \in V_h$. And the corresponding discrete eigenvalue problem reads as

$$P_h(H - EI)u_h = 0 \ . \tag{7}$$

The solution of the eigenvalue problem (5) can be approximated by well known numerical methods. Commonly used are Galerkin methods, collocation methods and finite differences.

The Galerkin scheme has the advantage to be variational, and therefore the numerically computed eigenvalues are always larger as or equal to the exact eigenvalues.

The matrix representations of the different parts of the Hamilton Fock operator are given by

$$\mathbf{T} = \left(-\frac{1}{2}\langle \Delta\phi_\lambda, \widetilde{\theta}_{\lambda'}\rangle\right)_{\lambda',\lambda \in \mathcal{I}_h} \quad , \quad \mathbf{V}_c = (\langle \mathcal{V}_{core}(\mathbf{x})\phi_\lambda, \widetilde{\theta}_{\lambda'}\rangle)_{\lambda',\lambda \in \mathcal{I}_h},$$

$$\mathbf{V}_H^{(n)} = (\langle \mathcal{V}_H(\mathbf{x})\phi_\lambda, \widetilde{\theta}_{\lambda'}\rangle)_{\lambda',\lambda \in \mathcal{I}_h}, \mathbf{W}^{(n)} = \left(\langle (-\frac{1}{2}\beta\mathcal{W} + \alpha\mathcal{V}_{xc})\phi_\lambda, \widetilde{\theta}_{\lambda'}\rangle\right)_{\lambda',\lambda \in \mathcal{I}_h},$$

with the Hartree potential $\mathcal{V}_H$ and the exchange energy term $\mathcal{W}$, which together give the *Hamilton Fock matrix* (or discrete Hamiltonian)

$$\mathbf{H}^{(n)} = \mathbf{H}_{\Phi^{(n)}} = \mathbf{T} + \mathbf{V}_c + \mathbf{V}_H^{(n)} + \mathbf{W}^{(n)}.$$

For the Galerkin scheme one uses $\widetilde{\theta}_{\lambda'} = \phi_{\lambda'}$ and for the collocation scheme $\langle f, \widetilde{\theta}_{\lambda'}\rangle = f(x_{\lambda'})$.

The self consistent field iteration works as follows. Once the Hamilton Fock matrix $\mathbf{H}^{(n)}$ is built, the invariant eigenspace for the N lowest eigenvalues must be computed. Usually this is done by computing the N eigenvectors $\mathbf{c}_i^{(n+1)}$, $i = 1, \ldots, N$, corresponding to the N lowest eigenvalues of $\mathbf{H}^{(n)}$, $\lambda_{1,h}^{(n+1)} \leq \ldots \leq \lambda_{N,h}^{(n+1)} < 0$. From these eigenvectors the approximate eigenfunctions $\varphi_{i,h}^{(n+1)}$, $i = 1, \ldots, N$, the density matrix and the approximate electron density can be computed. From the latter quantities one gets the Hamilton Fock matrix for the next iteration step. Let us remark that this procedure requires at least $\mathcal{O}(N^3)$ arithmetic operations due to the involved diagonalization. Recent methods can reduce this complexity to $\mathcal{O}(N)$.

3.5. *Multiresolution spaces*

We consider a *scaling function* ϕ satisfying the refinement equation

$$\phi(\mathbf{x}) = \sum_{\mathbf{k}\in\mathbb{Z}^3} a_{\mathbf{k}}\phi(2\mathbf{x} - \mathbf{k}) \ , \ \mathbf{x} \in \mathbb{R}^3. \tag{8}$$

For $j \in \mathbb{Z}$ we introduce the basis functions $\phi_{\mathbf{k}}^j := 2^{3j/2}\phi(2^j\mathbf{x} - \mathbf{k})$, $\mathbf{k} \in \mathbb{Z}^3$. These basis functions span the multiresolution spaces $V_j := \mathrm{span}\{\phi_{\mathbf{k}}^j : \mathbf{k} \in \mathbb{Z}^3 \ , \ |\mathbf{k}| \leq 2^j j^2\}$.

The scaling function ϕ is defined by the filter coefficients $a_{\mathbf{k}}$. In fact using the function $m_\phi(\xi) := m(\xi) := \sum_{\mathbf{k}\in\mathbb{Z}^3} a_{\mathbf{k}} e^{-i\mathbf{k}\cdot\xi}$ the Fourier transform of ϕ is given by (Ref. 32)

$$\widehat{\phi}(\xi) := \int e^{-i\xi\cdot\mathbf{x}}\phi(\mathbf{x})d\mathbf{x} = \prod_{j=1}^{\infty} \frac{m(2^{-j}\xi)}{8} \ .$$

It is known that if $m(\xi) = \prod_{l=1}^3 \frac{(1+e^{-i\xi_l})^d}{2^{d-1}}$ then the following approximation property holds for $s < t$, $s < \gamma$ and $t \leq d$,

$$\inf_{v_j\in V_j} \|u - v_j\|_{H^s} \lesssim 2^{j(s-t)}\|u\|_{H^t}, \tag{9}$$

additionally the inverse property holds

$$\|v_j\|_{H^t} \lesssim 2^{j(t-s)}\|v_j\|_{H^s} \ , \ \forall v_j \in V_j. \tag{10}$$

The regularity of ϕ is $\gamma = \sup\{s \in \mathbb{R} : \phi \in H^s(\mathbb{R}^3)\}$. It was shown in Ref. 42 how the regularity depends on the filter resp. the function m.

Typical examples of these spaces are splines, orthogonal and biorthogonal wavelets, interpolets. The interpolating scaling function satisfies $\phi(\mathbf{k}) = \delta_{\mathbf{0},\mathbf{k}}$, i.e. $\phi_{\mathbf{k}}^j$ is a nodal basis function on the Cartesian grid $\{2^{-j}\mathbf{k} : \mathbf{k} \in \mathbb{Z}^3\}$.

If ϕ_D is a Daubechies scaling function of order d, i.e. $\phi_D(. - \mathbf{k})$ are orthogonal and compactly supported, and m_{ϕ_D} the corresponding function given by the filter of ϕ_D, then the Fourier coefficients of $m_{\phi_I}(\xi) := |m_{\phi_D}(\xi)|^2/8$ form a filter for an interpolating scaling function ϕ_I with the order $2d$.

From the definition of the basis functions $\phi^j_{\mathbf{k}}$ it becomes obvious that the system matrix for a translation invariant operator, e.g. the Laplacian or a convolution operator, is a discrete convolution. E.g.

$$\langle -\Delta\phi^j_{\mathbf{k}}, \phi^j_{\mathbf{k}'}\rangle = \mathbf{T}_{\mathbf{k}-\mathbf{k}'} \ , \mathbf{k}, \mathbf{k}' \in \mathbb{Z}^3.$$

Moreover it is not difficult to show that

$$\langle -\Delta(\phi_D)^j_{\mathbf{k}}, (\phi_D)^j_{\mathbf{k}'}\rangle = -\Delta(\phi_I)^j_{\mathbf{k}}(2^{-j}\mathbf{k}'). \tag{11}$$

I.e., the Galerkin matrix using Daubechies scaling functions for representing the kinetic energy is the same as the collocation matrix using the corresponding interpolating scaling function. The collocation discretization of a local potential $\mathcal{V}(\mathbf{x})$ using interpolating scaling functions yields a diagonal matrix

$$\mathcal{V}(2^{-j}\mathbf{k})\phi^j_{\mathbf{k}'}(2^{-j}\mathbf{k}) = 2^{3j/2}\mathcal{V}(2^{-j}\mathbf{k})\delta_{\mathbf{k},\mathbf{k}'}.$$

This fact makes the collocation scheme with interpolating scaling functions extremely attractive for Schrödinger type equations. In general the system matrices of the collocation scheme may be not symmetric. But symmetry is retained for local potentials, and even it can be shown that the matrix for the nonlocal exchange term $\mathcal{W}$ remains symmetric. Many finite difference methods can be interpreted as collocation methods for a certain interpolating scaling function.

Multiresolution spaces are the starting point for wavelet decomposition. Wavelets can be very important for an efficient numerical treatment. However, the wavelet basis constitutes another basis in the spaces V_j or X. In the present paper we focus on approximation and convergence results of these spaces. In this respect it is not essential which basis is actually used for the discretization. Therefore we will not extend on wavelet bases in this paper.

4. Projection Methods for Eigenvalue Problems of Schrödinger Type

In the sequel we will consider a linear operator of the form

$$H = -\frac{1}{2}\Delta + V + W : H^2 \to L_2 \tag{12}$$

where V is a pseudo-potential $V \in C^\infty \cap (L_{3-\epsilon} + L_\infty)$, $L_{3-\epsilon} + L_\infty = \{V_1 + V_2 : V_1 \in L_p \text{ for all } p < 3, V_2 \in L_\infty\}$. It is also possible with minor changes to consider an optional exchange term W arising in the Hartree Fock model and in hybrid models. There holds for a certain $\delta \in (0, 1/2)$

$$V + W \in L(H^2) \quad \text{and } V + W \in K(H^2, H^{3/2+\delta}),$$

i.e. $V + W$ is bounded in H^2 and a compact operator from H^2 to $H^{3/2+\delta}$. It is worthwhile to mention that these properties are not valid for the original Coulomb potentials because of their singularities at the centers of the atoms. But if they are replaced by smooth pseudo-potentials, these assumptions hold even for nonlocal pseudo-potentials. Furthermore the correlation-exchange potentials arising from the models in Density Functional Theory are not known to have the required regularity. Since the underlying equations have a modeling error, possibly by applying some smoothing, it may be assumed that they satisfy also the above regularity requirements. In this respect the previous assumptions become reasonable.

If H is the Hamilton Fock operator in the self-consistent limit, it is known[31] that there exist countably many negative eigenvalues, λ_i, $i = 1, 2, 3, \ldots$, monotonically increasing and listed repeated according to their geometric multiplicity. We assume that this is valid also for the present operator H possibly arising from the Kohn Sham equations. Furthermore we assume that for $N \in \mathbb{N}$ (pairs of) electrons the difference $\lambda_{N+1} - \lambda_N$ is strictly positive, so that there exists $\mu > 0$ satisfying $\lambda_N < -\mu < \lambda_{N+1}$.

For every $i \in \mathbb{N}$, $\kappa_i = \lambda_i + \mu$ is an eigenvalue of the shifted operator $A = H + \mu I$ and the eigenfunctions $\{\phi_j\}_{j \in \{j \in \mathbb{N} : \kappa_j = \kappa_i\}}$ are supposed to form an L^2-orthonormal basis of $\ker(A - \kappa_i I)$.

To define the present setting for collocation methods as particular projection methods we consider the following spaces $X := H^2$, $Y := L_2$, $U := H^{3/2+\delta}$.

For each $h > 0$ let $V_h \subset X$ be a finite dimensional subspace with dimension $M_h = \dim V_h$, satisfying $V_h \supset V_{h'}$ for $h < h'$, and $\bigcup_{h>0} V_h$ is supposed to be dense in $X = H^2$. We also consider a finite set of collocation points $X_h \subset \mathbb{R}^3$ with cardinality $M_h = |X_h|$, such that for every $u \in U$ there exists a unique function $u_h \in V_h$ which satisfies $u_h(x_i) = u(x_i)\ \forall x_i \in X_h$. Let $P_h u := u_h$, then $P_h : U \to V_h$ defines the interpolation projector onto V_h with respect to X_h. Furthermore $P_h^\perp : X = H^2 \to V_h$ denotes the H^2-orthogonal projector onto V_h. We assume additionally the following uniformly boundedness with respect to $h > 0$: $\|P_h\|_{H^2 \to L^2} \le C$ and $\|P_h^\perp\|_{H^2 \to H^{7/2+\delta}} \le C$, and that there holds $P_h f \to f$ in L^2 for all $f \in U$

and $\|P_h - I\|_{H^{3/2+\delta}\to L^2} \to 0$. These properties are known to be valid for a large variety of spaces V_h and sets of collocation points.

Let $\Gamma \subset \{z \in \mathbb{C} : \mathrm{Re} z < 0\}$ be a positively orientated Jordan curve surrounding the set $\{\kappa_1, \ldots, \kappa_N\}$ and excluding all other points of the spectrum of A. Let I_Γ be the closure of the interior of Γ.

We assume further the uniformly boundedness of $\{P_h(A - zI)|_{V_h}\}$,

$$\exists C_{b,A} > 0 \forall z \in \Gamma \forall 0 < h < h_0 \forall u_h \in V_h : \|P_h(A - zI)u_h\|_{L^2} \leq C_{b,A}\|u_h\|_{H^2}.$$

We consider an auxiliary operator $B = -\frac{1}{2}\Delta + \mu I$. For this operator the equation $(B - zI)u = f$ is solved by the projection scheme $P_h(B - zI)u_h = P_h f$. In order to enable the application of the projector P_h we require $B(V_h) \subset U$. We assume uniformly boundedness, stability and consistency of this projection scheme uniformly with respect to z,

$$\exists C_{b,B} > 0 \forall z \in I_\Gamma \forall 0 < h < h_0 \forall u_h \in V_h : \|P_h(B - zI)u_h\|_{L^2} \leq C_{b,B}\|u_h\|_{H^2},$$
$$\exists C_{s,B} > 0 \forall z \in I_\Gamma \forall 0 < h < h_0 \forall u_h \in V_h : \|P_h(B - zI)u_h\|_{L^2} \geq C_{s,B}\|u_h\|_{H^2},$$
$$\forall u \in H^2 : \sup_{z \in I_\Gamma} \|(B - zI)u - P_h(B - zI)P_h^\perp u\|_{L^2} \to 0.$$

These crucial properties can be shown for many collocation schemes, for instance the collocation method using interpolating scaling functions of even order $2d$. This method yields the same system matrices for the operator $B - zI$ as the Galerkin method using Daubechies orthogonal scaling functions. For the latter the well known Lax Milgram Lemma applies yielding all required properties assumed above.

If the operator $(H + \mu I)^{-1}$ is compact, the approximation of the eigenvalue problem is reduced to the eigenvalue problem of compact operators, see e.g. Ref. 11. The corresponding results cannot be used for the previous Hamilton Fock operators, which admit a discrete as well as a continuous spectrum. Nevertheless the operator $B^{-1}V$ is compact in X and Y, and the eigenvalues of A are the poles of the meromorphic operator family $z \mapsto (I - (zI - B)^{-1}V)^{-1}$. Therefore many arguments from the Riesz-Schauder theory can be used also in the treatment of the present problem. However, a complete treatment requires a careful analysis.

Finally we assume that $H(V_h) \subset U$ and that $P_h A|_{V_h} : V_h \to V_h$ is diagonalisable, i.e. there is a basis $\{\phi_{i,h}\}_{i=1}^{M_h}$ of V_h with $P_h A|_{V_h}\phi_{i,h} = \kappa_{i,h}\phi_{i,h}$.

By $E = \mathrm{span}\{\phi_i : i = 1, \ldots, N\}$ we define an invariant subspace of the operator A. The mapping

$$P_E = \frac{-1}{2\pi i}\oint_\Gamma (A - zI)^{-1} dz$$

is the L^2-orthogonal projector $P_E : L^2 \to L^2$ from L^2 onto E. For all $u \in L^2$ there holds

$$P_E u = \sum_{i=1}^{N} \langle u, \phi_i \rangle \phi_i.$$

By $E_h = \text{span}\{\phi_{i,h} : i \in \{1, \ldots, M_h\}, \kappa_{i,h} \text{ is surrounded by } \Gamma\}$ we define an invariant subspace of the operator $P_h A|_{V_h}$. The mapping

$$P_{E_h} = \frac{-1}{2\pi i} \oint_\Gamma (P_h A - zI)|_{V_h}^{-1} P_h dz$$

is a projector $P_{E_h} : U \subset L^2 \to V_h \subset L^2$ from U onto E_h (in general oblique).

Based on the above properties of the discretization of B and due to the compactness of the set I_Γ we can show that the following pointwise convergence of sequences of operators is valid uniformly on I_Γ.

Lemma 4.1. *For every $f \in L^2$ there holds*

$$\sup_{z \in I_\Gamma} \|(I - P_h^\perp)(B - zI)^{-1} f\|_{H^2} \to 0.$$

For all $f \in H^{3/2+\delta}$ there holds

$$\sup_{z \in I_\Gamma} \|f - P_h(B - zI)P_h^\perp(B - zI)^{-1} f\|_{L^2} \to 0.$$

For all $u \in H^{7/2+\delta}$ there holds

$$\sup_{z \in I_\Gamma} \|u - [P_h(B - zI)]|_{V_h}^{-1} P_h(B - zI)u\|_{H^2} \to 0.$$

Furthermore for all $f \in H^{3/2+\delta}$

$$\sup_{z \in I_\Gamma} \|(B - zI)^{-1} f - [P_h(B - zI)]|_{V_h}^{-1} P_h f\|_{H^2} \to 0.$$

For all $z \in I_\Gamma$ and $0 < h < h_0$ we define the operators $K_z = (zI - B)^{-1}V$ and $\tilde{K}_{h,z} = (zI - P_h B)|_{V_h}^{-1} P_h V$. The following result establishes the crucial convergence with respect to the operator norm concluded from the compactness of V.

Lemma 4.2. *It holds*

$$\sup_{z \in I_\Gamma} \|\tilde{K}_{h,z} - K_z\|_{H^2 \to H^2} \to 0.$$

From $A = B + V$ it follows that $A - zI = (B - zI)(I - K_z)$ and $(P_h A - zI)|_{V_h} = (P_h B - zI)|_{V_h}(I - \tilde{K}_{h,z})|_{V_h}$. This fact and the preceding lemma allow us to deduce the stability of the operators $P_h(A - zI)|_{V_h}$ from those ones of the operators $P_h(B - zI)|_{V_h}$ and $(I - \tilde{K}_{h,z})|_{V_h}$. Together with the corresponding consistency we get the convergence of the projection scheme for the operator A.

Theorem 4.1. *There exist $C_{s,A} > 0$, $h_s > 0$, so that for all $z \in \Gamma$, $0 < h < h_s$, $u_h \in V_h$*

$$\|P_h(A - zI)u_h\|_{L^2} \geq C_{s,A}\|u_h\|_{H^2}.$$

Moreover, for all $u \in H^{7/2+\delta}$ there holds

$$\sup_{z\in\Gamma} \|(A - zI)u - P_h(A - zI)P_h^{\perp} u\|_{L^2} \to 0.$$

For all $u \in H^{7/2+\delta}$ there holds

$$\sup_{z\in\Gamma} \|u - [P_h(A - zI)]|_{V_h}^{-1} P_h(A - zI)u\|_{H^2} \to 0.$$

The following theorem establishes the convergence of the eigenspaces.

Theorem 4.2. *For all $0 < h < h_s$ there holds*

$$\|(P_E - P_{E_h})|_E\|_{L^2\to L^2} \leq C \sup_{z\in\Gamma} \|[I - (P_h A - zI)|_{V_h}^{-1} P_h(A - zI)]|_E\|_{L^2\to L^2}$$

with a constant C not depending on h. Consequently,

$$\|(P_E - P_{E_h})|_E\|_{L^2\to L^2} \to 0.$$

This theorem is of particular importance because the projection P_E is given by the density matrix and P_{E_h} is an approximation of the orthogonal projector defined by the discrete density matrix.

As an immediate consequence we can estimate the dimension of the discrete eigenspaces.

Corollary 4.1. *There exists $h_\epsilon \in (0, h_s)$ so that for all $h \in (0, h_\epsilon)$*

$$\dim E \leq \dim E_h.$$

The converse estimate has been more difficult to prove. We omit the details here presenting only the expected result.

Theorem 4.3. *For sufficiently small $h > 0$ there holds $\dim E_h = \dim E$.*

With these results at hand we can show the following convergence of eigenvalues and computed ground state energy.

Theorem 4.4. *Under the above assumptions there holds for the first N eigenvalues*

$$|\sum_{i=1}^{N}(\kappa_i - \kappa_{i,h})| \lesssim \sup_{i=1,\dots,N} \inf_{v_h\in V_h} \|\phi_i - v_h\|_{H^2}. \tag{13}$$

A similar estimate is valid for the Hartree Fock or Kohn Sham energy

$$|E_{HF/KS} - E_{HF/KS,h}| \lesssim \sup_{i=1,\dots,N} \inf_{v_h\in V_h} \|\phi_i - v_h\|_{H^2}.$$

Using the collocation scheme with interpolating scaling functions of order $2d$ we obtain a convergence rate

$$|E_{HF/KS} - E_{HF/KS,h}| \lesssim 2^{-j2(d-1)} .$$

The result of Theorem 4.4 is not optimal for the Galerkin scheme. It is worth to mention that with the Galerkin scheme one achieves higher convergence rates, namely twice the rate of the convergence of the eigenspaces with respect to the energy norm, i.e. the Sobolev H^1-norm. However, because of the identity (11), we have to compare the above collocation scheme with the Galerkin scheme using Daubechies scaling functions of order d which gives at most the same rate $|E_{HF/KS} - E_{HF/KS,h}| \lesssim 2^{-j2(d-1)}$. Therefore one obtains with the collocation scheme the same convergence speed at a considerably lower cost.

5. Numerical Experiments

The first three numerical experiments study the convergence of discretization methods in the case of the one-dimensional harmonic oscillator. The eigenvalue problem under consideration is $H\phi = E\phi$ with $H = -\frac{1}{2}\frac{d^2}{dx^2} + \frac{1}{2}x^2$.

The first discretization method is the collocation method with plane waves. Because the sought eigenfunctions are exponentially decaying it is justified to consider the corresponding L-periodic eigenvalue problem, $L > 0$ sufficiently large. Let N be a power of two. We choose the collocation points equidistant, $x_j = -L/2 + hj$, $j = 0, \dots, N-1$, $h = L/N$, and the trial space in the form $V_N = \text{span}\{\psi_{\frac{2\pi}{L}n}\}_{n=-N/2}^{N/2-1}$, $\psi_k(x) = \frac{1}{\sqrt{L}}e^{ikx}$.

The second discretization method is the collocation method with interpolating scaling functions. Let ϕ_I be the Deslaurier-Dubuc interpolating scaling function of order $2d$, $d \geq 3$, i.e. every polynomial of degree $\leq 2d-1$ can be written as a linear combination of the translates

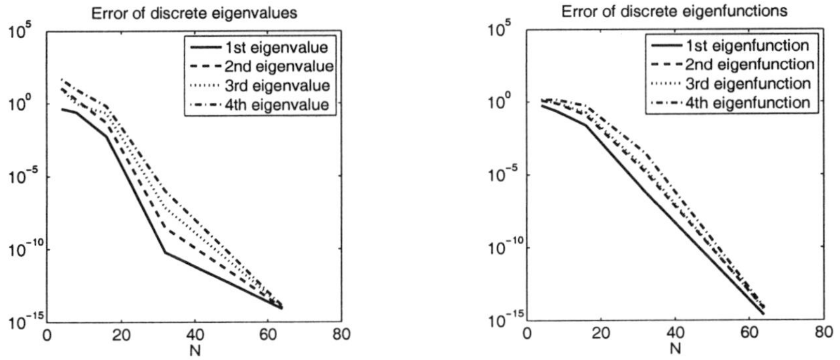

Fig. 1. Errors of collocation scheme with plane waves. We see a superalgebraic convergence.

$\phi_I(x-k)$, $k \in \mathbb{Z}$. Let $L \in 2\mathbb{N}, j \in \mathbb{N}$. We choose the collocation points as $x_{j,k} = k/2^j, k = -L2^{j-1}, \ldots, L2^{j-1}$. We set $(\phi_I)^j_k(x) = \phi_I(2^j x - k), k \in \mathbb{Z}$, and define the trial space as $V_j = \text{span}\{(\phi_I)^j_k\}_{k=-L2^{j-1}}^{L2^{j-1}}$ with dimension $N(j) = L2^j + 1$.

The last discretization method is the Galerkin method with Daubechies scaling functions. Let ϕ_D be the Daubechies scaling function of order $d \geq 3$, i.e. every polynomial of degree $\leq d-1$ can be written as a linear combination of the translates $\phi_D(x-k)$, $k \in \mathbb{Z}$. Let $L \in 2\mathbb{N}, j \in \mathbb{N}$. We define $(\phi_D)^j_k(x) = 2^{j/2}\phi_D(2^j x - k), k \in \mathbb{Z}$, and choose the trial space as $V_j = \text{span}\{(\phi_D)^j_k\}_{k=-L2^{j-1}}^{L2^{j-1}}$.

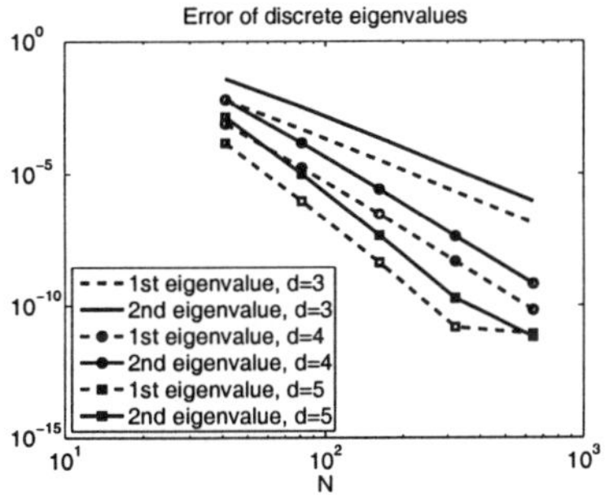

Fig. 2. Errors of collocation scheme with interpolating scaling functions. The slope of the error for $d = 3$ is about 4, that one for $d = 4$ is about 6 and that one for $d = 5$ is about 8.

$2d$	J	n	E_J	$\lvert E_J - E \rvert$
6	4	32	-1.83137	1.7e-02
6	5	64	-1.84042	7.5e-03
6	6	128	-1.84706	8.6e-04
8	4	32	-1.86363	1.6e-02
8	5	64	-1.84649	1.4e-03
8	6	128	-1.84777	1.4e-04
10	4	32	-1.87209	2.4e-02
10	5	64	-1.84743	4.8e-04
10	6	128	-1.84783	9.0e-05

Fig. 3. Errors of discrete ground state energy E_J for the Hartree-Fock model of H_2 using $[-7,7]^3$ as computational box.

Table 1. Errors of the discrete eigenvalues for the second and third discretization. E_n is the n-th exact eigenvalue, $E_{n,j,C}$ is the n-th discrete eigenvalue in the case of the collocation method, $E_{n,j,G}$ is the corresponding eigenvalue in the case of the Galerkin method. We see the approximative agreement of the corresponding errors.

d	j	$N(j)$	$\lvert E_{1,j,C}-E_1\rvert$	$\lvert E_{2,j,C}-E_2\rvert$	$\lvert E_{1,j,G}-E_1\rvert$	$\lvert E_{2,j,G}-E_2\rvert$
3	1	41	6.22e-03	3.88e-02	6.33e-03	3.92e-02
3	2	81	4.98e-04	3.40e-03	5.00e-04	3.41e-03
3	3	161	3.29e-05	2.29e-04	3.29e-05	2.29e-04
3	4	321	2.08e-06	1.45e-05	2.08e-06	1.45e-05
3	5	641	1.30e-07	9.13e-07	1.30e-07	9.13e-07
4	1	41	7.85e-04	6.29e-03	8.09e-04	6.44e-03
4	2	81	1.73e-05	1.52e-04	1.74e-05	1.53e-04
4	3	161	2.95e-07	2.64e-06	2.95e-07	2.64e-06
4	4	321	4.70e-09	4.23e-08	4.71e-09	4.23e-08
4	5	641	6.70e-11	6.58e-10	6.72e-11	6.58e-10
5	1	41	1.48e-04	1.43e-03	1.54e-04	1.48e-03
5	2	81	9.51e-07	1.01e-05	9.59e-07	1.02e-05
5	3	161	4.22e-09	4.61e-08	4.23e-09	4.62e-08
5	4	321	1.52e-11	1.85e-10	1.51e-11	1.85e-10
5	5	641	8.39e-12	6.45e-12	6.89e-12	6.55e-12

Now we present the results of two three-dimensional calculations. The first example is a Hartree-Fock calculation of the H_2 molecule which has a bond length of 1.4 atomic units. As reference value for the Hartree-Fock ground state energy (without nuclear repulsion energy) we choose that one of Mitin,[34] which is after rounding to six places $E = -1.847915$. The second example is a DFT calculation of a part of a LiH-crystal. Four hydrogen-atoms and four lithium-atoms form a cubic structure where the elements alternate. The bond-length is chosen as 3.836 atomic units. The used exchange-correlation functional is of the form $E_{xc}(n) = \int_{\mathbb{R}^3} \epsilon_{xc}(n(\mathbf{x}))n(\mathbf{x})d\mathbf{x}$, where $\epsilon_{xc} = \epsilon_x + \epsilon_c$. We use the exchange-part ϵ_x according to Slater and the correlation-part ϵ_c from Vosko, Wilk and Nusair Ref. 43. The exchange-part is $\epsilon_x(n(\mathbf{x})) = -\frac{9}{8}\alpha\left(\frac{3}{\pi}\right)^{1/3} n(\mathbf{x})^{1/3}, \alpha = 2/3$, the correlation-part is defined by more extensive formulas.

Before the corresponding eigenvalue problem is discretised by a collocation method we choose a sufficiently large computational box and map it onto the cube $[-1,1]^3$. Let $J \in \mathbb{N}$. The trial space is chosen as $V_J = \text{span}\{\phi_{J,k_1} \otimes \phi_{J,k_2} \otimes \phi_{J,k_3}\}_{k \in Q_J}$, where ϕ is the Deslaurier-Dubuc interpolating scaling function of order $2d$ and $Q_J = \{-2^J, \ldots, 2^J - 1\}^3$. The set of collocation points is $X_J = \{(\frac{k_1}{2^J}, \frac{k_2}{2^J}, \frac{k_3}{2^J})\}_{k \in Q_J}$. I.e. we consider a uniform grid on $[-1,1]^3$ with $n = 2^{J+1}$ points in every space direction.

In order to be able to apply the collocation method we replace the singular Coulomb-potential by a smooth pseudopotential from Goedecker, Teter and Hutter Ref. 18. This pseudopotential consists of a local and a nonlocal part. The local part $V_{loc}(\mathbf{x})$ has the form

$$V_{loc}(\mathbf{x}) = \frac{-Z_{ion}}{|\mathbf{x}|}\operatorname{erf}\left(\frac{|\mathbf{x}|}{\sqrt{2}r_{loc}}\right) + \exp\left(-\frac{1}{2}\left(\frac{|\mathbf{x}|}{r_{loc}}\right)^2\right)$$
$$\times\left(C_1 + C_2\left(\frac{|\mathbf{x}|}{r_{loc}}\right)^2 + C_3\left(\frac{|\mathbf{x}|}{r_{loc}}\right)^4 + C_4\left(\frac{|\mathbf{x}|}{r_{loc}}\right)^6\right),$$

where erf denotes the error function. Z_{ion} is the ionic charge (i.e., charge of the nucleus minus charge of the core electrons), and r_{loc} gives the range of the Gaussian ionic charge distribution leading to the erf potential. The parameters Z_{ion}, r_{loc}, C_1, ..., C_4 are listed for the elements of the first two rows of the Periodic Table in Ref. 18. The nonlocal part is zero for the elements up to Be and therefore not needed in our calculation examples.

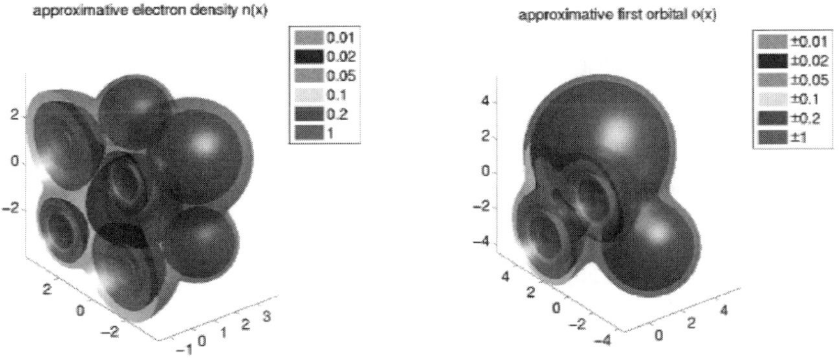

Fig. 4. Isosurface-plots of density and first orbital of LiH-crystal. The computational box is $[-7,7]^3$ and the resolution is 64 grid points in each space direction.

References

1. P.M. Anselone, *Collectively Compact Operator Approximation Theory* (Prentice-Hall, Englewood Cliffs, New Jersey, 1971).
2. T.A. Arias, *Rev. Mod. Phys.*, **71**, 267 (1999).
3. I. Babuska, J. Osborn, *Eigenvalue problems*, in *Handbook of Numerical Analysis*, Vol. 2 (North-Holland, Amsterdam, 1991).

4. T.L. Beck, *Rev. Mod. Phys.*, **72** 1041 (2000).
5. G. Beylkin, N. Coult, M.J. Mohlenkamp, *J. Comput. Phys.*, **152**, 32 (1999).
6. S. Börm, L. Grasedyck, W. Hackbusch, *Lecture notes, Max-Planck-Institut für Mathematik in den Naturwissenschaften* (Leipzig, 2003).
7. D. Bowler, M. Gillan, *Comp. Phys. Comm.*, **120**, 95 (1999).
8. H.J. Bungartz, M. Griebel, *Acta Numerica*, 1 (2004).
9. E. Cancès, C. Le Bris, *Int. J. Quant. Chem.*, **79** 82 (2000).
10. E. Cancès, M. Defranceschi, W. Kutzelnigg, C. Le Bris, Y. Maday, *Computational Chemistry: a primer*, in *Handbook of Numerical Analysis*, Vol. X (North-Holland, 2003).
11. F. Chatelin, *Spectral approximation of linear operators* (Academic Press, New York, 1983).
12. P.G. Ciarlet, *Handbook of Numerical Analysis*, Vol. X: Computational Chemistry (North-Holland, 2003).
13. A. Cohen, *Numerical analysis of wavelet methods* (North-Holland, 2003).
14. M. Defranceschi, P. Fischer, *SIAM J. Numer. Anal.*, **351** (1998).
15. R.M. Dreizler, E.K.U. Gross, *Density functional theory* (Springer, Berlin, 1990).
16. H.-J. Flad, W. Hackbusch, D. Kolb, R. Schneider, *J. Chem. Phys.*, **116**, 9641 (2002).
17. H.-J. Flad, W. Hackbusch, R. Schneider, *Best N term approximation for electronic wavefunctions* (Preprint 05-9, Mathematisches Seminar Kiel, 2005).
18. S. Goedecker, M. Teter, J. Hutter, *Phys. Rev. B*, **54**, 1703 (1996).
19. S. Goedecker, O. Ivanov, *Sol. State Comm.*, **105**, 665 (1998).
20. S. Goedecker, *Linear scaling methods for the solution of Schrödinger's equation*, in *Handbook of Numerical Analysis*, Vol. X (North-Holland, 2003).
21. R.D. Grigorieff, *Math. Nachr.* **69**, 253 (1975).
22. W. Hackbusch, *Multi-Grid Methods and Applications* (Springer, New York, 1985).
23. R.J. Harrison, G.I. Fann, T. Yanai, Z. Gan, G. Beylkin, *Preprint, University of Colorado at Boulder* (Department of Applied Mathematics, 2003).
24. T. Helgaker, P. Jørgensen, J. Olsen, *Molecular electronic-structure theory* (John Wiley & Sons, New York, 2002).
25. P.D. Hislop, I.M. Sigal, *Introduction to Spectral Theory* (Springer, 1996).
26. W. Hunziker, I.M. Sigal, *Journ. Math. Phys.*, **41**, 3448 (2000).
27. H.-C. Kaiser, J. Rehberg, *Math. Methods Appl. Sci.*, **20**, 1283 (1997).
28. W. Kutzelnigg, *Int. J. Quant. Chem.*, **51**, 447 (1994).
29. X.P. Li, R.W. Nunes, D. Vanderbilt, *Phys. Rev. B.*, **47**, 10891 (1993).
30. E.H. Lieb, B. Simon, *Comm. Math. Phys.*, **53**, 185 (1977).
31. P.-L. Lions, *Comm. Math. Phys.*, **109**, 33 (1987).
32. S. Mallat, *A wavelet tour of signal processing*, 2nd edn. (Academic Press, 1999).
33. R. McWeeny, *Methods of Molecular Quantum Mechanics*, 2nd edn. (Academic Press, San Diego, 1992).
34. A. Mitin, *Phys. Rev. A*, **6201**, 010501 (2000).
35. R.G. Parr, W. Yang, *Density functional theory of atoms and molecules* (Ox-

ford University Press, New York, 1989).
36. M. Reed, B. Simon, *Methods of modern mathematical physics*, Vol. 4 (Academic Press, New York, 1978).
37. M. A. Shubin, *Pseudodifferential Operators and Spectral Theory* (Springer, New York, 1987).
38. B. Simon, *Journ. Math. Phys.*, **41**, 3523 (2000).
39. A. Szabo, N.S. Ostlund, *Modern quantum chemistry* (Dover publications, New York, 1996).
40. M.E. Taylor, *Partial Differential Equations II, Qulitative Studies of Linear Equations* (Springer, New York, 1996).
41. G.M. Vainikko, *Funktionalanalysis der Diskretisierungsmethoden* (Teubner, Leipzig, 1976).
42. L. Villemoes, *SIAM J. Math. Anal.*, **25**, 1433 (1994).
43. S.H. Vosko, L. Wilk, M. Nusair, *Can. J. Phys.*, **58**, 1200 (1980).
44. H. Yserentant, *Numer. Math.*, **98**, 731 (2004).

Harmonic, Wavelet and p-Adic Analysis
Eds. N. M. Chuong *et al.* (pp. 83–109)

§5. LOCALIZATION OPERATORS AND TIME-FREQUENCY ANALYSIS

ELENA CORDERO and LUIGI RODINO

Department of Mathematics, University of Torino, Italy
E-mail: elena.cordero@unito.it; luigi.rodino@unito.it

KARLHEINZ GRÖCHENIG*

Department of Mathematics, University of Vienna, Austria
E-mail: karlheinz.groechenig@univie.ac.at

Localization operators have been object of study in quantum mechanics, in PDE and signal analysis recently. In engineering, a natural language is given by time-frequency analysis. Arguing from this point of view, we shall present the theory of these operators developed so far. Namely, regularity properties, composition formulae and their multilinear extension shall be highlighted. Time-frequency analysis will provide tools, techniques and function spaces. In particular, we shall use modulation spaces, which allow "optimal" results in terms of regularity properties for localization operators acting on $L^2(\mathbb{R}^d)$.

1991 Mathematics Subject Classification. 47G30,35S05,46E35,47B10.

Keywords: Localization operator, modulation space, Weyl calculus, convolution relations, Wigner distribution, short-time Fourier transform, Schatten class

1. Introduction and Definitions

The name "localization operators" goes back to 1988, when I. Daubechies[17] first used these operators as a mathematical tool to localize a signal on the frequency plane. Localization operators with Gaussian windows were already known in physics: they were introduced as a quantization rule by Berezin[4] in 1971 and called anti-Wick operators. Since their first appearance, they have been extensively studied as an important mathematical tool in signal analysis and other applications (see Refs. 18,37,44 and references therein). Beyond signal analysis and the anti-Wick quantization

*K. G. was supported by the Marie-Curie Excellence Grant MEXT-CT 2004-517154, E. C. and L. R. by the FIRB Grant RBAU01XCWT

procedure,[4,38] we recall their employment as approximation of operators ("wave packets").[16,27] Besides, in other branches of mathematics, localization operators are also named Toeplitz operators (see, e.g., Ref. 19) or short-time Fourier transform multipliers.[25]

The objective of this chapter is to report on recent progress on localization operators and to present the state-of-the-art. We complement the "First survey of Gabor multipliers"[25] by Feichtinger and Nowak. Since the appearance of their survey our understanding of localization operators has expanded considerably, and many open questions have since been resolved satisfactorily.

The very definition of localization operators is carried out by frequency tools and representations, see for example.[28] Indeed, we consider the linear operators of translation and modulation (so-called time-frequency shifts) given by

$$T_x f(t) = f(t-x) \quad \text{and} \quad M_\omega f(t) = e^{2\pi i \omega t} f(t)\,. \tag{1}$$

These occur in the following time-frequency representation. Let g be a non-zero window function in the Schwartz class $\mathcal{S}(\mathbb{R}^d)$, then the short-time Fourier transform (STFT) of a signal $f \in L^2(\mathbb{R}^d)$ with respect to the window g is given by

$$V_g f(x,\omega) = \langle f, M_\omega T_x g\rangle = \int_{\mathbb{R}^d} f(t)\, \overline{g(t-x)}\, e^{-2\pi i \omega t}\, dt\,. \tag{2}$$

We have $V_g f \in L^2(\mathbb{R}^{2d})$. This definition can be extended to every pair of dual topological vector spaces, whose duality, denoted by $\langle \cdot, \cdot \rangle$, extends the inner product on $L^2(\mathbb{R}^d)$. For instance, it may be suited to the framework of distributions and ultra-distributions.

Just few words to explain the meaning of the previous "time-frequency" representation. If $f(t)$ represents a signal varying in time, its Fourier tansform $\hat{f}(\omega)$ shows the distribution of its frequency ω, without any additional information about "when" these frequencies appear. To overcome this problem, one may choose a non-negative window function g well localized around the origin. Then, the information of the signal f at the instant x can be obtained by shifting the window g till the instant x under consideration, and by computing the Fourier transform of the product $f(x)g(t-x)$, that localizes f around the instant time x.

Once the analysis of the signal f is terminated, we can reconstruct the original signal f by a suitable inversion procedure. Namely, the reproducing formula related to the STFT, for every pairs of windows $\varphi_1, \varphi_2 \in \mathcal{S}(\mathbb{R}^d)$ with

$\langle \varphi_1, \varphi_2 \rangle \neq 0$, reads as follows

$$\int_{\mathbb{R}^{2d}} V_{\varphi_1} f(x,\omega) M_\omega T_x \varphi_2 \, dx d\omega = \langle \varphi_2, \varphi_1 \rangle f \,. \tag{3}$$

The function φ_1 is called the *analysis* window, because the STFT $V_{\varphi_1} f$ gives the frequency distribution of the signal f, whereas the window φ_2 permits to come back to the original f and, consequently, is called the *synthesis* window.

The signal analysis often requires to highlight some features of the frequency distribution of f. This is achieved by first multiplying the STFT $V_{\varphi_1} f$ by a suitable function $a(x,\omega)$ and secondly by constructing $\tilde{f}$ from the product the product $aV_{\varphi_2} f$. In other words, we recover a filtered version of the original signal f which we denote by $A_a^{\varphi_1,\varphi_2}$. This intuition motivates the definition of time-frequency localization operators.

Definition 1.1. The localization operator $A_a^{\varphi_1,\varphi_2}$ with symbol $a \in \mathcal{S}(\mathbb{R}^{2d})$ and windows $\varphi_1, \varphi_2 \in \mathcal{S}(\mathbb{R}^d)$ is defined to be

$$A_a^{\varphi_1,\varphi_2} f(t) = \int_{\mathbb{R}^{2d}} a(x,\omega) V_{\varphi_1} f(x,\omega) M_\omega T_x \varphi_2(t) \, dx d\omega \,, \quad f \in L^2(\mathbb{R}^d). \tag{4}$$

The preceding definition makes sense also if we assume $a \in L^\infty(\mathbb{R}^{2d})$, see below. In particular, if $a = \chi_\Omega$ for some compact set $\Omega \subseteq \mathbb{R}^{2d}$ and $\varphi_1 = \varphi_2$, then $A_a^{\varphi_1,\varphi_2}$ is interpreted as the part of f that "lives on the set Ω" in the time-frequency plane. This is why $A_a^{\varphi_1,\varphi_2}$ is called a *localization* operator.

Often it is more convenient to interpret the definition of $A_a^{\varphi_1,\varphi_2}$ in a weak sense, then (4) can be recast as

$$\langle A_a^{\varphi_1,\varphi_2} f, g \rangle = \langle a V_{\varphi_1} f, V_{\varphi_2} g \rangle = \langle a, \overline{V_{\varphi_1} f} \, V_{\varphi_2} g \rangle, \quad f, g \in \mathcal{S}(\mathbb{R}^d). \tag{5}$$

If we enlarge the class of symbols to the tempered distributions, i.e., we take $a \in \mathcal{S}'(\mathbb{R}^{2d})$ whereas $\varphi_1, \varphi_2 \in \mathcal{S}(\mathbb{R}^d)$, then (4) is a well-defined continuous operator from $\mathcal{S}(\mathbb{R}^d)$ to $\mathcal{S}'(\mathbb{R}^d)$. The previous assertion can be proven directly using the weak definition. For every window $\varphi_1 \in \mathcal{S}(\mathbb{R}^d)$ the STFT V_{φ_1} is a continuous mapping from $\mathcal{S}(\mathbb{R}^d)$ into $\mathcal{S}(\mathbb{R}^{2d})$ (see, e.g., Ref. 28, Thm. 11.2.5). Since also $V_{\varphi_2} g \in \mathcal{S}(\mathbb{R}^{2d})$, the brackets $\langle a, \overline{V_{\varphi_1} f} \, V_{\varphi_2} g \rangle$ are well-defined in the duality between $\mathcal{S}'(\mathbb{R}^{2d})$ and $\mathcal{S}(\mathbb{R}^{2d})$. Consequently, the left-hand side of (5) can be interpreted in the duality between $\mathcal{S}'(\mathbb{R}^d)$ and $\mathcal{S}(\mathbb{R}^d)$ and shows that $A_a^{\varphi_1,\varphi_2}$ is a continuous operator from $\mathcal{S}(\mathbb{R}^d)$ to $\mathcal{S}'(\mathbb{R}^d)$. The continuity of the mapping $A_a^{\varphi_1,\varphi_2}$ is achieved by using the continuity of both the STFT and the brackets $\langle \cdot, \cdot \rangle$. Similar arguments can be applied for tempered ultra-distributions, as we are going to see later on.

If $\varphi_1(t) = \varphi_2(t) = e^{-\pi t^2}$, then $A_a = A_a^{\varphi_1,\varphi_2}$ is the classical anti-Wick operator and the mapping $a \longrightarrow A_a^{\varphi_1,\varphi_2}$ is interpreted as a quantization rule.[4,38,44]

Note that the time-frequency shifts $(x, \omega, \tau) \mapsto \tau T_x M_\omega$, $(x, \omega) \in \mathbb{R}^{2d}$, $|\tau| = 1$, define the Schrödinger representation of the Heisenberg group; for a deeper understanding of localization operators it is therefore natural to use the mathematical tools associated to harmonic analysis and time-frequency shifts, see Refs. 27,28 and the next Sec. 2.

Localization operators can be viewed as a multilinear mapping

$$(a, \varphi_1, \varphi_2) \mapsto A_a^{\varphi_1,\varphi_2}, \tag{6}$$

acting on products of symbol and window spaces. The dependence of the localization operator $A_a^{\varphi_1,\varphi_2}$ on all three parameters has been widely studied in different functional frameworks. The start was given by subspaces of the tempered distributions. The basic subspace is $L^2(\mathbb{R}^d)$, but many other Banach and Hilbert spaces, as well as topological vector spaces, have been considered. We mention L^p spaces,[6,44] potential and Sobolev spaces,[7] modulation spaces[10,25,35,42,43] and Gelfand-Shilov spaces[15] (the last ones in the ultra-distribution environment) as samples of spaces either for choosing symbol and windows or for defining the action of the related localization operator. The outcomes are manifold. The continuity of the mapping in (6) can be expressed by an inequality of the form

$$\|A_a^{\varphi_1,\varphi_2}\|_{op} \le C \|a\|_{B_1} \|\varphi_1\|_{B_2} \|\varphi_2\|_{B_3}, \tag{7}$$

where B_1, B_2, B_3 are suitable spaces of symbols and windows. For example, if $a \in L^\infty(\mathbb{R}^d)$ and $\varphi_1, \varphi_2 \in L^2(\mathbb{R}^d)$, then

$$\begin{aligned}
\|A_a^{\varphi_1,\varphi_2}\|_{B(L^2)} &= \sup_{\|f\|_{L^2}=1} \sup_{\|g\|_{L^2}=1} |\langle A_a^{\varphi_1,\varphi_2} f, g\rangle| \\
&= \sup_{\|f\|_{L^2}=1} \sup_{\|g\|_{L^2}=1} |\langle a, \overline{V_{\varphi_1} f}\, V_{\varphi_2} g\rangle| \\
&\le \sup_{\|f\|_{L^2}=1} \sup_{\|g\|_{L^2}=1} \|a\|_{L^\infty} \|\overline{V_{\varphi_1} f}\, V_{\varphi_2} g\|_{L^1} \\
&\le \sup_{\|f\|_{L^2}=1} \sup_{\|g\|_{L^2}=1} \|a\|_{L^\infty} \|V_{\varphi_1} f\|_{L^2} \|V_{\varphi_2} g\|_{L^2} \\
&= \|a\|_{L^\infty} \|\varphi_1\|_{L^2} \|\varphi_2\|_{L^2},
\end{aligned}$$

where the last inequality is achieved by using the orthogonality relations for the STFT

$$\|V_\varphi f\|_{L^2(\mathbb{R}^{2d})} = \|\varphi\|_{L^2(\mathbb{R}^d)} \|f\|_{L^2(\mathbb{R}^d)}, \quad \forall \varphi, f \in L^2(\mathbb{R}^d).$$

Thus for this particular choice of symbol classes and window spaces we obtain the L^2 boundedness. The previous easy proof gives just a flavour of the boundedness results for localization operators, we shall see that the symbol class L^∞ can be enlarged significantly. Even a tempered distribution like δ may give the boundedness of the corresponding localization operator. Apart from continuity, estimates of the type (7) also supply Hilbert-Schmidt, Trace class and Schatten class properties for $A_a^{\varphi_1,\varphi_2}$.[11,15]

Among the many function/(ultra-)distribution spaces employed, modulation spaces reveal to be the *optimal choice* for handling localization operators, see Sec. 3 below. As special case we mention Feichtinger's algebra $M^1(\mathbb{R}^d)$ defined by the norm

$$\|f\|_{M^1} := \|V_g f\|_{L^1(\mathbb{R}^{2d})}$$

for some (hence all) non-zero $g \in \mathcal{S}(\mathbb{R}^d)$.[23,28] Its dual space $M^\infty(\mathbb{R}^{2d})$ is a very useful subspace of tempered distributions and possesses the norm

$$\|f\|_{M^\infty} := \sup_{(x,\omega)\in\mathbb{R}^{2d}} |V_g f(x,\omega)| \,.$$

With these spaces the estimate (7) reads as follows:

Theorem 1.1. *If $a \in M^\infty(\mathbb{R}^{2d})$, and $\varphi_1, \varphi_2 \in M^1(\mathbb{R}^d)$, then $A_a^{\varphi_1,\varphi_2}$ is bounded on $L^2(\mathbb{R}^d)$, with operator norm at most*

$$\|A_a^{\varphi_1,\varphi_2}\|_{B(L^2)} \le C\|a\|_{M^\infty}\|\varphi_1\|_{M^1}\|\varphi_2\|_{M^1}\,.$$

The striking fact is the converse of the preceding result.[10]

Theorem 1.2. *If $A_a^{\varphi_1,\varphi_2}$ is bounded on $L^2(\mathbb{R}^d)$ uniformly with respect to* all *windows $\varphi_1, \varphi_2 \in M^1$, i.e., if there exists a constant $C > 0$ depending only on the symbol a such that, for all $\varphi_1, \varphi_2 \in \mathcal{S}(\mathbb{R}^d)$,*

$$\|A_a^{\varphi_1,\varphi_2}\|_{B(L^2)} \le C\|\varphi_1\|_{M^1}\,\|\varphi_2\|_{M^1}\,, \tag{8}$$

then $a \in M^\infty$.

Similar statements hold true for Schatten class properties[11] and for weighted ultra-distributional modulation spaces.[15] A recent result in the study of localization operators[26] reveals the optimality of modulation spaces even for the compactness property. These topics shall be detailed in Secs. 4 and 5.

In Sec. 6 we shall treat the composition of localization operators. Whereas the product of two operators is again a pseudodifferential operator, in general the composition of two localization operators is no longer a localization operator. This additional difficulty has captured the interest of several authors, generating some remarkable ideas. An exact product formula for localization operators, obtained in Ref. 20, shall be presented. Notice, however, that it works only under very restrictive conditions and is unstable. In another direction, many authors have made resort to asymptotic expansions that realize the composition of two localization operators as a sum of localization operators and a controllable remainder.[1,14,33,40] These contributions are mainly motivated by applications to PDEs and energy estimates, and therefore use smooth symbols defined by differentiability properties, such as the traditional Hörmander or Shubin classes, and Gaussian windows. In the context of time-frequency analysis, where modulation spaces can be employed, much rougher symbols and more general window functions are allowed to be used for localization operators. Consequently, the product formula in Ref.14 has been extended to rougher spaces of symbols in Ref. 12, as we are going to show.

In the end (Sec. 7), we shall present a new framework for localization operators. Namely, the study of multilinear pseudodifferential operators[2,3] motivates the definition of multilinear localization operators. For them, we shall present the sufficient and necessary boundedness properties together with connection with Kohn-Nirenberg operators.[13]

Notation. We define $t^2 = t \cdot t$, for $t \in \mathbb{R}^d$, and $xy = x \cdot y$ is the scalar product on $\mathbb{R}^d$.

The Schwartz class is denoted by $\mathcal{S}(\mathbb{R}^d)$, the space of tempered distributions by $\mathcal{S}'(\mathbb{R}^d)$. We use the brackets $\langle f, g\rangle$ to denote the extension to $\mathcal{S}(\mathbb{R}^d) \times \mathcal{S}'(\mathbb{R}^d)$ of the inner product $\langle f, g\rangle = \int f(t)\overline{g(t)}dt$ on $L^2(\mathbb{R}^d)$. The Fourier transform is normalized to be $\hat{f}(\omega) = \mathcal{F}f(\omega) = \int f(t)e^{-2\pi i t\omega}dt$, the involution g^* is $g^*(t) = \overline{g(-t)}$.

The singular values $\{s_k(L)\}_{k=1}^{\infty}$ of a compact operator $L \in B(L^2(\mathbb{R}^d))$ are the eigenvalues of the positive self-adjoint operator $\sqrt{L^*L}$. Equivalently, for every $k \in \mathbb{N}$, the singular value $\{s_k(L)\}$ is given by

$$s_k(L) = \inf\{\|L - T\|_{L^2} : T \in B(L^2(\mathbb{R}^d)) \quad \text{and} \quad \dim \mathrm{Im}\,(T) \leq k\}.$$

For $1 \leq p < \infty$, the Schatten class S_p is the space of all compact operators whose singular values lie in l^p. For consistency, we define $S_\infty := B(L^2(\mathbb{R}^d))$ to be the space of bounded operators on $L^2(\mathbb{R}^d)$. In particular, S_2 is the space of Hilbert-Schmidt operators, and S_1 is the space of trace class operators.

Throughout the paper, we shall use the notation $A \lesssim B$ to indicate

$A \leq cB$ for a suitable constant $c > 0$, whereas $A \asymp B$ if $A \leq cB$ and $B \leq kA$, for suitable $c, k > 0$.

2. Time-Frequency Methods

First we summarize some concepts and tools of time-frequencya, for an extended exposition we refer to the textbooks.[27,28]

The time-frequencyr s required for localization operators and the Weyl calculus are the *short-time Fourier transform* and the *Wigner distribution.*

The short-time Fourier transform (STFT) is defined in (2). The *cross-Wigner distribution* $W(f,g)$ of $f, g \in L^2(\mathbb{R}^d)$ is given by

$$W(f,g)(x,\omega) = \int f(x + \frac{t}{2})\overline{g(x - \frac{t}{2})} e^{-2\pi i \omega t}\, dt. \tag{9}$$

The quadratic expression $Wf = W(f,f)$ is usually called the Wigner distribution of f.

Both the STFT $V_g f$ and the Wigner distribution $W(f,g)$ are defined for f, g in many possible pairs of Banach spaces. For instance, they both map $L^2(\mathbb{R}^d) \times L^2(\mathbb{R}^d)$ into $L^2(\mathbb{R}^{2d})$ and $\mathcal{S}(\mathbb{R}^d) \times \mathcal{S}(\mathbb{R}^d)$ into $\mathcal{S}(\mathbb{R}^{2d})$. Furthermore, they can be extended to a map from $\mathcal{S}'(\mathbb{R}^d) \times \mathcal{S}'(\mathbb{R}^d)$ into $\mathcal{S}'(\mathbb{R}^{2d})$.

For a non-zero $g \in L^2(\mathbb{R}^d)$, we write V_g^* for the adjoint of V_g, given by

$$\langle V_g^* F, f\rangle = \langle F, V_g f\rangle, \quad f \in L^2(\mathbb{R}^d),\ F \in L^2(\mathbb{R}^{2d}).$$

In particular, for $F \in \mathcal{S}(\mathbb{R}^{2d})$, $g \in \mathcal{S}(\mathbb{R}^d)$, we have

$$V_g^* F(t) = \int_{\mathbb{R}^{2d}} F(x,\omega) M_\omega T_x g(t)\, dx\, d\omega \in \mathcal{S}(\mathbb{R}^d). \tag{10}$$

Take $f \in \mathcal{S}(\mathbb{R}^d)$ and set $F = V_g f$, then

$$f(t) = \frac{1}{\|g\|_{L^2}^2} \int_{\mathbb{R}^{2d}} V_g f(x,\omega) M_\omega T_x g(t)\, dx\, d\omega \in \mathcal{S}(\mathbb{R}^d) = \frac{1}{\|g\|_{L^2}^2} V_g^* V_g f(t). \tag{11}$$

We refer to Ref. 28, Prop. 11.3.2 for a detailed treatment of the adjoint operator.

Representation of localization operators as Weyl/Kohn-Nirenberg operators. Let $W(g,f)$ be the cross-Wigner distribution as defined in (9). Then the Weyl operator L_σ of symbol $\sigma \in \mathcal{S}'(\mathbb{R}^{2d})$ is defined by

$$\langle L_\sigma f, g\rangle = \langle \sigma, W(g,f)\rangle, \qquad f, g \in \mathcal{S}(\mathbb{R}^d). \tag{12}$$

Every linear continuous operator from $\mathcal{S}(\mathbb{R}^d)$ to $\mathcal{S}'(\mathbb{R}^d)$ can be represented as a Weyl operator, and a calculation in Refs. 7,27,38 reveals that

$$A_a^{\varphi_1,\varphi_2} = L_{a*W(\varphi_2,\varphi_1)}, \tag{13}$$

so the (Weyl) symbol of $A_a^{\varphi_1,\varphi_2}$ is given by

$$\sigma = a * W(\varphi_2, \varphi_1). \tag{14}$$

To get boundedness results for a localization operator, it is sometimes convenient to write it in a different pseudodifferential form. Consider the *Kohn-Nirenberg* form of a pseudodifferential operator, given by

$$T_\tau f(x) = \int_{\mathbb{R}^d} \tau(x,\omega)\hat{f}(\omega)e^{2\pi i x\omega}\, d\omega, \quad f \in \mathcal{S}(\mathbb{R}^d), \tag{15}$$

where τ is a measurable function, or even a tempered distribution on $\mathbb{R}^{2d}$.

If we define the *rotation* operator $\mathcal{U}$ acting on a function F on $\mathbb{R}^{2d}$ by

$$\mathcal{U}\mathcal{F}(x,\omega) = \mathcal{F}(\omega,-x), \quad \forall\,(x,\omega) \in \mathbb{R}^{2d}, \tag{16}$$

then, the identity of operators below holds:[13]

$$A_a^{\varphi_1,\varphi_2} = T_\tau, \tag{17}$$

with the Kohn-Nirenberg symbol τ given by

$$\tau = a * \mathcal{U}\mathcal{F}(V_{\varphi_1}\varphi_2) \tag{18}$$

The expression $\mathcal{U}\mathcal{F}(V_{\varphi_1}\varphi_2)$ is usually called the Rihaczek distribution.

3. Function Spaces

Gelfand-Shilov spaces. The Gelfand-Shilov spaces were introduced by Gelfand and Shilov in Ref. 30. They have been applied by many authors in different contexts, see, e.g. Refs. 9,32,34,41. For the sake of completeness, we recall their definition and properties in more generality than required.

Definition 3.1. Let $\alpha, \beta \in \mathbb{R}_+^d$, and assume $A_1, \ldots, A_d$, $B_1, \ldots, B_d > 0$. Then the Gelfand-Shilov space $\mathcal{S}_{\beta,A}^{\alpha,B} = \mathcal{S}_{\beta,A}^{\alpha,B}(\mathbb{R}^d)$ is defined by

$$\begin{aligned}\mathcal{S}_{\beta,A}^{\alpha,B} =&\{f \in C^\infty(\mathbb{R}^d) \mid (\exists C > 0)\ \|x^p\partial^q f\|_{L^\infty} \\ &\leq CA^p(p!)^\beta B^q(q!)^\alpha,\ \forall p,q \in \mathbb{N}_0^d\}.\end{aligned}$$

We then consider projective and inductive limits denoted by

$$\Sigma_\beta^\alpha = \text{proj} \lim_{A>0,B>0} \mathcal{S}_{\beta,A}^{\alpha,B}; \quad \mathcal{S}_\beta^\alpha := \text{ind} \lim_{A>0,B>0} \mathcal{S}_{\beta,A}^{\alpha,B}.$$

For a comprehensive treatment of Gelfand-Shilov spaces we refer to Ref. 30. We limit ourselves to those features that will be useful for our study.

Proposition 3.1. *The next statements are equivalent:*[15,30,31]

- $f \in \mathcal{S}_\beta^\alpha(\mathbb{R}^d)$.
- $f \in \mathcal{C}^\infty(\mathbb{R}^d)$ *and there exist real constants* $h > 0, k > 0$ *such that:*

$$\|fe^{h|x|^{1/\beta}}\|_{L^\infty} < \infty \quad \textit{and} \quad \|\mathcal{F}fe^{k|\omega|^{1/\alpha}}\|_{L^\infty} < \infty, \tag{19}$$

where $|x|^{1/\beta} = |x_1|^{1/\beta_1} + \cdots + |x_d|^{1/\beta_d}$, $|\omega|^{1/\alpha} = |\omega_1|^{1/\alpha_1} + \cdots + |\omega_d|^{1/\alpha_d}$.
- $f \in \mathcal{C}^\infty(\mathbb{R}^d)$ *and there exists* $C > 0, h > 0$ *such that*

$$\|(\partial^q f)e^{h|x|^{1/\beta}}\|_{L^\infty} \le C^{|q|+1}(q!)^\alpha, \quad \forall q \in \mathbb{N}_0^d. \tag{20}$$

Gelfand-Shilov spaces enjoy the following embeddings:
(i) For $\alpha, \beta \ge 0$,[30]

$$\Sigma_\beta^\alpha \hookrightarrow \mathcal{S}_\beta^\alpha \hookrightarrow \mathcal{S}. \tag{21}$$

(ii) For every $0 \le \alpha_1 < \alpha_2$ and $0 \le \beta_1 < \beta_2$,[15]

$$\mathcal{S}_{\beta_1}^{\alpha_1} \hookrightarrow \Sigma_{\beta_2}^{\alpha_2}. \tag{22}$$

Furthermore, S_β^α is not trivial if and only if $\alpha + \beta > 1$ or $\alpha + \beta = 1$ and $\alpha\beta > 0$. The spaces Σ_α^α with $\alpha \ge 1/2$ are studied by Pilipović.[34] In particular, the case $\alpha = 1/2$ yields $\Sigma_{1/2}^{1/2} = \emptyset$.

The Fourier transform $\mathcal{F}$ is a topological isomorphism between S_α^β and S_β^α ($\mathcal{F}(S_\alpha^\beta) = S_\beta^\alpha$) and extends to a continuous linear transform from $(S_\alpha^\beta)'$ onto $(S_\beta^\alpha)'$. If $\alpha \ge 1/2$, then $\mathcal{F}(S_\alpha^\alpha) = S_\alpha^\alpha$. The Gelfand-Shilov spaces are invariant under time-frequency shifts:

$$T_x(S_\alpha^\beta) = S_\alpha^\beta \quad \text{and} \quad M_\omega(S_\alpha^\beta) = S_\alpha^\beta, \tag{23}$$

and similarly to the Σ_β^α.

Therefore the spaces S_α^α are a family of *Fourier transform and time-frequencys invariant spaces* which are contained in the Schwartz class $\mathcal{S}$. Among these S_α^α the *smallest* non-trivial Gelfand-Shilov space is given by $\mathcal{S}_{1/2}^{1/2}$. A basic example is given by $f(x) = e^{-\pi x^2} \in \mathcal{S}_{1/2}^{1/2}(\mathbb{R}^d)$.

Another useful characterization of the space $\mathcal{S}_\alpha^\alpha$ involves the STFT: $f \in \mathcal{S}_\alpha^\alpha(\mathbb{R}^d)$ if and only if $V_g f \in \mathcal{S}_\alpha^\alpha(\mathbb{R}^{2d})$ (see Ref. 32, Prop. 3.12 and reference therein). We will use the case $\alpha = 1/2$: for a non-zero window $g \in \mathcal{S}_{1/2}^{1/2}$ we have

$$V_g f \in \mathcal{S}_{1/2}^{1/2}(\mathbb{R}^{2d}) \Leftrightarrow f \in \mathcal{S}_{1/2}^{1/2}(\mathbb{R}^d). \tag{24}$$

The strong duals of Gelfand-Shilov classes $\mathcal{S}^\alpha_\beta$ and Σ^α_β are spaces of tempered ultra-distributions of Roumieu and Beurling type and will be denoted by $(\mathcal{S}^\alpha_\beta)'$ and $(\Sigma^\alpha_\beta)'$, respectively.

Modulation Spaces. The modulation space norms traditionally measure the joint time-frequency distribution of $f \in \mathcal{S}'$, we refer, for instance, to Refs. 21,28, Ch. 11-13 and the original literature quoted there for various properties and applications. In that setting it is sufficient to observe modulation spaces with weights which admit at most polynomial growth at infinity. However the study of ultra-distributions requires a more general approach that includes the weights of exponential growth.

Weight Functions. In the sequel v will always be a continuous, positive, even, submultiplicative function (submultiplicative weight), i.e., $v(0) = 1$, $v(z) = v(-z)$, and $v(z_1 + z_2) \leq v(z_1)v(z_2)$, for all $z, z_1, z_2 \in \mathbb{R}^{2d}$. Moreover, v is assumed to be even in each group of coordinates, that is, $v(\pm x, \pm\omega) = v(x,\omega)$, for all $(x,\omega) \in \mathbb{R}^{2d}$ and all choices of signs. Submultiplicativity implies that $v(z)$ is *dominated* by an exponential function, i.e.

$$\exists\, C, k > 0 \quad \text{such that} \quad v(z) \leq Ce^{k|z|}, \quad z \in \mathbb{R}^{2d}. \tag{25}$$

For example, every weight of the form $v(z) = e^{a|z|^b}(1+|z|)^s \log^r(e+|z|)$ for parameters $a, r, s \geq 0$, $0 \leq b \leq 1$ satisfies the above conditions.

Associated to every submultiplicative weight we consider the class of so-called *v-moderate* weights $\mathcal{M}_v$. A positive, even weight function m on $\mathbb{R}^{2d}$ belongs to $\mathcal{M}_v$ if it satisfies the condition

$$m(z_1 + z_2) \leq Cv(z_1)m(z_2) \quad \forall z_1, z_2 \in \mathbb{R}^{2d}.$$

We note that this definition implies that $\frac{1}{v} \lesssim m \lesssim v$, $m \neq 0$ everywhere, and that $1/m \in \mathcal{M}_v$.

For the investigation of localization operators the weights mostly used are defined by

$$v_s(z) = v_s(x,\omega) = \langle z\rangle^s = (1 + x^2 + \omega^2)^{s/2}, \quad z = (x,\omega) \in \mathbb{R}^{2d} \tag{26}$$
$$w_s(z) = w_s(x,\omega) = e^{s|(x,\omega)|}, \quad z = (x,\omega) \in \mathbb{R}^{2d}, \tag{27}$$
$$\tau_s(z) = \tau_s(x,\omega) = \langle\omega\rangle^s \tag{28}$$
$$\mu_s(z) = \mu_s(x,\omega) = e^{s|\omega|}. \tag{29}$$

Definition 3.2.

Let m be a weight in $\mathcal{M}_v$, and g a non-zero *window* function in $\mathcal{S}^{1/2}_{1/2}$. For $1 \leq p, q \leq \infty$ and $f \in \mathcal{S}^{1/2}_{1/2}$ we define the modulation space norm (on

$\mathcal{S}_{1/2}^{1/2}$) by

$$\|f\|_{M_m^{p,q}} = \|V_g f\|_{L_m^{p,q}} = \left(\int_{\mathbb{R}^d}\left(\int_{\mathbb{R}^d}|V_g f(x,\omega)|^p m(x,\omega)^p\,dx\right)^{q/p} d\omega\right)^{1/q},$$

(with obvious changes if either $p=\infty$ or $q=\infty$). If $p,q<\infty$, the modulation space $M_m^{p,q}$ is the norm completion of $\mathcal{S}_{1/2}^{1/2}$ in the $M_m^{p,q}$-norm. If $p=\infty$ or $q=\infty$, then $M_m^{p,q}$ is the completion of $\mathcal{S}_{1/2}^{1/2}$ in the weak* topology. If $p=q$, $M_m^p := M_m^{p,p}$, and, if $m\equiv 1$, then $M^{p,q}$ and M^p stand for $M_m^{p,q}$ and $M_m^{p,p}$, respectively.

Notice that:

- If $f,g\in\mathcal{S}_{1/2}^{1/2}(\mathbb{R}^d)$, the above integral is convergent thanks to (19) and (24). Namely, the constant h in (19) guarantees $\|V_g f e^{h|\cdot|^2}\|_{L^\infty}<\infty$ and, for $m\in\mathcal{M}_v$, we have

$$\int_{\mathbb{R}^d}\left(\int_{\mathbb{R}^d}|V_g f(x,\omega)|^p m(x,\omega)^p\,dx\right)^{q/p} d\omega$$
$$\leq C\,\|(V_g f)e^{h|\cdot|^2}\|_{L^\infty}\int_{\mathbb{R}^d}\left(\int_{\mathbb{R}^d}|m(x,\omega)|^p e^{-hp|(x,\omega)|^2}\,dx\right)^{q/p} d\omega<\infty\,.$$

- By definition, $M_m^{p,q}$ is a Banach space. Besides, it is proven for the subexponential case in Ref. 21 and for the exponential one in Ref. 15 that their definition does not depend on the choice of the window g, that can be enlarged to the modulation algebra M_v^1.
- For $m\in\mathcal{M}_v$ of at most polynomial growth, $M_m^{p,q}\subset\mathcal{S}'$ and the definition 3.2 reads as:[10,28]

$$M_m^{p,q}(\mathbb{R}^d)=\{f\in\mathcal{S}'(\mathbb{R}^d):\ V_g f\in L_m^{p,q}(\mathbb{R}^{2d})\}.$$

For every weight $m\in\mathcal{M}_v$, $M_m^{p,q}$ is the subspace of ultra-distribution $(\Sigma_1^1)'$ defined in Ref. 15, Def. 2.1.

(iv) If m belongs to $\mathcal{M}_v$ and fulfills the GRS-condition $\lim_{n\to\infty}v(nz)^{1/n}=1$, for all $z\in\mathbb{R}^{2d}$, the definition of modulation spaces is the same as in Ref. 12 (because the "space of special windows" $\mathcal{S}_\mathcal{C}$ is a subset of $\mathcal{S}_{1/2}^{1/2}$).
(v) For related constructions of modulation spaces, involving the theory of coorbit spaces, we refer to Refs. 22,24.

The class of modulation spaces contains the following well-known function spaces:
Weighted L^2-spaces: $M^2_{\langle x\rangle^s}(\mathbb{R}^d)=L^2_s(\mathbb{R}^d)=\{f\,:\,f(x)\langle x\rangle^s\in L^2(\mathbb{R}^d)\}$, $s\in$

$\mathbb{R}$.
Sobolev spaces: $M^2_{\langle\omega\rangle^s}(\mathbb{R}^d) = H^s(\mathbb{R}^d) = \{f : \hat{f}(\omega)\langle\omega\rangle^s \in L^2(\mathbb{R}^d)\}$, $s \in \mathbb{R}$.
Shubin-Sobolev spaces:[7,38] $M^2_{\langle(x,\omega)\rangle^s}(\mathbb{R}^d) = L^2_s(\mathbb{R}^d) \cap H^s(\mathbb{R}^d) = Q_s(\mathbb{R}^d)$.
Feichtinger's algebra: $M^1(\mathbb{R}^d) = S_0(\mathbb{R}^d)$.

The characterization of the Schwartz class of tempered distributions is given in Ref. 31: we have $\mathcal{S}(\mathbb{R}^d) = \bigcap_{s\geq 0} M^1_{\langle\cdot\rangle^s}(\mathbb{R}^d)$ and $\mathcal{S}'(\mathbb{R}^d) = \bigcup_{s\geq 0} M^\infty_{1/\langle\cdot\rangle^s}(\mathbb{R}^d)$. A similar characterization for Gelfand-Shilov spaces and tempered ultra-distributions was obtained in Ref. 15, Prop. 2.3: Let $1 \leq p, q \leq \infty$, and let w_s be given by (27), then,

$$\mathcal{S}^1_1 = \bigcap_{s\geq 0} M^{p,q}_{w_s}, \quad (\mathcal{S}^1_1)' = \bigcup_{s\geq 0} M^{p,q}_{1/w_s}. \tag{30}$$

$$\Sigma^1_1 = \bigcup_{s>0} M^{p,q}_{w_s}, \quad (\Sigma^1_1)' = \bigcap_{s>0} M^{p,q}_{1/w_s}. \tag{31}$$

Potential spaces. For $s \in \mathbb{R}$ the Bessel kernel is

$$G_s = \mathcal{F}^{-1}\{(1 + |\cdot|^2)^{-s/2}\}, \tag{32}$$

and the *potential space* [5] is defined by

$$W^p_s = G_s * L^p(\mathbb{R}^d) = \{f \in \mathcal{S}',\, f = G_s * g,\, g \in L^p\}$$

with norm $\|f\|_{W^p_s} = \|g\|_{L^p}$.

For comparison we list the following embeddings between potential and modulation spaces.[10]

Lemma 3.1. *We have*

- *If* $p_1 \leq p_2$ *and* $q_1 \leq q_2$, *then* $M^{p_1,q_1}_m \hookrightarrow M^{p_2,q_2}_m$.
- *For* $1 \leq p \leq \infty$ *and* $s \in \mathbb{R}$

$$W^p_s(\mathbb{R}^d) \hookrightarrow M^{p,\infty}_{\tau_s}(\mathbb{R}^d).$$

Consequently, $L^p \subseteq M^{p,\infty}$, and in particular, $L^\infty \subseteq M^\infty$. But M^∞ contains all bounded measures on $\mathbb{R}^d$ and other tempered distributions. For instance, the point measure δ belongs to M^∞, because for $g \in \mathcal{S}$ we have

$$|V_g\delta(x,\omega)| = |\langle\delta, M_\omega T_x g\rangle| = |\bar{g}(-x)| \leq \|g\|_{L^\infty}, \quad \forall (x,\omega) \in \mathbb{R}^{2d}.$$

Convolution Relations and Wigner Estimates. In view of the relation between the multiplier a and the Weyl symbol (14), we need to understand the convolution relations between modulation spaces and some properties of the Wigner distribution.

We first state a *convolution relation for modulation spaces* proven in Ref. 10, in the style of Young's theorem. Let v be an arbitrary submultiplicative weight on $\mathbb{R}^{2d}$ and m a v-moderate weight. We write $m_1(x) = m(x,0)$ and $m_2(\omega) = m(0,\omega)$ for the restrictions to $\mathbb{R}^d \times \{0\}$ and $\{0\} \times \mathbb{R}^d$, and likewise for v.

Proposition 3.2. *Let $\nu(\omega) > 0$ be an arbitrary weight function on $\mathbb{R}^d$ and $1 \leq p, q, r, s, t \leq \infty$. If*

$$\frac{1}{p} + \frac{1}{q} - 1 = \frac{1}{r}, \qquad \text{and} \qquad \frac{1}{t} + \frac{1}{t'} = 1\,,$$

then

$$M^{p,st}_{m_1 \otimes \nu}(\mathbb{R}^d) * M^{q,st'}_{v_1 \otimes v_2 \nu^{-1}}(\mathbb{R}^d) \hookrightarrow M^{r,s}_m(\mathbb{R}^d) \tag{33}$$

with norm inequality $\|f * h\|_{M^{r,s}_m} \lesssim \|f\|_{M^{p,st}_{m_1 \otimes \nu}} \|h\|_{M^{q,st'}_{v_1 \otimes v_2 \nu^{-1}}}$.

1. Despite the large number of indices, the statement of this proposition has some intuitive meaning: a function $f \in M^{p,q}$ behaves like $f \in L^p$ and $\hat{f} \in L^q$; so the parameters related to the x-variable behave like those in Young's theorem for convolution, whereas the parameters related to ω behave like Hölder's inequality for pointwise multiplication.
2. A special case of Proposition 3.2 with a different proof is contained in Ref. 42.

The modulation space norm of a cross-Wigner distribution may be controlled by the window norms, as taken from Refs. 10,15.

Proposition 3.3. *Let $1 \leq p \leq \infty$ and $s \geq 0$.*

- *If $\varphi_1 \in M^1_{v_s}(\mathbb{R}^d)$ and $\varphi_2 \in M^p_{v_s}(\mathbb{R}^d)$, then $W(\varphi_2, \varphi_1) \in M^{1,p}_{\tau_s}(\mathbb{R}^{2d})$, with*
$$\|W(\varphi_2, \varphi_1)\|_{M^{1,p}_{\tau_s}} \lesssim \|\varphi_1\|_{M^1_{v_s}} \|\varphi_2\|_{M^p_{v_s}}. \tag{34}$$
- *If $\varphi_1 \in M^1_{w_s}(\mathbb{R}^d)$ and $\varphi_2 \in M^p_{w_s}(\mathbb{R}^d)$, then $W(\varphi_2, \varphi_1) \in M^{1,p}_{\mu_s}(\mathbb{R}^{2d})$ with*
$$\|W(\varphi_2, \varphi_1)\|_{M^{1,p}_{\mu_s}} \lesssim \|\varphi_1\|_{M^1_{w_s}} \|\varphi_2\|_{M^p_{w_s}}. \tag{35}$$

4. Regularity Results

In this section, we first give general sufficient conditions for boundedness and Schatten classes of localization operators. Then we treat ultradistributions with compact support as symbols, and finally we shall state a compactness result.

4.1. *Sufficient conditions for boundedness and schatten class*

Using the tools of time-frequencya in Sec. 3, we can now obtain the properties of localization operators with symbols in modulation spaces, by reducing the problem to the corresponding one for the Weyl calculus.

First, we recall a boundedness and trace class result for the Weyl operators in terms of modulation spaces.

Theorem 4.1.

- *If* $\sigma \in M^{\infty,1}(\mathbb{R}^{2d})$, *then* L_σ *is bounded on* $M^{p,q}(\mathbb{R}^d)$, $1 \leq p,q \leq \infty$, *with a uniform estimate* $\|L_\sigma\|_{S_\infty} \lesssim \|\sigma\|_{M^{\infty,1}}$ *for the operator norm. In particular,* L_σ *is bounded on* $L^2(\mathbb{R}^d)$.
- *If* $\sigma \in M^1(\mathbb{R}^{2d})$, *then* $L_\sigma \in S_1$ *and* $\|L_\sigma\|_{S^1} \lesssim \|\sigma\|_{M^1}$.
- *If* $1 \leq p \leq 2$ *and* $\sigma \in M^p(\mathbb{R}^{2d})$, *then* $L_\sigma \in S_p$ *and* $\|L_\sigma\|_{S_p} \lesssim \|\sigma\|_{M^p}$.
- *If* $2 \leq p \leq \infty$ *and* $\sigma \in M^{p,p'}(\mathbb{R}^{2d})$, *then* $L_\sigma \in S_p$ *and* $\|L_\sigma\|_{S_p} \lesssim \|\sigma\|_{M^{p,p'}}$.

One of many proofs of (i) can be found in Ref. 28, Thm. 14.5.2, the L^2-boundedness was first discovered by Sjöstrand.[39] The trace class property (ii) is proved in Ref. 29, whereas (iii) and (iv) follow by interpolation from the first two statements, since $[M^1, M^2]_\theta = M^p$ for $1 \leq p \leq 2$, and $[M^{\infty,1}, M^{2,2}]_\theta = M^{p,p'}$ for $2 \leq p \leq \infty$.

Based on the Thm. 4.1 and Prop. 3.2, we present the most general boundedness results for localization operators obtained so far. We detail the polynomial weight case, the exponential one is stated and proved by replacing the weight v_s by w_s and τ_s by μ_s (see Ref. 15, Thm. 3.2).

Theorem 4.2. *Let* $s \geq 0$, $a \in M^\infty_{1/\tau_s}(\mathbb{R}^{2d})$, $\varphi_1, \varphi_2 \in M^1_{v_s}(\mathbb{R}^d)$. *Then* $A_a^{\varphi_1,\varphi_2}$ *is bounded on* $M^{p,q}(\mathbb{R}^d)$ *for all* $1 \leq p,q \leq \infty$, *and the operator norm satisfies the uniform estimate*

$$\|A_a^{\varphi_1,\varphi_2}\|_{S_\infty} \lesssim \|a\|_{M^\infty_{1/\tau_s}} \|\varphi_1\|_{M^1_{v_s}} \|\varphi_2\|_{M^1_{v_s}}.$$

Proof. See Ref. 10, Thm. 3.2. To highlight the role of time-frequency analysis, we sketch the proof. An appropriate convolution relation is employed to show that the Weyl symbol $a*W(\varphi_2,\varphi_1)$ of $A_a^{\varphi_1,\varphi_2}$ is in $M^{\infty,1}$. Namely, if $\varphi_1, \varphi_2 \in M^1_{v_s}(\mathbb{R}^d)$, then by (34) we have $W(\varphi_2,\varphi_1) \in M^1_{\tau_s}(\mathbb{R}^{2d})$. Applying Proposition 3.2 in the form $M^\infty_{1/\tau_s} * M^1_{\tau_s} \subseteq M^{\infty,1}$, we obtain that the Weyl symbol $\sigma = a * W(\varphi_2,\varphi_1) \in M^{\infty,1}$. The result now follows from Theorem 4.1 (*i*). □

To compare Theorem 4.2 to existing results, we recall that the standard condition for $A_a^{\varphi_1,\varphi_2}$ to be bounded is $a \in L^\infty(\mathbb{R}^{2d})$, see Ref. 44. A more subtle result of Feichtinger and Nowak[25] shows that the condition a in the Wiener amalgam space $W(M, L^\infty)$ is sufficient for boundedness. Since we have the proper embeddings $L^\infty \subset W(M, L^\infty) \subset M^\infty \subset M^\infty_{1/\tau_s}$ for $s \geq 0$, Theorem 4.2 appears as a significant improvement. A special case of Theorem 4.2 follows also from Toft's work.[43]

Since $\tau_s(z, \zeta) = \langle \zeta \rangle^s$ depends only on the frequency variable, the condition $a \in M^\infty_{1/\tau_s}$ describes the admissible roughness of a, while in some sense a remains bounded in z. On the other hand, if we allow the symbol a to grow in both time and frequency by choosing the "full" weight $v_s = \langle (z, \zeta) \rangle^s$, then we obtain a negative result Ref. 10, Prop. 3.3:

Proposition 4.1. *For any $s > 0$ there exist symbols $a \in M^\infty_{1/v_s}(\mathbb{R}^{2d})$ and windows $\varphi_1, \varphi_2 \in \mathcal{S}(\mathbb{R}^d)$ such that $A_a^{\varphi_1,\varphi_2}$ is unbounded on $L^2(\mathbb{R}^d)$.*

These results demonstrate that bounded symbols with negative smoothness may still yield bounded localization operators, provided that the roughness of a is compensated by a suitable time-frequency localization of the windows. On the other hand, a smooth unbounded symbol cannot, in general, yield a bounded operator.

The Schatten class properties of localization operators with symbols in modulation spaces are achieved accordingly. Combining Proposition 3.2 with Theorem 4.1, almost optimal conditions for $A_a^{\varphi_1,\varphi_2} \in S_p$ are derived in Refs. 10,15. Again, we state the earliest result for weights of polynomial growth Ref. 10, Thm. 3.4.

Theorem 4.3.

- *If $1 \leq p \leq 2$, then the mapping $(a, \varphi_1, \varphi_2) \mapsto A_a^{\varphi_1,\varphi_2}$ is bounded from $M^{p,\infty}_{1/\tau_s}(\mathbb{R}^{2d}) \times M^1_{v_s}(\mathbb{R}^d) \times M^p_{v_s}(\mathbb{R}^d)$ into S_p, in other words,*

$$\|A_a^{\varphi_1,\varphi_2}\|_{S_p} \lesssim \|a\|_{M^{p,\infty}_{1/\tau_s}} \|\varphi_1\|_{M^1_{v_s}} \|\varphi_2\|_{M^p_{v_s}} .$$

- *If $2 \leq p \leq \infty$, then the mapping $(a, \varphi_1, \varphi_2) \mapsto A_a^{\varphi_1,\varphi_2}$ is bounded from $M^{p,\infty}_{1/\tau_s} \times M^1_{v_s} \times M^{p'}_{v_s}$ into S_p, and*

$$\|A_a^{\varphi_1,\varphi_2}\|_{S_p} \lesssim \|a\|_{M^{p,\infty}_{1/\tau_s}} \|\varphi_1\|_{M^1_{v_s}} \|\varphi_2\|_{M^{p'}_{v_s}} .$$

Using the embeddings $W^p_{-s} \hookrightarrow M^{p,\infty}_{1/\tau_s}$ (Lemma 3.1) and $M^1_{v_s} \hookrightarrow M^p_{v_s}$, one obtains a slightly weaker statement for symbols in potential spaces. This result was already derived in Ref. 7, Thm. 4.7.

Corollary 4.1. *Let $a \in W^p_{-s}(\mathbb{R}^{2d})$ for some $s \geq 0$, $1 \leq p \leq \infty$, and $\varphi_1, \varphi_2 \in M^1_{v_s}(\mathbb{R}^d)$. Then*

$$\|A_a^{\varphi_1,\varphi_2}\|_{S_p} \lesssim \|a\|_{W^p_{-s}} \|\varphi_1\|_{M^1_{v_s}} \|\varphi_2\|_{M^1_{v_s}} .$$

By using other convolution relations provided by Proposition 3.2, interpolation and embedding properties of modulation spaces, one may derive many variations of Theorem 4.3. We only mention two small modifications that might be of interest.
(a) If $a \in M^{1,p}_{1/\tau_s}$ and $\varphi_1 \in M^1_{v_s}$, $\varphi_2 \in M^{p'}_{v_s}$, then $A_a^{\varphi_1,\varphi_2}$ is of trace class, because $M^{1,p}_{1/\tau_s} * M^{1,p'}_{\tau_s} \subseteq M^1$. Comparing to Theorem 4.3(i), we see that this result allows us to use a window φ_2 with less time-frequency concentration, however, at the price of a slightly smaller symbol class.
(b) If $(a, \varphi_1, \varphi_2) \in M^{p,\infty}_{1/\tau_s} \times M^q_{v_s} \times M^r_{v_s}$, where $1/q + 1/r - 1 = 1/p$ and $1 \leq p \leq 2$, then $A_a^{\varphi_1,\varphi_2} \in S_p$. To see this, we observe that Theorem 4.3(i) also holds with the role of the windows reversed, i.e., for $(\varphi_1, \varphi_2) \in M^p_{v_s} \times M^1_{v_s}$. The result then follows from the interpolation property $[M^1_{v_s} \times M^p_{v_s}, M^p_{v_s} \times M^1_{v_s}]_\theta = M^q_{v_s} \times M^r_{v_s}$ with $1/q + 1/r - 1 = 1/p$.

4.2. *Ultra-distributions with compact support as symbols*

As an application we present the result shown in Ref. 15, Sec. 4, in terms of ultra-distributions with compact support, denoted by $\mathcal{E}_t$ $t > 1$. Recall the embeddings:

$$\mathcal{E}'_t \subset (\mathcal{S}^t_t)' \subset (\Sigma^t_t)', \quad t > 1.$$

We skip the precise definition of $\mathcal{E}'_t$, which can be found in many places, see e.g. Ref. 36, Def. 1.5.5 and subsequent anisotropic generalization. The following structure theorem, obtained by a slight generalization of Ref. 36, Thm. 1.5.6 to the anisotropic case, will be sufficient for our purposes.

Theorem 4.4. *Let $t \in \mathbb{R}^d$, $t > 1$, i.e. $t = (t_1, \ldots, t_d)$, with $t_1 > 1, \ldots, t_d > 1$. Every $u \in \mathcal{E}'_t$ can be represented as*

$$u = \sum_{\alpha \in \mathbb{N}_0^d} \partial^\alpha \mu_\alpha, \tag{36}$$

where μ_α is a measure satisfying

$$\int_K |d\mu_\alpha| \leq C_\epsilon \epsilon^{|\alpha|} (\alpha!)^{-t}, \tag{37}$$

for every $\epsilon > 0$ and a suitable compact set $K \subset \mathbb{R}^d$, independent of α.

Using the preceding characterization, the STFT of an ultra-distribution with compact support is estimated as follows. Ref. 15, Prop. 4.2.

Proposition 4.2. *Let $t \in \mathbb{R}^d$, $t > 1$, and $a \in \mathcal{E}'_t(\mathbb{R}^d)$. Then its STFT with respect to any window $g \in \Sigma_1^1$ satisfies the estimate*

$$|V_g a(x, \omega)| \lesssim e^{-h|x|} e^{t|2\pi\omega|^{1/t}},$$

for every $h > 0$, cf. (19) *and below for the vectorial notation.*

The STFT estimate given in Proposition 4.2, is the key of the following trace class result for localization operators:

Corollary 4.2. *Let $t \in \mathbb{R}^d$, $t > 1$. If $a \in \mathcal{E}'_t(\mathbb{R}^{2d})$ and $\varphi_1, \varphi_2 \in \mathcal{S}_1^1(\mathbb{R}^d)$, then $A_a^{\varphi_1,\varphi_2}$ is a trace class operator.*

Proof. See Ref. 15, Cor. 4.3; for sake of clarity we sketch the proof. If $\varphi_1, \varphi_2 \in \mathcal{S}_1^1(\mathbb{R}^d)$, the characterization in (30) with $p = q = 1$ implies that $\varphi_1, \varphi_2 \in M^1_{w_\epsilon}(\mathbb{R}^d)$ for some (all) $\epsilon > 0$. Since, for $|\omega| > C_\epsilon$ (where C_ϵ is a suitable positive constant depending on ϵ) we can write

$$t \cdot |2\pi\omega|^{1/t} = \sum_{i=1}^{d} t_i |2\pi\omega_i|^{1/t_i} \le \epsilon|\omega|,$$

then the estimate of Proposition 4.2 gives $a \in M^{1,\infty}_{1/\mu_\epsilon}(\mathbb{R}^{2d})$. Finally, since $\varphi_1, \varphi_2 \in M^1_{w_\epsilon}(\mathbb{R}^d)$ and $a \in M^{1,\infty}_{1/\mu_\epsilon}(\mathbb{R}^{2d})$, Theorem 4.3 (i), written for the case $p = 1$ with τ_s replaced by μ_s, and v_s by w_s, implies that the operator $A_a^{\varphi_1,\varphi_2}$ is trace class. □

Similar results show that tempered distributions with compact support give trace class operators, see Ref. 10, Cor. 3.7.

4.3. *Compactness of localization operators*

Localization operators with symbols and windows in the Schwartz class are compact.[7] If we define by M^0 the closed subspace of M^∞, consisting of all $f \in \mathcal{S}'$ such that its STFT $V_g f$ (with respect to a non-zero Schwartz window g) vanishes at infinity, it is easy to show that localization operators with symbols in M^0 and Schwartz windows are compact. Namely, let $a \in M^0(\mathbb{R}^{2d})$ and $g \in \mathcal{S}(\mathbb{R}^{2d})$, for simplicity normalized to be $\|g\|_{L^2} = 1$; consider then $V_g a$. The Schwartz class is dense in M^0, hence there exists a sequence F_n of Schwartz functions on $\mathbb{R}^{4d}$ that converge to $V_g a$ in the

L^∞-norm. Define the sequence $a_n := V_g^* F_n$, $n \in \mathbb{N}$, where V_g^* is the adjoint operator defined in (10). Then $a_n \in \mathcal{S}(\mathbb{R}^{2d})$ and $a_n \to a$ in the M^∞-norm, since by (11)

$$\|a - a_n\|_{M^\infty} = \|V_g a - V_g V_g^* F_n\|_{L^\infty} = \|V_g a - F_n\|_{L^\infty} \to 0,$$

for $n \to \infty$. From Theorem 4.2 we have

$$\|A_{a_n}^{\varphi_1,\varphi_2} - A_a^{\varphi_1,\varphi_2}\|_{B(L^2)} = \|A_{(a_n-a)}^{\varphi_1,\varphi_2}\|_{B(L^2)} \leq C\|a-a_n\|_{M^\infty}\|\varphi_1\|_{M^1}\|\varphi_2\|_{M^1} \to 0.$$

Since compact operators are a closed subspace of the space of all bounded operators $B(L^2)$, then the localization operator $A_a^{\varphi_1,\varphi_2}$ is compact.

The symbol class $M^0(\mathbb{R}^{2d})$ is not optimal as the next simple example shows. Consider $a = \delta \notin M^0(\mathbb{R}^{2d})$. Since $V_g\delta(z,\zeta) = \bar{g}(z)$, it does not tend to zero when $z \in \mathbb{R}^{2d}$ is fixed and $|\zeta|$ goes to infinity. Hence $\delta \notin M^0(\mathbb{R}^{2d})$. However $A_\delta^{\varphi_1,\varphi_2}$ is a trace class operator for every $\varphi_1, \varphi_2 \in \mathcal{S}(\mathbb{R}^d)$, in fact, a rank-one operator, and therefore it is compact.

The example just mentioned has been the inspiration for the following compactness result Ref. 26, Prop. 3.6:

Proposition 4.3. *Let $g \in \mathcal{S}(\mathbb{R}^{2d})$ be given and $a \in M^\infty(\mathbb{R}^{2d})$. If $\varphi_1, \varphi_2 \in \mathcal{S}(\mathbb{R}^d)$ and*

$$\lim_{|z|\to\infty} \sup_{|\zeta|\leq R} |V_g a(z,\zeta)| = 0, \qquad \forall R > 0, \tag{38}$$

then $A_a^{\varphi_1,\varphi_2}$ is a compact operator.

5. Necessary Conditions

In this section we show that the sufficient conditions obtained so far are essentially optimal. This investigation requires different techniques and we limit ourselves to state the main results. A first attempt is done in Theorems 4.3, 4.4 of Ref. 10, where a converse for bounded and Hilbert-Schmidt operators is obtained for modulation spaces with polynomial weights:

Theorem 5.1.

- *Let $a \in \mathcal{S}'(\mathbb{R}^{2d})$ and fix $s \geq 0$. If there exists a constant $C = C(a) > 0$ depending only on a such that*

$$\|A_a^{\varphi_1,\varphi_2}\|_{S_\infty} \leq C\, \|\varphi_1\|_{M^1_{v_s}} \|\varphi_2\|_{M^1_{v_s}}$$

for all $\varphi_1, \varphi_2 \in \mathcal{S}(\mathbb{R}^d)$, then $a \in M^\infty_{1/\tau_s}$.

- *Let* $a \in \mathcal{S}'(\mathbb{R}^{2d})$. *If there exists a constant* $C = C(a) > 0$ *depending only on* a *such that*

$$\|A_a^{\varphi_1,\varphi_2}\|_{S_2} \leq C\, \|\varphi_1\|_{M^1} \|\varphi_2\|_{M^1}$$

for all $\varphi_1, \varphi_2 \in \mathcal{S}(\mathbb{R}^d)$, *then* $a \in M^{2,\infty}$.

Next, an extension for the boundedness necessary condition is given in Ref. 15, Thm. 3.3:

Theorem 5.2. *Let* $a \in (\Sigma_1^1)'(\mathbb{R}^{2d})$ *and fix* $s \geq 0$. *If there exists a constant* $C = C(a) > 0$ *depending only on* a *such that*

$$\|A_a^{\varphi_1,\varphi_2}\|_{S_\infty} \leq C\, \|\varphi_1\|_{M^1_{w_s}} \|\varphi_2\|_{M^1_{w_s}}$$

for all $\varphi_1, \varphi_2 \in \Sigma_1^1(\mathbb{R}^d)$, *then* $a \in M^\infty_{1/\mu_s}$.

Necessary conditions for localization operators belonging to the Schatten class S_p have been obtained for unweighted modulation spaces in Ref. 11:

Theorem 5.3. *Let* $a \in \mathcal{S}'(\mathbb{R}^{2d})$ *and* $1 \leq p \leq \infty$. *Assume that* $A_a^{\varphi_1,\varphi_2} \in S_p$ *for* all *windows* $\varphi_1, \varphi_2 \in \mathcal{S}(\mathbb{R}^d)$ *and that there exists a constant* $B > 0$ *depending only on the symbol* a *such that*

$$\|A_a^{\varphi_1,\varphi_2}\|_{S_p} \leq B\, \|\varphi_1\|_{M^1}\, \|\varphi_2\|_{M^1}, \qquad \forall \varphi_1, \varphi_2 \in \mathcal{S}(\mathbb{R}^d)\,,$$

then $a \in M^{p,\infty}$.

The techniques employed for the converse results are thoroughly different from the techniques for the sufficient conditions. Gabor frames and equivalent norms for modulation spaces are some of the crucial ingredients in the proofs. For the sake of completeness, we shall sketch the main features. First, by using the Gabor frame of the form

$$\{M_{\beta n} T_{\alpha k} \Phi\}_{k,n \in \mathbb{Z}^{2d}}, \quad 0 < \alpha, \beta < 1,$$

with the Gaussian window $\Phi(x,\omega) = 2^{-d}\, e^{-\pi(x^2+\omega^2)}$, the $M^{p,\infty}(\mathbb{R}^{2d})$-norm of a can be expressed by the equivalent norms

$$\|a\|_{M^{p,\infty}(\mathbb{R}^{2d})} \asymp \|\langle a, M_{\beta n} T_{\alpha k} \Phi\rangle_{n,k \in \mathbb{Z}^{2d}}\|_{\ell^{p,\infty}(\mathbb{Z}^{4d})}. \tag{39}$$

Then one relates the action of the localization operator on certain time-frequencys of the Gaussian φ to the Gabor coefficients, and for a diligent choice of (x,ξ) and (u,η) one obtains that $\langle A_a^{\varphi_1,\varphi_2} M_\xi T_x \varphi, M_\eta T_u \varphi\rangle = \langle a, M_{\beta n} T_{\alpha k} \Phi\rangle$. The result is then obtained by using (39).

In view of the sufficient Schatten class results known so far, it is left as an exercise to show that the necessary conditions for the Schatten class can also be formulated for weighted modulation spaces.

We end up with the compactness necessary result of Ref. 26, Thm. 3.15.

Theorem 5.4. *Let $a \in M^\infty(\mathbb{R}^{2d})$ and $g \in \mathcal{S}(\mathbb{R}^{2d})$ be given. Then, the following conditions are equivalent:*

(a) The localization operator $A_a^{\varphi_1,\varphi_2} : L^2(\mathbb{R}^d) \to L^2(\mathbb{R}^d)$ is compact for every pair $\varphi_1, \varphi_2 \in \mathcal{S}(\mathbb{R}^d)$.

(b) The symbol a satisfies condition (38).

6. Composition Formula

Given two localization operators, we want to compute their product and develop a symbolic calculus. It would be useful to express it in terms of localization operators. We shall present two different product formulae. The first one is an exact formula that expresses the composition of two localization operators again as a localization operator. However, the formula holds only for Gaussian windows and very special symbols. The second formula is much more general, but in this case the product of two localization operators is a sum of localization operators plus a remainder term, which can be expressed in either the Weyl or integral operator form.

6.1. *Exact product*

We reformulate the result of Ref. 20 in the notation of Refs. 27,28.

We consider the window functions $\varphi_1(t) = \varphi_2(t) = \varphi(t) = 2^{d/4}e^{-\pi t^2}$, $t \in \mathbb{R}^d$. In this case, the Wigner distribution of the Gaussian φ is a Gaussian as well: $W(\varphi,\varphi)(z) = 2^{d/2}e^{-2\pi z^2}$, $z \in \mathbb{R}^{2d}$. According to (14) the Weyl symbol σ of the operator $A_a^{\varphi,\varphi}$ is $\sigma(\zeta) = 2^{d/2}(a * e^{-2\pi z^2})(\zeta)$, $z, \zeta \in \mathbb{R}^{2d}$. We first recall the well-known composition of Weyl transforms from Ref. 27, Chp. 3.2 and then make the transition to localization operators. Let $[\cdot,\cdot]$ be the standard *symplectic form* on $\mathbb{R}^{2d}$ defined by

$$[(z_1,z_2),(\zeta_1,\zeta_2)] = z_1\zeta_2 - z_2\zeta_1, \quad \text{with} \quad z = (z_1,z_2),\, \zeta = (\zeta_1,\zeta_2)\,,$$

and let the twisted convolution $\natural$ on $\mathbb{R}^{2d}$ be given by

$$F \natural G(\zeta) = \iint_{\mathbb{R}^{2d}} F(z)\, G(\zeta - z) e^{\pi i [z,\zeta]}\, dz\,. \tag{40}$$

Then the composition of two Weyl transforms with symbols σ and τ can be written formally as

$$L_\sigma L_\tau = L_{\mathcal{F}^{-1}(\hat{\sigma}\natural\hat{\tau})}\,. \tag{41}$$

For any $f, g \in \mathcal{S}(\mathbb{R}^{2d})$, we define the $\natural^b$ product by

$$f\natural^b g(\zeta) = \iint_{\mathbb{R}^{2d}} f(z)g(z-\zeta)e^{\pi(z\zeta+i[z,\zeta])}\, e^{-\pi z^2}\, dz\,. \tag{42}$$

Then the product of localization operators is given by the following formula.

Theorem 6.1. *Let $a, b \in \mathcal{S}(\mathbb{R}^{2d})$. If there exists a symbol $c \in \mathcal{S}'(\mathbb{R}^{2d})$ such that*

$$\hat{c} = 2^{-d/2}\hat{a}\,\natural^b\,\hat{b}, \tag{43}$$

then we have

$$A_a^{\varphi,\varphi}A_b^{\varphi,\varphi} = A_c^{\varphi,\varphi}.$$

The proof is a straightforward consequence of relations (40) and (42). Indeed, one rewrites $A_a^{\varphi,\varphi}A_b^{\varphi,\varphi}$ in the Weyl form and uses relation (40) for the Weyl product. The result is the Weyl operator L_μ, where the Fourier transform of μ is given by

$$\begin{aligned}\hat{\mu}(\zeta) &= [\mathcal{F}(a * W(\varphi,\varphi))\,\natural\,\mathcal{F}(b * W(\varphi,\varphi))](\zeta)\\ &= 2^{-d}\iint_{\mathbb{R}^{2d}} \hat{a}(z)\hat{b}(z-\zeta)e^{-(\pi/2)z^2}e^{-(\pi/2)(\zeta-z)^2}e^{\pi i[z,\zeta]}\,dz\\ &= 2^{-d}e^{-(\pi/2)\zeta^2}\iint_{\mathbb{R}^{2d}} \hat{a}(z)\hat{b}(z-\zeta)e^{\pi(z\zeta+i[z,\zeta])}\,e^{-\pi z^2}\,dz.\end{aligned}$$

Hence, we have

$$\hat{\mu}(\zeta) = \hat{c}(\zeta)(2^{-d/2}e^{-(\pi/2)\zeta^2}) = \mathcal{F}(c * W(\varphi,\varphi))(\zeta),$$

where $\hat{c}$ is given by relation (43).

6.2. *Asymptotic product*

A second approach to the composition of two localization operators derives asymptotic expansions.[1,12,14,33,40] These realize the product as a sum of localization operators plus a controllable remainder. Most of these expansions were motivated by PDEs and energy estimates, and therefore use smooth symbols that are defined by differentiability properties, such as the Hörmander or Shubin classes. For applications in quantum mechanics and signal analysis, alternative notions of smoothness — "smoothness in phase-space" or quantitative measures of "time-frequency concentration" — have turned out to be useful. This point of view is pursued in Ref. 12, and we shall present the corresponding results.

The starting point is the following composition formula for two localization operators derived in Ref. 14:

$$A_a^{\varphi_1,\varphi_2} A_b^{\varphi_3,\varphi_4} = \sum_{|\alpha|=0}^{N-1} \frac{(-1)^{|\alpha|}}{\alpha!} A_{a\partial^\alpha b}^{\Phi_\alpha,\varphi_2} + E_N. \tag{44}$$

The essence of this formula is that the product of two localization operators can be written as a sum of localization operators with suitably defined, new windows Φ_α and a remainder term E_N, which is "small".

In the spirit of the classical symbolic calculus, this formula was derived in Ref. 14, Thm. 1.1 for *smooth* symbols belonging to some Shubin class $S^m(\mathbb{R}^{2d})$ and for windows in the Schwartz class $\mathcal{S}(\mathbb{R}^d)$.

In Ref. 12 the validity of (44) is established on the modulation spaces. The innovative features of this extension are highlighted below. Since the results are very technical, we do not give the detailed statements and proofs, but refer the reader to Ref. 12.

(i) *Rough symbols.* While in (44) the symbol b must be N-times differentiable, the symbol a only needs to be locally bounded. The classical results in symbolic calculus require both symbols to be smooth.

(ii) *Growth conditions on symbols.* The symbolic calculus in (44) can handle symbols with ultra-rapid growth (as long as it is compensated by a decay of b or vice versa). For instance, a may grow subexponentially as $a(z) \sim e^{\alpha|z|^\beta}$ for $\alpha > 0$ and $0 < \beta < 1$. This goes far beyond the usual polynomial growth and decay conditions.

(iii) *General window classes.* A precise description of the admissible windows φ_j in (44) is provided. Usually only the Gaussian $e^{-\pi x^2}$ or Schwartz functions are considered as windows.

(iv) *Size of the remainder term.* Norm estimates for the size of the remainder term E_N are derived. They depend explicitly on the symbols a, b and the windows φ_j.

(v) *The Fredholm Property of Localization Operators.* By choosing $N = 1$, $\varphi_1 = \varphi_2 = \varphi$ with $\|\varphi\|_2 = 1$, $a(z) \neq 0$ for all $z \in \mathbb{R}^{2d}$, and $b = 1/a$, the composition formula (44) yields the following important special case:

$$A_a^{\varphi,\varphi} A_{1/a}^{\varphi,\varphi} = A_1^{\varphi,\varphi} + R = \mathrm{I} + R\,. \tag{45}$$

Under the following conditions on a:

(i) $|a| \asymp 1/m$ (in particular, $a \in L^\infty_m(\mathbb{R}^{2d})$,) where $m \in \mathcal{M}_v$,
(ii) $(\partial_j a)m \in L^\infty$ and vanishes at infinity for $j = 1, \ldots, 2d$;

the remainder R is shown to be compact, and as a consequence, $A_a^{\varphi,\varphi}$ is a Fredholm operator between the two modulation spaces $M^{p,q}$ and $M^{p,q}_m$ (with different weights). This result works even for ultra-rapidly growing symbols such as $a(z) = e^{\alpha|z|^\beta}$ for $\alpha > 0$ and $0 < \beta < 1$. For comparison, the reduction of localization operators to standard pseudodifferential calculus requires elliptic or hypo-elliptic symbols, and the proof of the Fredholm property works only under severe restrictions, see Ref. 8.

7. Multilinear Localization Operators

Multilinear localization operators are introduced in Ref. 13; they not only generalize the linear case but also yield a subclass of multilinear pseudodifferential operators. To understand their meaning, one can think of localizing m-fold products of functions. For the sake of clarity, we shall first introduce the bilinear case and show how the construction arises naturally from the framework of reproducing formulae and linear localization operators. The general case can be treated similarly.

Bilinear localization operators. Let $f_1, f_2 \in \mathcal{S}(\mathbb{R}^d)$, then the tensor product $(f_1 \otimes f_2)(x_1, x_2) = f_1(x_1)f_2(x_2)$ is a function in $\mathcal{S}(\mathbb{R}^{2d})$. Given four window functions $\varphi_i \in \mathcal{S}(\mathbb{R}^d)$, $i = 1, \ldots, 4$, with $\langle \varphi_1, \varphi_3\rangle = \langle \varphi_2, \varphi_4\rangle = 1$, the usual reproducing formula for the functions f_1, f_2 stated in (3) reads as follows:

$$f_1 = \int_{\mathbb{R}^{2d}} V_{\varphi_1} f_1(z_1, \zeta_1)\, M_{\zeta_1} T_{z_1} \varphi_3 \, d\zeta_1 \, dz_1. \tag{46}$$

$$f_2 = \int_{\mathbb{R}^{2d}} V_{\varphi_2} f_2(z_2, \zeta_2)\, M_{\zeta_2} T_{z_2} \varphi_4 \, d\zeta_2 \, dz_2. \tag{47}$$

The product of both sides of equalities (46) and (47) yields

$$\begin{aligned}(f_1 \otimes f_2)(x_1, x_2) &= \int_{\mathbb{R}^{2d}} \int_{\mathbb{R}^{2d}} V_{\varphi_1} f_1(z_1, \zeta_1) V_{\varphi_2} f_2(z_2, \zeta_2) M_{\zeta_1} T_{z_1} \varphi_3(x_1) \\ &\quad \times M_{\zeta_2} T_{z_2} \varphi_4(x_2) d\zeta dz \\ &= \int_{\mathbb{R}^{4d}} V_{\varphi_1 \otimes \varphi_2}(f_1 \otimes f_2)(z, \zeta)\, M_\zeta T_z (\varphi_3 \otimes \varphi_4)(x_1, x_2)\, d\zeta\, dz\end{aligned}$$

with $z = (z_1, z_2)$, $\zeta = (\zeta_1, \zeta_2) \in \mathbb{R}^{2d}$.

The previous reproducing formula for the function $f_1 \otimes f_2$ can be localized in the time-frequency plane yielding a localization operator $A_a^{\varphi_1 \otimes \varphi_2, \varphi_3 \otimes \varphi_4}$ with symbol a (defined on $\mathbb{R}^{4d}$) and windows $\varphi_1 \otimes \varphi_2$, $\varphi_3 \otimes \varphi_4$. Formally, the action of the operator on the function $f_1 \otimes f_2$ is given by

$$A_a^{\varphi_1 \otimes \varphi_2, \varphi_3 \otimes \varphi_4}(f_1 \otimes f_2)(x_1, x_2)$$
$$= \int_{\mathbb{R}^{4d}} a(z,\zeta) V_{\varphi_1} f_1(z_1,\zeta_1) V_{\varphi_2} f_2(z_2,\zeta_2) M_{\zeta_1} T_{z_1} \varphi_3(x_1) M_{\zeta_2} T_{z_2} \varphi_4(x_2) dz d\zeta.$$

For any symbol $a \in \mathcal{S}'(\mathbb{R}^{4d})$, and window functions φ_j on $\mathcal{S}(\mathbb{R}^d)$, the operator $A_a^{\varphi_1 \otimes \varphi_2, \varphi_3 \otimes \varphi_4}$ can be seen as a bilinear mapping from the 2-fold product of Schwartz spaces $\mathcal{S}(\mathbb{R}^d) \times \mathcal{S}(\mathbb{R}^d)$ into the space $\mathcal{S}'(\mathbb{R}^{2d})$ of tempered distributions. Moreover, if we restrict now our attention to a smoother symbol $a \in \mathcal{S}(\mathbb{R}^{4d})$, we obtain a multilinear mapping from $\mathcal{S}(\mathbb{R}^d) \times \mathcal{S}(\mathbb{R}^d)$ into $\mathcal{S}(\mathbb{R}^{2d})$.

In Ref. 13 the boundedness properties of the trace of $A_a^{\varphi_1 \otimes \varphi_2, \varphi_3 \otimes \varphi_4}$ on the diagonal $x_1 = x_2$ are studied. This restriction leads to a new kind of localization operator.

Definition 7.1. Let $f_1, f_2 \in \mathcal{S}(\mathbb{R}^d)$. Given a symbol $a \in \mathcal{S}'(\mathbb{R}^{4d})$ and window functions $\varphi_i \in \mathcal{S}(\mathbb{R}^d)$, with $i = 1, \dots, 4$, the *bilinear localization operator* A_a is given by

$$A_a(f_1, f_2)(x)$$
$$= \int_{\mathbb{R}^{4d}} a(z,\zeta)\, V_{\varphi_1} f_1(z_1,\zeta_1)\, V_{\varphi_2} f_2(z_2,\zeta_2)\, M_{\zeta_1} T_{z_1} \varphi_3(x)\, M_{\zeta_2} T_{z_2} \varphi_4(x)\, dz\, d\zeta, \tag{48}$$

where $x \in \mathbb{R}^d$.

Notice that if the symbol $a \in \mathcal{S}'(\mathbb{R}^{4d})$ then the corresponding operator A_a maps $\mathcal{S}(\mathbb{R}^d) \times \mathcal{S}(\mathbb{R}^d)$ into $\mathcal{S}'(\mathbb{R}^d)$.

In order to give a weak definition of the bilinear localization operator A_a, we first introduce the following time-frequency representation. For $\varphi_3, \varphi_4 \in \mathcal{S}(\mathbb{R}^d) \setminus \{0\}$, $z = (z_1, z_2)$, $\zeta = (\zeta_1, \zeta_2) \in \mathbb{R}^{2d}$, we define $\mathcal{V}_{\varphi_3,\varphi_4}$ by

$$\mathcal{V}_{\varphi_3,\varphi_4} g(z,\zeta) = \int_{\mathbb{R}^d} g(t)\, \overline{M_{\zeta_1} T_{z_1} \varphi_3(t) M_{\zeta_2} T_{z_2} \varphi_4(t)}\, dt, \quad g \in \mathcal{S}(\mathbb{R}^d). \tag{49}$$

Thus, for $f_1, f_2, g \in \mathcal{S}(\mathbb{R}^d)$ the weak definition of (48) is given by

$$\langle A_a(f_1, f_2), g\rangle = \langle a, \overline{V_{\varphi_1 \otimes \varphi_2}(f_1 \otimes f_2)} \mathcal{V}_{\varphi_3,\varphi_4} g\rangle. \tag{50}$$

Multilinear localization operators. Without any further work — just some extra notation — it is straightforward to generalize the above definition of multilinear localization operators and relate it to a multilinear pseudodifferential operator. Thus we are led to make the following definition. Fix $m \in \mathbb{N}$. For every symbol $a \in \mathcal{S}'(\mathbb{R}^{2md})$ and windows φ_i, $i = 1, \dots, 2m$, in the Schwartz class $\mathcal{S}(\mathbb{R}^d)$, we introduce the analysis, synthesis window functions $\phi_1, \phi_2 : \mathbb{R}^{md} \to \mathbb{C}$, defined respectively as tensor products of the m analysis, and m synthesis windows, i.e.,

$$\phi_1(t_1, \dots, t_m) := \varphi_1(t_1) \cdots \varphi_m(t_m), \tag{51}$$

and

$$\phi_2(t_1, \dots, t_m) := \varphi_{m+1}(t_1) \cdots \varphi_{2m}(t_m). \tag{52}$$

Let $\mathcal{R}$ be the trace mapping that assigns to each function defined on $\mathbb{R}^{md}$ a function defined on $\mathbb{R}^d$ by the following formula:

$$\mathcal{R}F(t) := F_{|\{t_1=t_2=\cdots=t_m=t\}}(t_1, \dots, t_m) = F(t, \dots, t), \tag{53}$$

for any $t \in \mathbb{R}^d$.

Definition 7.2. The multilinear localization operator A_a with symbol $a \in \mathcal{S}'(\mathbb{R}^{2md})$ and windows $\varphi_j \in \mathcal{S}(\mathbb{R}^d)$, $j = 1, \dots, 2m$ is the multilinear mapping defined on the m-fold product of $\mathcal{S}(\mathbb{R}^d)$ into $\mathcal{S}'(\mathbb{R}^d)$ by

$$\begin{aligned} A_a(\overrightarrow{f})(x) : &= \int_{\mathbb{R}^{2md}} a(z,\zeta)\, V_{\otimes_{j=1}^m \varphi_j}\left(\otimes_{j=1}^m f_j\right)(z,\zeta) \prod_{j=1}^m M_{\zeta_j} T_{z_j} \varphi_{m+j}(x)\, d\zeta\, dz \\ &= \int_{\mathbb{R}^{2md}} a(z,\zeta)\, V_{\phi_1}\left(\otimes_{j=1}^m f_j\right)(z,\zeta)\ \mathcal{R} M_\zeta T_z \phi_2(x)\, d\zeta\, dz, \end{aligned} \tag{54}$$

where $(z,\zeta) \in \mathbb{R}^{md} \times \mathbb{R}^{md}$, $x \in \mathbb{R}^d$, and $\overrightarrow{f} = (f_1, \dots, f_m) \in \mathcal{S}(\mathbb{R}^d) \times \cdots \times \mathcal{S}(\mathbb{R}^d)$.

If $m = 1$ we are back to the linear localization operator $A_a^{\varphi_1,\varphi_2}$, whereas the case $m = 2$ gives the bilinear localization operator introduced in (48).

One of the results of Ref. 13 is related to the boundedness properties of multilinear localization operators on products of modulations spaces. To this end, these operators are represented as bilinear (or, in general, as multilinear) pseudodifferential operators and known results on boundedness of multilinear pseudodifferential operators on products of modulation spaces (Refs. 2,3) lead to boundedness results of these multilinear localization operators. In analogy to the linear case, it is worth detailing their connection with multilinear pseudodifferential operators.

Proposition 7.1. *Let $a \in \mathcal{S}'(\mathbb{R}^{2md})$ and $\varphi_j \in \mathcal{S}(\mathbb{R}^d)$, $j = 1, \ldots, 2m$. Then the multilinear localization operator A_a is the multilinear pseudodifferential operator T_τ defined on $\overrightarrow{f} = (f_j)_{j=1}^m \in \mathcal{S}(\mathbb{R}^d) \times \ldots \times \mathcal{S}(\mathbb{R}^d)$ by*

$$A_a(\overrightarrow{f})(x) = T_\tau(\overrightarrow{f})(x) := \int_{\mathbb{R}^{md}} \tau(x,\xi) \prod_{j=1}^m \hat{f}_j(\xi_j)\, e^{2\pi i x \sum_{j=1}^m \xi_j}\, d\xi\,. \tag{55}$$

The symbol τ is given as

$$\tau(x,\xi) = a * \Phi(X,\xi) \tag{56}$$

with $x \in \mathbb{R}^d$, $X = (x, \ldots, x)$, $\xi = (\xi_1, \ldots, \xi_m) \in \mathbb{R}^{md}$, and

$$\Phi(z,\xi) = \prod_{j=1}^m \mathcal{U}\mathcal{F}(V_{\varphi_j}\varphi_{j+m})(z_j,\xi_j), \tag{57}$$

for $z = (z_1, \ldots, z_m) \in \mathbb{R}^{md}$.

According to what happens for linear localization operators, we shall provide both sufficient and necessary conditions for boundedness on products of modulation spaces.

Theorem 7.1.
(a) ***Sufficient conditions.*** *Let $m \in \mathbb{N}$, a symbol $a \in M^\infty(\mathbb{R}^{2md})$, and window functions $\varphi_j \in M^1(\mathbb{R}^d)$, $j = 1, \ldots, 2m$, be given. Then the m-linear localization operator A_a defined by (54) extends to a bounded operator from $M^{p_1,q_1}(\mathbb{R}^d) \times \cdots \times M^{p_m,q_m}(\mathbb{R}^d)$ into $M^{p_0,q_0}(\mathbb{R}^d)$, when*

$$\frac{1}{p_1} + \cdots + \frac{1}{p_m} = \frac{1}{p_0}, \quad \frac{1}{q_1} + \cdots + \frac{1}{q_m} = m - 1 + \frac{1}{q_0},$$

and $1 \leq p_j, q_j \leq \infty$, for $j = 0, \ldots, m$. Moreover, we have the following norm estimate

$$\|A_a\| \leq C \|a\|_{M^\infty(\mathbb{R}^{2md})} \prod_{i=1}^{2m} \|\varphi_i\|_{M^1(\mathbb{R}^d)}, \tag{58}$$

where the positive constant C is independent of a and of φ_j, $j = 1, \ldots, 2m$.

(b) ***Necessary conditions.*** *Let $m \in \mathbb{N}$, and $a \in \mathcal{S}'(\mathbb{R}^{2md})$ be given. Assume that*

- *the m-linear localization operator A_a is bounded from $M^{p_1,q_1}(\mathbb{R}^d) \times \cdots \times M^{p_m,q_m}(\mathbb{R}^d)$ into $M^{p_0,q_0}(\mathbb{R}^d)$, where*

$$\frac{1}{p_1} + \cdots + \frac{1}{p_m} = \frac{1}{p_0}, \quad \frac{1}{q_1} + \cdots + \frac{1}{q_m} = m - 1 + \frac{1}{q_0},$$

and $1 \leq p_j, q_j \leq \infty$, *for* $j = 0, \ldots, m$, *and moreover that*

- A_a *satisfies the following norm estimate*

$$\|A_a\| \leq C(a) \prod_{i=1}^{2m} \|\varphi_i\|_{M^1(\mathbb{R}^d)}, \quad \forall \varphi_i \in \mathcal{S}(\mathbb{R}^d),\ i = 1, \ldots, 2m, \tag{59}$$

with a positive constant $C(a)$ *depending only on* a. *Then the symbol* a *belongs necessarily to* $M^\infty(\mathbb{R}^{2md})$.

An application of this theory is that it provides symbols for multilinear bounded Kohn-Nirenberg operators. Two steps are needed to construct symbols in $M^{\infty,1}$: first, suitable windows φ_i, $i = 1, \ldots, d$ are chosen for computing the function Φ defined in (57). Secondly, symbols a are provided explicitly, the convolution with Φ in (56) is computed, yielding the Kohn-Nirenberg symbols τ desired. We refer to Ref. 13, Sec. 7 for concrete examples.

References

1. A. Bényi and K. A. Okoudjou, *Studia Math.*, **172**, 169 (2006).
2. E. Cordero and K. Gröchenig, *J. Fourier Anal. Appl.*, **12**, 371 (2006).
3. E. Cordero adm K. Okoudjou, *J. Math. Anal.* (to appear).
4. E. Cordero, S. Pilipović, L. Rodine and N. Teofanov, *Mediterranean J. Math.*, **2**, 381 (2005).
5. J. Bergh and J. Löfström, *Grundlehren der Mathematischen Wissenschaften*, No. 223 (Springer-Verlag, Berlin, 1976).
6. P. Boggiatto and E. Cordero, *Proc. Amer. Math. Soc.* **130**, 2679 (2002).
7. P. Boggiatto, E. Cordero, and K. Gröchenig, *Integral Equations and Operator Theory*, **48**, 427 (2004).
8. P. Boggiatto, J. Toft, *Appl. Anal.* **84**, 269 (2005).
9. M. Cappiello and L. Rodino, *Rocky Mountain J. Math.* (to appear).
10. E. Cordero and K. Gröchenig, *J. Funct. Anal.* **205** 107 (2003).
11. E. Cordero and K. Gröchenig, *Proc. Amer. Math. Soc.* **133** 3573 (2005).
12. E. Cordero and K. Gröchenig, *Symbolic calculus and Fredholm property for localization operators* (Preprint, 2005).
13. E. Cordero and K. Okoudjou, *Multilinear localization operators* (Preprint, 2005).
14. E. Cordero and L. Rodino, *Osaka J. Math.* **42**43 (2005).
15. E. Cordero, S. Pilipović, L. Rodino, and N. Teofanov, *Mediterranean J. Math.* (to appear).
16. A. Córdoba and C. Fefferman, *Comm. Partial Differential Equations* **3**, 979 (1978).
17. I. Daubechies, *IEEE Trans. Inform. Theory* **34**, 605 (1988).

18. F. De Mari, H. G. Feichtinger, and K. Nowak, *J. London Math. Soc. (2)* **65**, 720 (2002).
19. F. De Mari and K. Nowak, *J. Geom. Anal.* **12**, 9 (2002).
20. J. Du and M. W. Wong, *Bull. Korean Math. Soc.* **37** 77 (2000).
21. H. G. Feichtinger, *Technical Report, University Vienna, 1983.* and also in *Wavelets and Their Applications*, Eds., M. Krishna, R. Radha, S. Thangavelu, (Allied Publishers, 2003) pp. 99–140.
22. H. G. Feichtinger and K. Gröchenig, *J. Funct. Anal.* **86**, 307 (1989).
23. H. G. Feichtinger, *Monatsh. Math.* **92**, 269 (1981).
24. H. G. Feichtinger and K. H. Gröchenig, *Monatsh. f. Math.* **108**, 129 (1989).
25. H. G. Feichtinger and K. Nowak. *A First Survey of Gabor Multipliers*, in *Advances in Gabor Analysis*, Eds., H. G. Feichtinger and T. Strohmer (Birkhäuser, Boston, 2002).
26. C. Fernández and A. Galbis, *J. Funct. Anal.* (2005)(to appear).
27. G. B. Folland, *Harmonic Analysis in Phase Space* (Princeton Univ. Press, Princeton, NJ, 1989).
28. K. Gröchenig, *Foundations of Time-Frequency Analysis*, (Birkhäuser, Boston, 2001).
29. K. Gröchenig, *Studia Math.* **121**, 87 (1996).
30. I. M. Gelfand and G. E. Shilov, *Generalized Functions II* (Academic Press, 1968).
31. K. Gröchenig and G. Zimmermann, *J. London Math. Soc.* **63**, 205 (2001).
32. K. Gröchenig, G. Zimmermann, *Journal of Function Spaces and Applications* **2**, 25 (2004).
33. N. Lerner, *The Wick calculus of pseudo-differential operators and energy estimates*, in *New trends in microlocal analysis (Tokyo, 1995)*(Springer, Tokyo, 1997) pp. 23–37.
34. S. Pilipović, *Boll. Un. Mat. Ital.* **7**, 235 (1988).
35. S. Pilipović and N. Teofanov, *J. Funct. Anal.* **208** 194 (2004).
36. L. Rodino, *Linear Partial Differential Operators in Gevrey Spaces* (World Scientific, 1993).
37. J. Ramanathan and P. Topiwala, *SIAM J. Math. Anal.* **24**, 1378 (1993).
38. M. A. Shubin, *Pseudodifferential Operators and Spectral Theory*, in *Translated from the 1978 Russian original*, Ed., Stig I. Andersson, (Springer-Verlag, Berlin, second edition, 2001).
39. J. Sjöstrand, *Math. Res. Lett.* **1**185 (1994).
40. D. Tataru, *Comm. Partial Differential Equations* **27**, 2101 (2002).
41. N. Teofanov, *Ultradistributions and time-frequency analysis,* in *Pseudo-differential Operators and Related Topics, Operator Theory: Advances and Applications*, Eds., P. Boggiatto, L. Rodino, J. Toft, M.W. Wong, Vol. 164 (Birkhäuser, 2006), pp. 173–191.
42. J. Toft, *J. Funct. Anal.* **207**, 399 (2004).
43. J. Toft, *Ann. Global Anal. Geom.* **26**, 73 (2004).
44. M. W. Wong, *Wavelets Transforms and Localization Operators*, in *Operator Theory Advances and Applications*, Vol. 136 (Birkhäuser, 2002).

Chapter II

HARMONIC ANALYSIS

Harmonic, Wavelet and p-Adic Analysis
Eds. N. M. Chuong *et al.* (pp. 113–139)

§6. ON MULTIPLE SOLUTIONS FOR ELLIPTIC BOUNDARY VALUE PROBLEM WITH TWO CRITICAL EXPONENTS

YU. V. EGOROV

Université Paul Sabatier, Toulouse, France
E-mail: egorov@mipups-tlse.fr

YAVDAT IL'YASOV*

Bashkir State University, Ufa, Russia
E-mail: IlyasovYS@ic.bashedu.ru

We study a semilinear elliptic boundary value problem with critical exponents both in the equation and in the boundary condition. We don't suppose that the energy functional is always positive and prove the existence of two positive solutions.

Keywords: AMS Subject Classifications: 35J70, 35J65, 47H17

1. Introduction and Main Results

Let (M, g) be a compact Riemannian manifold of dimension $n > 2$ with smooth boundary ∂M. We study the following problem:

$$\begin{cases} -\Delta_g u + r(x)u = R(x)u^{2^*-1} & \text{in } M, \\ \frac{\partial u}{\partial N} + h(x)u = H(x)u^{2^{**}-1} & \text{on } \partial M, \end{cases} \tag{1}$$

where Δ_g, ∇_g denotes the Laplace–Beltrami operator and the gradient in the metric g, respectively. N is the direction of the outward normal on ∂M in the metric g. Here $2^* = \frac{2n}{n-2}$, $2^{**} = \frac{2(n-1)}{(n-2)}$ are the critical Sobolev exponents. Further we will always suppose that $r, R \in C(M)$, $h, H \in C(\partial M)$.

This problem arises in differential geometry with r, R playing role of the scalar curvatures of M and h, H being mean curvatures of ∂M for the

*The author was supported in part by grants INTAS 03-51-5007, RFBR 05-01-00370, 05-01-00515.

Riemann metrics g, g' such that $g' = gu^{4/(n-2)}$. It is called Yamabe problem on manifolds with boundary.

It is important that the problem can be stated in the variational form, i.e., its solution corresponds to a critical point of the Euler functional

$$I(u) = \frac{1}{2}E(u) - \frac{1}{2^{**}}B(u) - \frac{1}{2^*}F(u). \tag{2}$$

Here

$$E(u) = \int_M |\nabla u|^2 dv_g + \int_M r(x)|u|^2 dv_g + \int_{\partial M} h(x)|u|^2 d\sigma_g \tag{3}$$

is the energy functional and we denote

$$F(u) = \int_M R(x)|u|^{2^*} dv_g, \quad B(u) = \int_{\partial M} H(x)|u|^{2^{**}} d\sigma_g, \tag{4}$$

where dv_g and $d\sigma_g$ are the Riemannian measures (induced by the metric g) on M and ∂M, respectively.

Note that there is always a function $u \in W_2^1(M)$ such that $E(u) > 0$. Our principal hypothese is the following

A. *the sign of* E *is indefinite, i.e. there exists* $u \in W_2^1(M)$ *such that* $E(u) < 0$.

It is an open problem to find the necessary and sufficient conditions for existence of positive solution to (1) in the case **A**.

Consider the problem

$$\begin{cases} -\Delta_g u + r(x)u = 0 \ \text{ in } \ M, \\ \frac{\partial u}{\partial N} + h(x)u = \lambda_g u \ \text{ on } \ \partial M. \end{cases} \tag{5}$$

Condition **A** implies that $\lambda_g < 0$.

The homogeneous cases with definite signs of nonlinearities $R = const, H = 0$ and $H = const, R = 0$, have been considered by Escobar in.[6,7] The case when $R = 0$ and $E(u) > 0$, has been considered by Escobar in Ref. 8.

We will use the following conditions introduced by Escobar in Ref. 8

B. $n > 5$, (M^n, g) be an n-dimensional compact Riemannian manifold with boundary, that has a nonumbilic point on ∂M.

Recall that a point of ∂M is *umbilic* if the tensor $\pi - hg$ vanishes at it, where π is the second fundamental form of ∂M.

C. $H(x)$ achieves a global maximum at a nonumbilic point of the boundary $0 \in \partial M$ where $\nabla H(0) = 0$, the second derivatives $\partial^2 H(0)/\partial x_i \partial x_j$ are defined and $|\Delta H(0)| \leq c(n)||\pi(0) - h(0)g(0)||$, where $c(n)$ is a suitable constant.

Let us state our main results. First we consider a homogeneous case when

D. $R(x) \equiv 0$;

E. $H(x) = \mu H^+(x) - H^-(x)$ where $H^+(x) \geq 0$, $H^-(x) > 0$ as $x \in \partial M$, $\mu \geq 0$; and the set $\{x \in M; H^+(x) > 0\}$ is non-empty.

We introduce the following characteristic value

$$\mu_1 = \sup\{\mu \in \text{Real}^+ |\ \mu B^+(\phi) - B^-(\phi) < 0, E(\phi) \leq 0,\ \forall \phi \in C^\infty(M)\}, \quad (6)$$

where $B^\pm(u) = \int_{\partial M} H^\pm(x)|u|^{2^{**}} d\sigma_g$. It is easy to see that $0 \leq \mu_1 \leq \infty$, where

$$\mu_1 = \infty \quad \text{if and only if} \quad \partial M^+ := \{x \in \partial M :\ H(x) > 0\} = \emptyset.$$

Remark, that in the case when $\partial M^+ \neq \emptyset$

$$\mu_1 = \inf\{\frac{B^-(\phi)}{B^+(\phi)} : E(\phi) \leq 0,\ \phi \in C^\infty(M)\}. \quad (7)$$

Our first main result is the following

Theorem 1.1. *Suppose that $n \geq 3$ and **A,D,E** hold. Then*

1) *If $\mu_1 = 0$ then problem (1) has no positive solutions for any $\mu \geq 0$.*
2) *If $\mu_1 > 0$ then*

2.1) *For every $\mu \in (0, \mu_1]$ problem (1) has a positive solution u^1_μ such that $I_\mu(u^1_\mu) < 0$.*

2.2) *If conditions **B,C** hold too, then there exists $\varepsilon > 0$ such that for every $\mu \in (\mu_1 - \varepsilon, \mu_1]$ problem (1) has an other positive solution u^2_μ such that $I_\mu(u^2_\mu) > 0$ if $\mu < \mu_1$; $I_{\mu_1}(u^2_{\mu_1}) = 0$.*

Now we state our main result for the nonhomogeneous case when the boundary value problem has two critical exponents. Suppose that

F. $R(x) \geq 0$ on M, $\int_M R(x)dx > 0$.

We will study the problem with parameters $\lambda \geq 0$ at $R(x)$ and $\mu \geq 0$ at $H^+(x)$, i.e., we consider the critical points of

$$I_{\lambda,\mu}(u) = \frac{1}{2}E(u) - \frac{1}{2^{**}}B_\mu(u) - \frac{1}{2^*}\lambda F(u). \quad (8)$$

Let $\mu > 0$. Introduce the following characteristic value

$$\Lambda^1_\mu = \inf\{\frac{-B^2_\mu(\phi)}{4E(\phi)F(\phi)} : E(\phi) < 0,\ \phi \in C^\infty(\bar{M})\}. \tag{9}$$

Our main result is the following

Theorem 1.2. *Suppose that* $n \geq 3$ *and* ***A,D,E,F*** *hold. Assume that* $\mu_1 > 0$. *Then for every* $\mu \in (0, \mu_1]$
1) *for every* $\lambda \in [0, \Lambda^1_\mu]$ *problem* (1) *has a positive solution* $u^1_{\lambda,\mu}$ *such that* $I_{\lambda,\mu}(u^1_{\lambda,\mu}) < 0$.
2) *If conditions* ***B, C*** *hold too, then there exists* $\varepsilon > 0$ *such that for every* $\mu \in (\mu_1 - \varepsilon, \mu_1]$ *there exists* $\lambda_\mu > 0$ *such that for every* $\lambda \in (0, \lambda_\mu)$ *problem* (1) *has a second positive solution* $u^2_{\lambda,\mu}$ *for which* $I_{\lambda,\mu}(u^2_{\lambda,\mu}) > 0$.

In order to construct a solution it is important to find the limiting Palais-Smale levels C_{P-S}. Note that $p = 2^* = \frac{2n}{n-2}$, $q = 2^{**} = \frac{2(n-1)}{(n-2)}$ are the critical Sobolev exponents for the embedding $W^1_2(M) \subset L_p(M)$ and for the trace-embedding $W^1_2(M) \subset L_q(\partial M)$, respectively. The functional I does not satisfy the Palais-Smale condition on all levels $I = \beta$. Usually the value $\hat{I}^\infty < +\infty$, expressed in terms of the Sobolev quotients $Q(B^n)$ and $Q(S^{n,+})$ for the ball B^n and the semi-sphere $S^{n,+}$, respectively, is determined (see Refs. 1,2,4,6–8,10,13). Then the levels $I = \beta$ where the functional I satisfies the Palais-Smale condition are defined by the inequality $\beta < \hat{I}^\infty$ and thus $C_{P-S} = \hat{I}^\infty$. It turns out that generally it is necessary to take into account also the ground state level $\hat{I}^{gr}$ that is the point with the least level of I among all critical points of I and $\hat{I}^{gr} = I(u_g)$ (cf. Ref. 5). We show $C_{P-S} = \hat{I}^\infty + \hat{I}^{gr}$ and that the levels $I = \beta$, where the functional I satisfies the Palais-Smale condition, are defined by the inequalities

$$\hat{I}^{gr} \leq \beta < \hat{I}^\infty + \hat{I}^{gr}.$$

Note that $u = 0$ is a critical point of I so that $I^{gr} \leq 0$. If E has a definite positive sign then $\hat{I}^{gr} = 0$, i.e. $u_{gr} = 0$. In our case we deal with $u_{gr} = u^1$, $\hat{I}^{gr} < 0$.

Remark also that in order to verify the Palais-Smale condition one has to show that any Palais-Smale sequence is bounded, and thereafter to prove the strong convergence of the sequence. If E has a definite positive sign, i.e. $\hat{I}^{gr} = 0$, then the first step is rather easy (see Refs. 1,2,4,6–8,10,13). But in the case (A) a Palais-Smale sequence can be unbounded. To overcome this difficulty we use the characteristic values μ_1 and Λ^1_μ.

The paper is organized as follows. In Sec. 2 using the fibering scheme (see Ref. 11) we introduce the basic variational formulations related to the problem (1). In Sec. 3, we prove the existence of the ground state of (1). In Sec. 4, we study the Palais-Smale property. In Sec. 5, we obtain some subcritical auxiliary results. Finally in Sec. 6, we conclude the proof of our main theorems.

Remark 1.1. Following our scheme and the Escobar articles Refs.7, 8 it is possible to replace the condition **B** by one of the following:
$\mathbf{B_1}$ $n \geq 3$, M is locally conformally flat and the maximal value of H is attained at an umbilic point of the boundary;
$\mathbf{B_2}$ $n = 3$, the maximal value of H is attained at an umbilic point on the boundary and M is not conformally diffeomorphic to B^3;
$\mathbf{B_3}$ $n = 3$ and the maximal value of H is attained at a nonumbilic point on the boundary;
$\mathbf{B_4}$ $n = 4$, the maximal value of H is attained at an umbilic point of ∂M and M is not conformally diffeomorphic to B^4;
$\mathbf{B_5}$ $n = 5$, the maximal value of H is attained at an umbilic point of ∂M conformally diffeomorphic to B^5.

2. The Basic Variational Problems

Let $g_{i,j}$ be the components of a given metric tensor $g = (g_{ij})$ with the inverse matrix $(g^{i,j})$, and let $|g| = \det(g_{i,j})$. Let (x^i) be a local system of coordinates on M. By definition, $\div_g X = \dfrac{1}{\sqrt{|g|}} \sum_i \dfrac{\partial}{\partial x_i}(\sqrt{|g|}X^i)$ is the divergence operator on the C^1-vector field $X = (X^i)$; $\nabla = \sum_i g^{i,j} \dfrac{\partial}{\partial x_i}$ is the gradient vector field; $\Delta u = \div_g(\nabla u)$ is the Laplace–Beltrami operator. We denote by dv_g and $d\sigma_g$ respectively the Riemannian measures (induced by the metric g) on M and ∂M.

We are working within the framework of the Sobolev space $W = W_2^1(M)$ equipped with the norm

$$||u|| = \left(\int_M |u|^2 dv_g + \int_M |\nabla u|^2 dv_g\right)^{1/2}. \tag{10}$$

Introduce the following notation

$$B_\mu(w) = \mu B^+(w) - B^-(w),\ w \in W,$$
$$B^\pm(w) = \int_{\partial M^\pm} H(x)|w|^{2^{**}} d\sigma_g,\ F(w) = \int_M R(x)|w|^{2^*} dv_g,$$
$$\partial M^+ = \{x \in \partial M :\ H(x) \geq 0\},\ \partial M^- = \{x \in \partial M :\ H(x) < 0\}.$$

Using the fibering scheme (see Refs. 9,11), we introduce constrained minimization problems for the functional $I_{\lambda,\mu}$. Namely, we consider the fibering functional

$$\tilde{I}_{\lambda,\mu}(t,v) = I_{\lambda,\mu}(tv)$$
$$= \frac{1}{2}t^2 E(v) - \frac{1}{2^{**}}t^{2^{**}} B_\mu(v) - \lambda \frac{1}{2^*}t^{2^*} F(v),\ (t,v)\ \in \text{Real} \times W \setminus \{0\} \quad (11)$$

For $v \in W \setminus \{0\}$ we consider the equation

$$Q_{\lambda,\mu}(t,v) := \frac{\partial}{\partial t}\tilde{I}_{\lambda,\mu}(t,v) = tE(v) - t^{\frac{n}{n-2}} B_\mu(v) - \lambda t^{\frac{n+2}{n-2}} F(v) = 0 \quad (12)$$

with $t \in \text{Real}^+$. To separate two positive solutions t of this equation we will consider also the functional

$$L_{\lambda,\mu}(t,v) = \frac{\partial^2}{\partial t^2}\tilde{I}_{\lambda,\mu}(t,v) \quad (13)$$
$$= E(v) - \frac{n}{n-2}t^{2/(n-2)} B(v) - \frac{n+2}{n-2}\lambda t^{4/(n-2)} F(v).$$

If $\lambda = 0$ and $\mu_1 > 0$ then from (7) we see that for every $\mu \in [0, \mu_1[$ and $w \in W \setminus \{0\}$

$$\text{if} \quad E(w) \leq 0 \text{ then } \ B_\mu(w) < 0, \quad (14)$$
$$\text{if} \quad B_\mu(w) \geq 0 \text{ then } \ E(w) > 0. \quad (15)$$

Let us define the following subsets of $W \setminus \{0\}$

$$\Theta^- = \{v \in W \setminus \{0\} :\ E(v) < 0\},\ \Theta^+ = \{v \in W \setminus \{0\} :\ B_\mu(v) > 0\}.$$

Note that (14),(15) imply that $\Theta^- \cap \Theta^+ = \emptyset$.

If $\lambda > 0$ then for every $v \in \Theta^-$ we can find the value

$$\lambda(v) = -\frac{B_\mu^2(v)}{4E(v)F(v)} \quad (16)$$

such that for every $\lambda \in (0, \lambda(v))$ equation (12) has two positive solutions $t^1_{\lambda,\mu}(v), t^2_{\lambda,\mu}(v)$. Hence for every $\mu \in (0, \mu_1[$ we can introduce (see (9))

$$\Lambda^1_\mu = \inf\{-\frac{B_\mu^2(v)}{4E(v)F(v)} : v \in \Theta^-\}. \quad (17)$$

First we have to prove that $\Lambda^1_\mu > 0$.

Proof of assertion 1) of Theorem 1.2.

Lemma 2.1. *Assume $\mu_1 > 0$ and $\mu \in]0, \mu_1[$. Suppose that **A**, **E** hold. Then $\Lambda^1_\mu > 0$.*

Proof. Let $\mu \in]0, \mu_1[$. Consider the sets:

$$\mathcal{F}_1 = \{u \in W,\ \|u\| = 1,\ E(u) \le 0\},\ \mathcal{F}_2 = \{u \in W,\ \|u\| = 1,\ B_\mu(u) \ge 0\}.$$

They are closed sets on the unit sphere S^1 in W which intersection is empty by (14), (15). Therefore, the distance between the frontiers of $\mathcal{F}_1$ and $\mathcal{F}_2$ is positive. Moreover, there exists a positive c_1 such that $B(u) \le -c_1$ if $u \in \mathcal{F}_1$. Thus

$$B(u) \le -c_1 \|u\|^{2^{**}} \quad \text{if } E(u) \le 0.$$

Since $|E(u)| \le C_1 \|u\|^2$ and $|F(u)| \le C_2 \|u\|^{2^*}$ we see that $\Lambda^1_\mu \ge 4c_1^2/C_1 C_2 > 0$. □

From the above constructions we can conclude the following

Claim 2.1. *Assume $0 < \mu_1$ and $\mu \in]0, \mu_1[$.*
1) If $\lambda = 0$ then equation (12) has exactly one positive solution $t_{\mu_1}(w)$ if $w \in \Theta^-$ and exactly one positive solution $t^2_\mu(w) > 0$ if $w \in \Theta^+$. These solutions are separated by the sign of $L_\mu(t, v)$, i.e. $L_\mu(t^1_\mu(v), v) > 0$, $L_\mu(t^2_\mu(v), v) < 0$.
2) If $0 < \lambda < \Lambda^1_\mu$ then for every $v \in \Theta^-$ the equation $Q_{\lambda,\mu}(t, v) = 0$ has a positive solution $t^1_{\lambda,\mu}(v) > 0$ such that $L_{\lambda,\mu}(t^1_{\lambda,\mu}(v), v) > 0$.
3) If $0 < \lambda < \infty$ then for every $v \in \Theta^+$ the equation $Q_{\lambda,\mu}(t, v) = 0$ has exactly one positive solution $t^2_{\lambda,\mu}(v) > 0$ such that $L_{\lambda,\mu}(t^2_{\lambda,\mu}(v), v) < 0$.

Here and in what follows in case $\lambda = 0$ we use the abridged notations $I_\mu := I^{0,\mu}$, $Q_\mu := Q^{0,\mu}$, $t^j_\mu := t^j_{0,\mu}$, etc.
It is not hard to prove that

Claim 2.2. *If $\mu \in]0, \mu_1[$ and $\lambda \in [0, \Lambda^1_\mu[$ then the solutions $t^1_{\lambda,\mu}(v)$, $t^2_{\lambda,\mu}(v)$ are C^1-functions of $v \in S^1$.*

If $\lambda = 0$ then $t^1_{\lambda,\mu}(v)$, $t^2_{\lambda,\mu}(v)$ are C-function of $\mu \in (0, \mu_1)$.

If $\mu \in]0, \mu_1[$ then the function $t^1_{\lambda,\mu}(v)$ is a C-function of $\lambda \in [0, \Lambda^1_\mu[$ and $t^2_{\lambda,\mu}(v)$ is a C-function of $\lambda \in [0, +\infty[$.

Let us define the fibering functionals

$$J^1_{\lambda,\mu}(v) = \tilde{I}_{\lambda,\mu}(t^1_{\lambda,\mu}(v), v),\ v \in \Theta^-,\ \mu \in (0,\mu_1), \lambda \in [0, \Lambda^1_\mu) \tag{18}$$

and

$$J^2_{\lambda,\mu}(v) = \tilde{I}_{\lambda,\mu}(t^2_{\lambda,\mu}(v), v),\ v \in \Theta^+,\ \mu \in (0,\mu_1), \lambda \in [0, +\infty). \tag{19}$$

Proposition 2.2 implies

Claim 2.3. *The functionals* $J^1_{\lambda,\mu}(v)$, $J^2_{\lambda,\mu}(v)$ *are* C^1*-functionals of* $v \in \Theta^-, \Theta^+$. *They are continuous functions of* μ *and* λ.

Observe that the functions $J^j_{\lambda,\mu}(v)$ are 0-homogeneous, i.e. $J^j_{\lambda,\mu}(sv) = J^j_{\lambda,\mu}(v)$ for any $s \neq 0$, $j = 1, 2$.

Thus we have the following two basic variational problems

$$\hat{I}^j_{\lambda,\mu} = inf\{J^j_{\lambda,\mu}(v) | v \in \Theta\},\quad j = 1, 2. \tag{20}$$

Observe that in the case $\lambda = 0$ variational problems (20) are equivalent to the following ones:

$$\hat{I}^1_\mu = \inf\{J^1_\mu(w) : w \in W \setminus \{0\},\ E(w) < 0\}, \tag{21}$$

$$\hat{I}^2_\mu = \inf\{J^2_\mu(w) :\ w \in W \setminus \{0\},\ B_\mu(w) > 0\}, \tag{22}$$

where

$$J^1_\mu = \frac{1}{2(n-1)} \frac{E(w)^{n-1}}{B_\mu(w)^{n-2}},\ J^2_\mu = -\frac{1}{2(n-1)} \frac{|E(w)|^{n-1}}{|B_\mu(w)|^{n-2}}. \tag{23}$$

It is not hard to prove (see Refs. 9, 11)

Lemma 2.2. *Let* $j = 1, 2$. *Suppose that* $v^j_{\lambda,\mu}$ *is a solution of problem (20). Then* $u^j_{\lambda,\mu} = t^j_{\lambda,\mu}(v^j_{\lambda,\mu})v^j_{\lambda,\mu}$ *is a nonzero critical point of the functional* $I_{\lambda,\mu}$.

Since the functionals $J^1_{\lambda,\mu}$, $J^2_{\lambda,\mu}$ are 0-homogeneous, they are uniformly continuous with respect to μ and λ on $S^1 = \{w : ||w||_W = 1\}$. Using these facts it can be shown the following

Claim 2.4. 1) *Let* $\lambda = 0$, $j = 1, 2$. *Then* $\hat{I}^j_\mu$ *are continuous monotone non-increasing functions on* $\mu \in]0, \mu_1]$.
2) *Let* $\mu \in]0, \mu_1[$, $j = 1, 2$. *Then* $\hat{I}^j_{\lambda,\mu}$ *are continuous monotone nonincreasing functions of* λ.
3) *If* $v \in \Theta^-$ *then* $J^1_{\lambda,\mu}(v) < 0$ *for* $\mu \in]0, \mu_1[$ *and* $\lambda \in]0, \Lambda^1_\mu[$;
4) *If* $v \in \Theta^+$ *then* $J^2_{\lambda,\mu}(v) > 0$ *for* $\mu \in]0, \mu_1[$ *and* $\lambda \in]0, +\infty[$.

Corollary 2.1. 1) *If* $\lambda = 0$ *then* $\hat{I}^1_\mu < 0$, $\hat{I}^2_\mu > 0$ *for every* $\mu \in]0, \mu_1[$.

2) *If* $\mu \in]0, \mu_1[$ *and* $\lambda \in [0, \Lambda^1_\mu)$ *then* $\hat{I}^1_{\lambda,\mu} < 0$ *and* $\hat{I}^2_{\lambda,\mu} > 0$.

2.1. *The level of ground state and the Sobolev level*

To study the critical points of $I_{\lambda,\mu}$ it is important to know the ground state and the Sobolev level of $I_{\lambda,\mu}$.

By definition, the critical point $u_g \in W$ of $I_{\lambda,\mu}$ is said to be *the ground state* if it is a point with the least level of $I_{\lambda,\mu}$ among all the critical points Z (see Ref. 5), i.e.

$$\min\{I_{\lambda,\mu}(u) : u \in Z\} = I_{\lambda,\mu}(u_g). \tag{24}$$

In this case the value $\hat{I}^{gr}_{\lambda,\mu} := I_{\lambda,\mu}(u_g)$ is called *the level of ground state.*

Our main lemma on the level of ground state is the following

Lemma 2.3. *Let* $\mu \in]0, \mu_1]$, $\lambda = 0$ *or* $\mu \in]0, \mu_1[$, $\lambda \in]0, \Lambda^1_\mu[$. *Assume that there exists a solution* $v_0 \in \Theta^-$ *of variational problem* (20), $j = 1$. *Then* $u_0 = t^1_{\lambda,\mu}(v_0)v_0 \in W \setminus 0$ *is a ground state and* $\hat{I}^1_{\lambda,\mu}$ *is a level of ground state for* $I_{\lambda,\mu}$.

Proof. Let $u \in W \setminus \{0\}$ be a critical point of $I_{\lambda,\mu}$. Put $t(u) = ||u|| > 0$ and $v(u) = u/||u|| \in S^1$. Then $(t(u), v(u))$ satisfies equation (12). Consider the case $\mu \in]0, \mu_1]$ and $\lambda = 0$. By direct analysis of (12) and using (15) it can be deduced that $(t(u), v(u)) \in \Theta^- \cup \Theta^+$. The Proposition 2.4 yields that

$$\hat{I}^2_\mu > \hat{I}^1_\mu = J^1_\mu(t^1_\mu(v_0)v_0) = I_\mu(u_0).$$

Hence we obtain that $I_\mu(u) \geq I_\mu(u_0)$ and therefore u_0 is a ground state and $\hat{I}^1_\mu$ is the level of ground state for I_μ. The case $\mu \in]0, \mu_1[$, $\lambda \in [0, \Lambda^1_\mu[$ is considered using the same arguments.

Let us now introduce a conception of the Sobolev level. We adapt the arguments of P.L. Lions.[10]

Along with the functional $I_{\mu,\lambda}$ we consider the functional $I^{[y],\infty}_{\lambda,\mu}(u)$ defined in the following way:

i) if $y \in M$ then

$$I^{[y],\infty}_{\lambda,\mu}(u) = \frac{1}{2}\int_{\mathrm{Real}^n} |\nabla u|^2 dx - \lambda \frac{1}{2^*} R(y) \int_{\mathrm{Real}^n} |u|^{2^*} dx, \; u \in W^1_2(\mathrm{Real}^n)\setminus\{0\};$$

ii) if $y \in \partial M$ then for $u \in W_2^1(\mathrm{Real}_+^n) \setminus \{0\}$

$$I_{\lambda,\mu}^{[y],\infty}(u) = \frac{1}{2}\int_{x_n \geq 0} |\nabla u|^2 dx - \lambda \frac{1}{2^*} R(y) \int_{x_n \geq 0} |u|^{2^*} dx$$
$$-\mu \frac{1}{2^{**}} H(y) \int_{x_n = 0} |u|^{2^{**}} dx'.$$

As above we can introduce the corresponding functionals $\tilde{I}_{\lambda,\mu}^{[y],\infty}(t,v)$, $Q_{\lambda,\mu}^{[y],\infty}(t,v)$ and consider the equations
iq) if $y \in M$

$$Q_{\lambda,\mu}^{[y],\infty}(t,v) = t^2 \int_{\mathrm{Real}^n} |\nabla v|^2 dx - \lambda t^{2^*} R(y) \int_{\mathrm{Real}^n} |v|^{2^*} dx = 0,$$

for $t > 0$ and as $v \in W_2^1(\mathrm{Real}^n) \setminus \{0\}$;
iiq) if $y \in \partial M$

$$Q_{\lambda,\mu}^{[y],\infty}(t,v) = t^2 \int_{x_n \geq 0} |\nabla v|^2 dx - \lambda t^{2^*} R(y) \int_{x_n \geq 0} |v|^{2^*} dx$$
$$-\mu t^{2^{**}} H(y) \int_{x_n = 0} |v|^{2^{**}} dx' = 0,$$

for $t > 0$ and $v \in W_2^1(\mathrm{Real}_+^n) \setminus \{0\}$.

It is easy to see that these equations can have at most one positive solution $t_\mu^{[y],\infty}(v) := t_{\lambda,\mu}^2(v)$. Moreover, the solution is absent if and only if $y \in N$, where

$$N = \{y \in M, R(y) = 0 \text{ or } y \in \partial M, R(y) = 0, H(y) \leq 0\}. \tag{25}$$

Thus as above we can introduce the fibering functional

$$J_{\lambda,\mu}^{[y],\infty}(v) := I_{\lambda,\mu}^{[y]}(t_{\lambda,\mu}^{[y],\infty}(v)v), \;\; v \in W_2^1 \setminus \{0\},$$

where we put for $y \in N$ by definition

$$J_{\lambda,\mu}^{[y],\infty}(v) := +\infty.$$

By the fibering scheme we have the following constrained minimization problem

$$\hat{I}_{\lambda,\mu}^{[y],\infty} = \inf\{J_{\lambda,\mu}^{[y],\infty}(v) : \; v \in W_2^1 \setminus \{0\}\}. \tag{26}$$

We call the *Sobolev level* the following number

$$\hat{I}_{\lambda,\mu}^{\infty} = \inf_{y \in \overline{M}} \hat{I}_{\lambda,\mu}^{[y],\infty}. \tag{27}$$

□

Claim 2.5. *If $\mu \in]0, +\infty[, 0 \leq \lambda < \infty$, then $\hat{I}_{\lambda,\mu}^{\infty} > 0$. Furthermore, $\hat{I}_{\lambda,\mu}^{\infty}$ is a continuous monotone non-increasing function of μ and λ.*

Proof. Consider for example the case $y \in \partial M$. Let $v \in W_2^1(\mathrm{Real}_+^n) \setminus \{0\}$. By 0-homogeneity of $J_{\lambda,\mu}^{[y],\infty}(v)$ we may assume $\int_{x_n \geq 0} |\nabla v|^2 dx = 1$. Then the Sobolev inequality implies $0 < \int_{x_n \geq 0} |v|^{2^*} dx \leq C$, $0 < \int_{x_n = 0} |v|^{2^{**}} dx' \leq C$ with some $0 < C < +\infty$ that does not depend on $v \in S^1$. Hence by iiq) and since $\int_{x_n \geq 0} |\nabla v|^2 dx = 1$ we conclude that $t_{\lambda,\mu}^{[y],\infty}(v) > c_0$ where $0 < c_0 < +\infty$ does not depend on $v \in W_2^1(\mathrm{Real}_+^n) \setminus \{0\}$. Using this fact in case $E(v) = \int_{x_n \geq 0} |\nabla v|^2 dx = 1$ we deduce that

$$J_{\lambda,\mu}^{[y],\infty}(v) > \frac{1}{n(n-1)} (t_{\lambda,\mu}^2(v))^2 E(v) > \frac{1}{n(n-1)} (c_0)^2 > 0, \ \forall v \in W_2^1(\mathrm{Real}_+^n) \setminus \{0\}.$$

Hence we obtain the first statement. The second statement follows by monotonicity and uniform continuity with respect to $v \in W_2^1(\mathrm{Real}_+^n) \setminus \{0\}$ of the function $J_{\lambda,\mu}^{[y],\infty}(v)$ of μ and λ. □

3. Existence of the Ground State

In this section we prove the assertions 2.1) of Theorem 1.1 and 1) of Theorem 1.2. Note that by Proposition 2.3 it gives us also the ground state of $I_{\lambda,\mu}$. Both these assertions will be obtained simultaneously.

But first we consider the cases $\mu \in]0, \mu_1[$, $\lambda \in [0, \Lambda_\mu^1[$ and thereafter the case $\mu = \mu_1$ and $\lambda = 0$.

Lemma 3.1. *Suppose that $\mu_1 > 0$ and $\mu \in]0, \mu_1[$, $\lambda \in [0, \Lambda_\mu^1[$. Then there exists a ground state $u_{\lambda,\mu}^1 \in W \setminus \{0\}$ of $I_{\lambda,\mu}$. Furthermore, $u_{\lambda,\mu}^1 \in C^{1,\alpha}(M)$ for some $\alpha \in]0,1[$, $u_{\lambda,\mu}^1 > 0$.*

Proof. Let us prove that problem (20), $j = 1$ has a solution $v_{\lambda,\mu}^1$.

Let $v_m \in \Theta^-$ be a minimizing sequence for this problem, i.e. $J_\mu^1(v_m) \to \hat{I}_{\lambda,\mu}^1$. Since $J_{\lambda,\mu}^1$ is 0-homogeneous, we may assume that $||v_m|| = 1$. Thus v_m is bounded in W. Since W is reflexive, we may assume that $v_m \rightharpoonup \bar{v} \in W$ weakly in W and strongly in $L_l(M)$, for $2 \leq l < 2^*$, and in $L_s(\partial M)$, for $2 \leq s < 2^{**}$.

Let us show that $\bar{v} \neq 0$. If $v_m \to 0$ in W then $v_m \to 0$ in $L_2(M)$, $L_2(\partial M)$ and $v_m \in S^1$, so that $E(\psi_m) \to 1$ as $m \to \infty$. But this contradicts to the assumption $E(v_m) < 0$. Thus $\bar{v} \neq 0$ and $\bar{v} \in \Theta^-$.

Let us show that $lim_{m\to\infty} t_{\lambda,\mu}^1(v_m) = \bar{t} < \infty$. Indeed, if $\bar{t} = \infty$ then the contradiction follows directly from equality (12) since $E(v_m)$ are bounded and from (7) in case $\mu < \mu_1$. Moreover, it follows that $lim_{m\to\infty} B_\mu(v_m) = \bar{B}_\mu < 0$.

Remark also that $\hat{I}_{\lambda,\mu}^1 > -\infty$.

Let us show now that $\bar{t} > 0$. Suppose contrary $\bar{t} = 0$. Then since $E(v_m)$, $B(v_m)$, $F(v_m)$ are bounded, it follows from (12) that $J^1_{\lambda,\mu}(v_m) \to 0$. However by Corollary 2.1 we have $\hat{I}^1_{\lambda,\mu} < 0$. Hence we get a contradiction and therefore $\bar{t} > 0$.

So we have

$$\bar{J}^1_{\lambda,\mu}(v_m) = \frac{1}{2}\bar{t}^2 E(v_m) - \lambda\frac{1}{2^*}\bar{t}^{2^*} F(v_m) - \frac{1}{2^{**}}\bar{t}^{2^{**}} B(v_m) \to \hat{I}^1_{\lambda,\mu}.$$

It is not hard to prove that

$$\hat{I}^1_{\lambda,\mu} = \inf_{w\in\Theta^-} \bar{J}^1_{\lambda,\mu}(w). \tag{28}$$

Using Vitali's convergence theorem we have (see Ref. 18, p. 174):

$$\int_M |\nabla v_m|^2 dx - \int_M |\nabla(v_m - v)|^2 dx = \int_M \int_0^1 \frac{d}{dt}|\nabla(v_m + (t-1)v)|^2 dt dx$$

$$= 2\int_0^1 \int_M (\nabla(v_m + (t-1)v), \nabla v) dx dt$$

$$\to 2\int_0^1 \int_M t|\nabla v|^2 dx dt = \int_M |\nabla v|^2 dx$$

as $m \to \infty$. Similarly,

$$E(v_m) = E(w) + E(v_m - w) + o(1),\ B(v_m) = B(v) + B(v_m - v) + o(1),$$

$$F(v_m) = F(v) + F(v_m - v) + o(1),$$

where $o(1) \to 0$ as $\gamma \to 2^{**}$ and so

$$\bar{J}^1_{\lambda,\mu}(v_m) = \bar{J}^1_{\lambda,\mu}(v) + \bar{J}^1_{\lambda,\mu}(v_m - v) + o(1).$$

The last equality implies that $\bar{J}^1_{\lambda,\mu}(v_m - v)$ converges to some a. Since

$$\bar{t}^2 E(v_m) - \lambda\bar{t}^{2^*} F(v_m) - \bar{t}^{2^{**}} B(v_m) \to 0,$$

it means that

$$(\frac{1}{2} - \frac{1}{2^*})\bar{t}^2 E(v_m - v) + \lambda(\frac{1}{2^{**}} - \frac{1}{2^*})\bar{t}^{2^*} F(v_m - v) \to a,$$

and since the both terms are positive, $a \geq 0$.

However,

$$\bar{J}^1_{\lambda,\mu}(v) = \hat{I}^1_{\lambda,\mu} - a,$$

and therefore, $a \leq 0$. Indeed, above we have shown that $v \in \Theta^-$ and therefore by (28) we have $\bar{J}^1_{\lambda,\mu}(v) \geq \hat{I}^1_{\lambda,\mu}$. Thus $a = 0$, the sequence $E(v_m -$

v) converges to 0, the sequence v_m converges to v in W and $J^1_{\lambda,\mu}(v) = \hat{I}^1_{\lambda,\mu}$. This completes the proof. □

Let us conclude the proof of assertions 2.1) in Theorem 1.1 and 1) in Theorem 1.2 supposing that $\mu \in]0, \mu_1[$, $\lambda \in [0, \Lambda^1_\mu[$.

It follows from Proposition 2.2 that $u_{\lambda,\mu} = t^1_{\lambda,\mu}(v_{\lambda,\mu})v_{\lambda,\mu}$ is a weak solution of (1).

Observe that all functionals in variational problem (20), $j = 1$ and Θ^- are even. Therefore, one can suppose that the minimizing point is non-negative, i.e. $v_{\lambda,\mu} \geq 0$. It implies that $u_{\lambda,\mu} = t^1_{\lambda,\mu}(v_{\lambda,\mu})v_{\lambda,\mu} > 0$ on M. Indeed, applying the regularity results from Ref. 4 we derive that $u_{\lambda,\mu} \in C^{1,\alpha}(M)$ for some $\alpha \in]0,1[$. Hence by the Harnack inequality[12] it follows that $u_{\lambda,\mu} > 0$.

Let us now consider the case $\mu = \mu_1$, $\lambda = 0$.

Lemma 3.2. *Suppose that $\mu_1 > 0$ and $\mu = \mu_1$, $\lambda = 0$. Then there exists a ground state $u^1_{\mu_1} \in W \setminus \{0\}$ of I_{μ_1}. Furthermore, $u_{\mu_1} \in C^{1,\alpha}(M)$ for some $\alpha \in]0,1[$, $u_{\mu_1} > 0$.*

Proof. By the above proof of Lemma 3.1 there exists a family of points $u_\mu = t^1_\mu(v_\mu)v_\mu$, $\mu \in]0, \mu_1[$. Let us show that $lim_{\mu \to \mu_1} t^1_\mu(v^\mu) = \bar{t} < \infty$. Assume the converse $lim_{\mu_i \to \mu_1} t^1_{\mu_1}(v_{\mu_i}) = \infty$ for some subsequence $\mu_i \uparrow \mu_1$. Since $||v_{\mu_i}|| = 1$, the set v_{μ_i} is bounded in W. Since W is reflexive, we may assume that $v_{\mu_i} \rightharpoonup \bar{V} \in W$ weakly in W and strongly in $L_l(M)$, for $2 \leq l < 2^*$, and in $L_s(\partial M)$, for $2 \leq s < 2^{**}$.

Reasoning as in the proof of Lemma 3.1 we can show that $\bar{V} \neq 0$. Since $u_\mu = t^1_\mu(v_\mu)v_\mu$ is a weak solution of (1) we have

$$\frac{1}{2t^1_\mu(v_\mu)^{2/(n-2)}} DE(v_\mu)(\psi) = \frac{1}{2^{**}} DB_\mu(u_\mu)(\psi), \ \forall \psi \in W^*.$$

If $lim_{\mu_i \to \mu_1} t^1_{\mu_1}(v_{\mu_i}) = \infty$ then we obtain $B'_{\mu_1}(\bar{V})(\psi) = 0$, $\forall \psi \in W^*$. This is possible if and only if $supp(\bar{V}) \cap \partial M \subset \partial M^0 = \{x \in \partial M | \ H(x) = 0\}$. Since E is a weakly lower semi-continuous functional on W, it follows that $E(\bar{V}) \leq lim_{\mu_i \to \mu_1} E(v_{\mu_i}) \leq 0$. On the other hand, since $B_{\mu_1}(\bar{V}) = 0$ and (7) holds, we have $E(\bar{V}) \geq 0$. Thus $E(\bar{V}) = 0$. However, $supp(\bar{V}) \cap \partial M \subset \partial M^0$ so that $B_\mu(\bar{V}) = 0$ for every $\mu \geq 0$. But it contradicts to the assumption $\mu_1 > 0$.

Thus we have proved that $lim_{\mu \to \mu_1} t^1_\mu(v^\mu) = \bar{t} < \infty$. It implies that $\hat{I}^1_{\mu_1} > -\infty$. As above in the proof of Lemma 3.1 it can be shown that $\bar{t} > 0$. Reasoning as above, we obtain that $u_\mu = t^1_\mu(v_\mu)v_\mu \to u_{\mu_1}$ as $\mu \to \mu_1$

strongly in W. From here we deduce that u_{μ_1} is a ground state of I_{μ_1}, $u_{\mu_1} \in C^{1,\alpha}(M)$ for some $\alpha \in (0,1)$, $u_{\mu_1} > 0$. □

4. The Palais-Smale Property

In this section we study the Palais-Smale property of $I_{\lambda,\mu}$.

Let $\mu > 0$, $\lambda > 0$. We say that a sequence $u_m \in W$ is *a Palais-Smale sequence* on W at a level β of $I_{\lambda,\mu}$ if

$$I_{\lambda,\mu}(u_m) \to \beta, \quad ||DI_{\lambda,\mu}(u_m)|| \to 0, \tag{29}$$

and we say that $I_{\lambda,\mu}$ satisfies on W *the Palais-Smale (P.-S.) condition at the level* β if any Palais-Smale sequence at the level β of $I_{\lambda,\mu}$ contains a strongly convergent subsequence in W.

Set

$$\mathcal{B}^+ = \{v \in W \setminus 0 | \ B_\mu(v) = 1\}.$$

Let $\mu \neq \mu_1$. We will say that a sequence $v_m \in \mathcal{B}^+$ is a *Palais-Smale sequence* on $\mathcal{B}^+$ at a level ν for $J^2_{\lambda,\mu}(v)$ if

$$J^2_{\lambda,\mu}(v_m) \to \nu, \quad ||DJ^2_{\lambda,\mu}(v_m)|| \to 0. \tag{30}$$

We say that $J^2_{\lambda,\mu}(v)$ satisfies on $\mathcal{B}^+$ the Palais-Smale (P.-S.) condition at the level ν if any Palais-Smale sequence $(v_m) \in \mathcal{B}^+$ at the level ν has a strongly convergent subsequence in $\mathcal{B}^+$.

Observe that if (v_m) in $\mathcal{B}^+$ is a Palais-Smale sequence at ν for $J^2_{\lambda,\mu}(v)$ then $u_m = t_m v_m$, where t_m is a solution of (12), is a Palais-Smale sequence at $\beta = \frac{1}{2(n-1)}\nu$ for $I_{\lambda,\mu}$, i.e.,

$$I_{\lambda,\mu}(t_m v_m) = \frac{t_m^2}{2}E(v_m) - \frac{t_m^{2^{**}}}{2^{**}}B_\mu(v_m) - \frac{t_m^{2^*}}{2^*}F_\mu(v_m) \to \beta, \tag{31}$$

$$||DI_\mu(t_m v_m)|| = ||t_m DE(v_m) - t_m^{2^{**}-1}DB_\mu(v_m) - t_m^{2^*-1}DF_\mu(v_m)|| \to 0. \tag{32}$$

Now we prove that $J^2_{\lambda,\mu}$ satisfies on $\mathcal{B}^+$ to the Palais-Smale (P.-S.) condition locally at the levels $\nu = 2(n-1)\beta$ with

$$0 \leq \beta < \hat{I}^{\infty}_{\lambda,\mu} + \hat{I}^{gr}_{\lambda,\mu}. \tag{33}$$

Lemma 4.1. *Suppose that* $0 < \mu_1$, $\mu \in]0, \mu_1[$, $\lambda \in [0, +\infty[$.
1) If $\beta > 0$ *and* (33) *holds, then the function* $J^2_{\lambda,\mu}$ *satisfies on* $\mathcal{B}^+$ *the Palais-Smale (P.-S.) condition at the level* $\nu = 2(n-1)\beta > 0$.
2) The function $I_{\mu_1}(u)$ *satisfies on* W *the Palais-Smale (P.-S.) condition at the level* $\beta = 0$.

Proof. Let us prove that the Palais-Smale sequences are bounded. First we prove

Claim 4.1. *Let* $0 < \mu_1$, $\mu \in]0, \mu_1[$, $\lambda \in [0, +\infty[$. *If* v_m *in* $\mathcal{B}^+$ *is a Palais-Smale sequence at* $\nu > 0$ *of* $J^2_{\lambda,\mu}(v)$, *then the sequence* $u_m = t^2_{\lambda,\mu}(v_m)v_m$, *where* $t^2_{\lambda,\mu}(v_m)$ *is the solution of* (12), *is bounded in* W.

Proof. Let (v_m) in $\mathcal{B}^+$ be a Palais-Smale sequence at $\nu > 0$ for $J^2_{\lambda,\mu}(v)$. Consider the following sets:

$$M_1 = \{u \in W,\ \|u\| = 1,\ E(u) \leq 0\},$$

$$M_2 = \{u \in W,\ \|u\| = 1,\ B_{\mu_1}(u) \leq 0\},\ M_3 = \{u \in W,\ \|u\| = 1,\ B_\mu(u) \leq 0\}.$$

They are closed sets on the unit sphere S^1 in W and $M_1 \subset M_2 \subset M_3$. Moreover, since $\mu \in]0, \mu_1[$, by (15) the distance between the frontiers of M_1 and M_3 is positive. Therefore, there exists a positive c_1 such that $E(u) \geq c_1$ if $u \in S^1 \setminus M_3$. Thus

$$E(u) \geq c_1\|u\|^2 \quad \text{if } B_\mu(u) \geq 0.$$

In particular, $E(u) \geq c_1\|u\|^2$ if $B_\mu(u) = 1$. Thus if

$$B_\mu(v_m) = 1,\ J^2_{\lambda,\mu}(v_m) \to \nu < +\infty \tag{34}$$

as $m \to \infty$, then the set $E(v_m)$ is bounded and therefore, the norms $\|v_m\|$ are uniformly bounded. From (34) and (12) it follows also that $t^2_{\lambda,\mu}(v_m)$ are bounded and separated from zero, i.e. there are two positive constants $0 < c_1, c_2 < \infty$ such that $c_1 \leq t^2_{\lambda,\mu}(v_m) \leq c_2$. Therefore, $u_m = t^2_{\lambda,\mu}(v_m)v_m$ are bounded and the proof of the lemma is complete. □

Now we prove

Claim 4.2. *Assume that* $\mu_1 > 0$, $\lambda = 0$. *Then any Palais-Smale sequence* (u_m) *in* W *of the function* I_{μ_1} *at the level* $\beta = 0$ *is bounded in* W.

Proof. Let $u_m \in W$ be a Palais-Smale sequence at the level $\beta = 0$ of I_{μ_1}. Then

$$I_{\mu_1}(t_m v_m) = \frac{t_m^2}{2}E(v_m) - \frac{t_m^{2^{**}}}{2^{**}}B_{\mu_1}(v_m) \to 0, \tag{35}$$

$$||DI_{\mu_1}(t_m v_m)|| = ||t_m DE(v_m) - t_m^{2^{**}-1}DB_{\mu_1}(v_m)|| \to 0, \tag{36}$$

where $u_m = t_m v_m$ and $||v_m||_W = 1$. Hence to prove the Proposition it is sufficient to show that the sequence t_m is bounded.

We may assume that there exists a weak limit: $v_m \rightharpoonup v$ in W as $m \to \infty$. It follows easily from (35), (36) that $E(v_m) \to 0$ as $m \to \infty$. Hence reasoning as above, in the proof of Lemma 3.1, we can prove that $v \neq 0$. Suppose that $t_m \to \infty$. From (36) we have

$$\frac{1}{t_m^{2^{**}-2}} DE(v_m)(\xi) = DB_{\mu_1}(v_m)(\xi) + o(1), \ \forall \, \xi \in W^*.$$

If $t_m \to \infty$ then we have $DB_{\mu_1}(v)(\xi) = 0$ for all $\xi \in W^*$. This is possible only if $supp\ v \subset \partial M^0 := \{x \in \partial M : H(x) = 0\}$. Thus $v \not\equiv 0$ on $\overline{M}$ and $B_{\mu_1}(v) = 0$. On the other hand, since E is weakly lower semi-continuous on W we have that $E(v) \leq liminf_{m\to\infty} E(v_m) = 0$. But in this case by definition (6) we have $\mu_1 = 0$, what contradicts to our assumption $\mu_1 > 0$. The proof is complete. □

From Proposition 4.1 it follows that if (v_m) in $\mathcal{B}^+$ is a Palais-Smale sequence for $J^2_{\lambda,\mu}(v)$ at the level $\nu = 2(n-1)\beta$, where $\beta > 0$ satisfies (33), then $u_m = t^2_{\lambda,\mu}(v_m)v_m$ is a bounded Palais-Smale sequence of $I_{\lambda,\mu}$ at the level $\beta > 0$ with (33). Hence and by Proposition 4.2 we see that in order to prove Lemma 4.1 it remains to prove the following

Claim 4.3. *Any bounded in W Palais-Smale sequence $\{u_m\}$ for $I_{\lambda,\mu}$ at a level β, satisfying* (33), *contains a subsequence strongly convergent in W.*

Proof. Let $u_m \in W$ be a Palais-Smale sequence at the level β of $I_{\lambda,\mu}$. By assumption the sequence$\{u_m\}$ is bounded in W. Therefore we can assume that $\{u_m\}$ is a weakly convergent sequence in W with the weak limit u, and using[10] we have

$$u_m \rightharpoonup u \text{ weakly in } W,$$

$$u_m \to u \text{ in } L^r(M),\ 1 < r < 2^*, \quad u_m \to u \text{ in } L^s(\partial M),\ 1 < s < 2^{**},$$

$$|\nabla u_m|^2 \rightharpoonup \xi \geq |\nabla u|^2 + \textstyle\sum_{k\in L} \xi_k \delta_{x_k}, \xi_k > 0, \tag{37}$$

$$|u_m|^{2^*} \rightharpoonup \eta = |u|^{2^*} + \textstyle\sum_{k\in L} \eta_k \delta_{x_k},\ \eta_k \geq 0,$$

$$|u_m|^{2^{**}} \rightharpoonup \zeta = |u|^{2^{**}} + \textstyle\sum_{k\in L} \zeta_k \delta_{x_k},\ \zeta_k \geq 0.$$

If the set L is empty then (37) implies that $u_m \to u$ strongly in W. Let us prove that L cannot be non-empty.

Since $||dI_\lambda(u_m)||_{W^*} \to 0, dI_\lambda(u) = 0$ and (37) holds we deduce the following

$$0 \geq \sum_{k\in L}(\xi_k - \lambda R(x_k)\eta_k - \mu H^+(x_k)\zeta_k).$$

By definition (26), (27) it follows easily that

$$\begin{aligned}\hat{I}^\infty_{\lambda,\mu} &\leq \sum_{k\in L}(\frac{1}{2}\xi_k - \frac{1}{2^*}\lambda R(x_k)\eta_k) - \frac{1}{2^{**}}\sum_{k\in L'} H(x_k)\zeta_k \\ &= \lambda\frac{1}{n}\sum_{k\in L} R(x_k)\eta_k + \frac{1}{2(n-1)}\sum_{k\in L'}\mu H^+(x_k)\zeta_k. \qquad (38)\end{aligned}$$

Using this fact and (37) we deduce from our assumptions (33) that

$$\begin{aligned}\hat{I}^\infty_{\lambda,\mu} + \hat{I}^{gr}_{\lambda,\mu} &> \beta = \lim_{m\to\infty} I_{\lambda,\mu}(u_m) \\ &= \lim_{m\to\infty}\{I_{\lambda,\mu}(u_m) - \frac{1}{2} < dI_{\lambda,\mu}(u_m), u_m >\} \\ &= \lim_{m\to\infty}\left\{\lambda\frac{1}{n}\int R|u_m|^{2^*}dv_g + \frac{1}{2(n-1)}\int H|u_m|^{2^{**}}d\sigma_g\right\} \\ &\geq \lambda\frac{1}{n}\int R|u|^{2^*}dv_g + \frac{1}{2(n-1)}\int H|u|^{2^{**}}d\sigma_g \\ &+ \lambda\frac{1}{n}\sum_{k\in L} R(x_k)\eta_k + \mu\frac{1}{2(n-1)}\sum_{k\in L'} H^+(x_k)\zeta_k \\ &\geq \lambda\frac{1}{n}\int R|u|^{2^*}dv_g + \frac{1}{2(n-1)}\int H|u|^{2^{**}}d\sigma_g + \hat{I}^\infty_{\lambda,\mu}. \qquad (39)\end{aligned}$$

Since by Lemma 2.3 $\hat{I}^{gr}_{\lambda,\mu} = \hat{I}^1_{\lambda,\mu} < 0$, this implies that $u \neq 0$. Hence we deduce that

$$\hat{I}^{gr}_{\lambda,\mu} > \lambda\frac{1}{n}\int R|u|^{2^*}dv_g + \frac{1}{2(n-1)}\int H|u|^{2^{**}}d\sigma_g = I_{\lambda,\mu}(u).$$

On the other hand, since u is a weak solution of (1) and $\hat{I}^{gr}_{\lambda,\mu} = \hat{I}^1_{\lambda,\mu}$ is the level of the ground state, we have $\hat{I}^{gr}_{\lambda,\mu} \leq I_{\lambda,\mu}(u)$. Thus we get a contradiction. The proof is complete. □

It is important for us to know when the level $\beta = \hat{I}^2_{\lambda,\mu}$ satisfies (33). The next lemma gives some conditions sufficient for that.

Lemma 4.2. *Suppose that* ***B*** *holds and* $0 < \mu_1$. *Then there exists* $\epsilon > 0$ *such that for every* $\mu \in]\mu_1 - \epsilon, \mu_1]$, $\lambda \in [0, \epsilon[$, *the inequality*

$$\hat{I}^2_{\lambda,\mu} + |\hat{I}^{gr}_{\lambda,\mu}| < \hat{I}^\infty_{\lambda,\mu} \qquad (40)$$

holds.

Proof. Since the functions $\hat{I}^2_{\lambda,\mu}$, $|\hat{I}^{gr}_{\lambda,\mu}|$, $\hat{I}^{\infty}_{\lambda,\mu}$ are continuous with respect to μ and λ it is sufficient to check (40) only for $\mu = \mu_1$, $\lambda = 0$, i.e. to show that

$$|\hat{I}^{gr}_{\mu_1}| < \hat{I}^{\infty}_{\mu_1} = \frac{1}{2(n-1)} \max_{x\in\partial M} (\mu_1 H(x))^{-(n-2)} S_{\partial}^{n-1}. \tag{41}$$

Let $v^1_{\mu_1} \in W$ be a ground state of I_{μ_1} such that $B_{\mu_1}(v^1_{\mu_1}) = -1$. Then by Lemma 2.3 and (23) we get

$$|\hat{I}^{gr}_{\mu_1}| = |\hat{I}^1_{\mu_1}| = -\frac{1}{2(n-1)}|E(v^1_{\mu_1})|^{n-1}. \tag{42}$$

Thus to prove the statement it is sufficient to show that

$$-E(v^1_{\mu_1}) < \max_{x\in\partial M} (\mu_1 H(x))^{-\frac{(n-2)}{(n-1)}} S_{\partial}. \tag{43}$$

By Lemma 3.3 in Ref. 7 we can assume that the metric g satisfies:

$$\begin{cases} (1)\ h = 0 \text{ on } \partial M; \\ (2)\ \bar{R}_{ij}(0) = 0; \\ (3)\ Ric(\eta)(0) = 0; \\ (4)\ R(0) = ||\pi||^2, \end{cases} \tag{44}$$

where π is the second fundamental form, h is its trace and $\bar{R}_{ij}$ are the coefficients of the Ricci tensor of ∂M.

Let $(x_1, ..., x_{n-1}, t)$ be the Fermi coordinates at $0 \in \partial M$ and the second fundamental form π has a diagonal form at 0. Let $\rho^2 = x_1^2 + ... + x_{n-1}^2 + t^2$ and ρ_0 be a small positive real number.

Using Lemma 3.1 in Ref. 7 it is derived the following equality

$$\sqrt{g} = 1 - \frac{1}{2}||\pi||^2 t^2 + O(|(x,t)|^3). \tag{45}$$

Let $\rho > 0$ and $B^n_\rho = \{y \in \text{Real}^n |\ |y| < \rho\}$. We denote $B = B^n_1$. Let us denote by $Q(B, \partial B)$ the Sobolev quotient of B, where B is endowed with the euclidean metric.

In Refs. 3,6 it has been proved that

$$Q(B,\partial B) = S_{\partial} = \frac{\int_{\text{Real}^n_+} |\nabla v_\varepsilon|^2 dx}{(\int_{x_n=0} |v_\varepsilon|^{2^{**}} d\sigma)^{(n-2)/(n-1)}}$$

with

$$v_\varepsilon(x,t) = \left(\frac{\varepsilon}{(\varepsilon + t)^2 + |x|^2}\right)^{(n-2)/2}, \tag{46}$$

where $\text{Real}^n_+ = \{(x,t)|\ x \in \text{Real}^{n-1},\ t > 0\}$.

Let $\psi_{\rho_0}(\rho)$ be a piecewise smooth, decreasing function of ρ, which satisfies $0 \leq \psi_{\rho_0}(\rho) \leq 1$, $\psi_{\rho_0}(\rho) = 1$ for $\rho < \rho_0$, $\psi_{\rho_0}(\rho) = 0$ for $\rho \geq 2\rho_0$, and $|\psi'_{\rho_0}(\rho)| \leq \rho_0^{-1}$ for $\rho_0 \leq \rho \leq 2\rho_0$.

Let $\delta \in$ Real. We will use the test function of Escobar[8] $\phi_\varepsilon = v_{\varepsilon,\delta}\psi_{\rho_0}$ where

$$v_{\varepsilon,\delta} = \left(\frac{\varepsilon}{(\varepsilon+t)^2 + |x|^2 - \delta x_1^2}\right)^{(n-2)/2}. \tag{47}$$

We may assume that for all sufficiently small $\varepsilon > 0$

$$\operatorname{supp}(\phi_\varepsilon) \cap \partial M \subset \partial M^+. \tag{48}$$

As in Ref. 8 using the asymptotic expansion (45) it is proved the following

Lemma 4.3. *Suppose $n > 5$ and the conditions **B**, **C** hold. Then there exist $a_0 > 0$, $\epsilon_0 > 0$, $\delta = \pm\varepsilon$, and a conformal metric $\tilde{g} = \tau g$ with some $\tau > 0$ such that for every $\rho_0 < a_0$ the following inequality*

$$E(\phi_\varepsilon) \leq (\mu_1 H(0))^{-\frac{n-2}{n-1}} Q(B, \partial B) B_{\mu_1}(\phi_\varepsilon)^{\frac{n-2}{n-1}} - c\varepsilon^2 + o(\varepsilon^2) \tag{49}$$

holds for every $\varepsilon < \epsilon_0$, where $c > 0$ does not depend on ε and ρ_0.

Introduce

$$\theta^\varepsilon = \frac{\phi_\varepsilon}{B_{\mu_1}(\phi_\varepsilon)^{\frac{(n-2)}{2(n-1)}}}.$$

Then $B_{\mu_1}(\theta^\varepsilon) = 1$. Hence and since $B_{\mu_1}(v^1_{\mu_1}) = -1$ it follows that

$$B_{\mu_1}(\theta^\varepsilon + v^1_{\mu_1}) = B_{\mu_1}(\theta^\varepsilon) + B_{\mu_1}(v^1_{\mu_1}) + \Theta(\varepsilon) = 1 - 1 + \Theta(\varepsilon) = \Theta(\varepsilon), \tag{50}$$

where $\Theta(\varepsilon) \to 0$ as $\varepsilon \to 0$. Moreover, it follows from (48) that $\Theta(\varepsilon) \geq 0$. By direct calculation we deduce that

$$E(\theta^\varepsilon + v^1_{\mu_1}) = E(\theta^\varepsilon) + E(v^1_{\mu_1}) + DE(v^1_{\mu_1})(\theta^\varepsilon). \tag{51}$$

Since $B_{\mu_1}(\theta^\varepsilon + v^1_{\mu_1}) = \Theta(\varepsilon) \geq 0$, we have by (7) that $E(\theta^\varepsilon + v^1_{\mu_1}) \geq 0$.

Therefore by (49),

$$\begin{aligned} -E(v^1_{\mu_1}) &\leq E(\theta^\varepsilon + v^1_{\mu_1}) - E(v^1_{\mu_1}) = E(\theta^\varepsilon) + DE(v^1_{\mu_1})(\theta^\varepsilon) \\ &\leq (\mu_1 H(0))^{-(n-2)/(n-1)} Q(B, \partial B) + DE(v^1_{\mu_1})(\theta^\varepsilon) - c\varepsilon^2 + o(\varepsilon^2). \end{aligned} \tag{52}$$

Observe that

$$|DE(v^1_{\mu_1})(\theta^\varepsilon)| \leq C\varepsilon^{(n-2)/2}\rho_0. \tag{53}$$

Indeed, since $u_{\mu_1} = t^2_{\mu_1}(v_{\mu_1})v_{\mu_1}$ is a weak solution of (1) then

$$DE(v^1_{\mu_1})(\theta^\varepsilon) = \kappa \int_{\partial M} H^+(v^1_{\mu_1})^{n/(n-2)}\phi_\varepsilon d\sigma_g / B_{\mu_1}(\phi_\varepsilon)^{\frac{(n-2)}{2(n-1)}}, \tag{54}$$

with $0 < \kappa < \infty$ which does not depend on ε.

Since $v_{\mu_1} \in C(\overline{M})$ by (46), (45) we have

$$|\int_{\partial M} H(x)(v^1_{\mu_1})^{2(n-1)/(n-2)}\phi_\varepsilon d\sigma_g| \le C_1[\int_{B^{n-1}_{2\rho_0}} \left(\frac{\varepsilon}{\varepsilon^2 + |x|^2}\right)^{(n-2)/2} dx$$
$$+ \int_{B^{n-1}_{\rho_0}} \left(\frac{\varepsilon}{\varepsilon^2 + |x|^2}\right)^{(n-2)/2} |x|^3 dx], \tag{55}$$

where $C_1 < \infty$ does not depend from ε.

Putting $x = \varepsilon y$ we deduce that

$$|\int_{\partial M} H^+(x)(v^1_{\mu_1})^{2(n-1)/(n-2)}\phi_\varepsilon d\sigma_g| \le C_2\varepsilon^{n/2} \int_{B^{n-1}_{2\rho_0/\varepsilon}} \left(\frac{1}{1 + |y|^2}\right)^{(n-2)/2} dy$$
$$+\varepsilon^{(n+6)/2} C_3 \int_{B^{n-1}_{2\rho_0/\varepsilon}} \frac{|y|^3}{(1 + |y|^2)^{(n-2)/2}} dy. \tag{56}$$

Since

$$\int_{B^{n-1}_{2\rho_0/\varepsilon}} \left(\frac{1}{1 + |y|^2}\right)^{(n-2)/2} dy \le C_4 \frac{\rho_0}{\varepsilon}, \quad \int_{B^{n-1}_{2\rho_0/\varepsilon}} \frac{|y|^3}{(1 + |y|^2)^{(n-2)/2}} dy \le C_4 (\frac{\rho_0}{\varepsilon})^4,$$

we get

$$|\int_{B^{n-1}_{\rho_0}} H^+(x)(v^1_{\mu_1})^{2(n-1)/(n-2)}\phi_\varepsilon d\sigma_g| \le C_5 \varepsilon^{(n-2)/2}\rho_0. \tag{57}$$

Hence we obtain

$$|DB_{\mu_1}(v^1_{\mu_1})(\phi^\varepsilon)| \le C_6 \varepsilon^{(n-2)/2}\rho_0. \tag{58}$$

It is easy to see that

$$0 < b_1 \le B_{\mu_1}(\phi_\varepsilon) < \infty, \tag{59}$$

with some b_1 which does not depend on $\varepsilon > 0$. This and (54), (58) imply the estimation (53).

Applying (53) in (52) we get (43), and therefore (40). If $n > 6$, we have $(n-2)/2 > 2$; if $n = 6$ we use that ρ_0 can be taken so small that $C\rho_0 < c/4$. The proof is complete. □

5. Subcritical Auxiliary Results

In this Section we prove auxiliary results in the subcritical cases $2 \le \gamma < 2^{**}$. Put

$$\mu_1(\gamma) = \inf\{\frac{B^{(-),\gamma}(\phi)}{B^{(+),\gamma}(\phi)} : E(\phi) \le 0,\ \phi \in C^\infty(M)\}, \tag{60}$$

where $B^{(\pm),\gamma}(u) = \int_{\partial M} H^{\pm}(x)|u|^\gamma d\sigma_g$. Since the function $\dfrac{B^{(-),\gamma}(\phi)}{B^{(+),\gamma}(\phi)}$ is 0-homogeneous, we may assume that the minimum in (60) is taken over the set

$$\{\phi : \max_M |\phi| \le 1,\ \phi \in C^\infty(M),\ E(\phi) \le 0\}.$$

Observe that $B^{(\pm),\gamma}(\phi) \to B^{(\pm),2^{**}}(\phi)$ as $\gamma \to 2^{**}$ uniformly on $\{\phi : \max_M |\phi| \le 1,\ \phi \in C^\infty(M)\}$. Hence we derive that

$$\mu_1(\gamma) \to \mu_1(2^{**}) \quad \text{as } \gamma \to 2^{**}. \tag{61}$$

Put

$$I_{\mu_1(\gamma)}(u) = \frac{1}{2}E(u) - \frac{1}{\gamma}[\mu_1(\gamma)B^{(+),\gamma}(u) - B^{(-),\gamma}(u)]. \tag{62}$$

Let us prove the following

Claim 5.1. *Let $2 \le \gamma < 2^{**}$. Then there exists a nonnegative solution $\phi_\gamma \in W$ of (60). Furthermore,*

1) If $\mu_1(\gamma) = 0$ then $\phi^\gamma(x) \equiv 0$ on ∂M^-.

2) If $\mu_1(\gamma) > 0$ then there exists a constant $t_\gamma > 0$ such that the function $u_\gamma = t_\gamma\phi_\gamma$ is a positive ($u_\gamma > 0$) critical point of $I_{\mu_1(\gamma)}(u_\gamma)$, i.e. $DI_{\mu_1(\gamma)}(u_\gamma) = 0$ and

$$I_{\mu_1(\gamma)}(u_\gamma) = 0, \quad \mu_1(\gamma)B^{(+),\gamma}(u_\gamma) - B^{(-),\gamma}(u_\gamma) = E(u_\gamma) = 0. \tag{63}$$

Proof. Let $\{\psi_m\}$ be a minimizing sequence for problem (60) such that

$$\mathcal{B}_{\mu_1(\gamma)}(\psi_m) = \frac{B^{(-),\gamma}(\psi_m)}{B^{(+),\gamma}(\psi_m)} \to \mu_1(\gamma) \quad \text{as } m \to \infty,$$

and $E(\psi_m) \le 0$, $m = 1, 2, \ldots$. The functional $\mathcal{B}(\cdot)$ is 0-homogeneous. Therefore we may assume without loss of generality that the sequence $\{\psi_m\}$ is bounded in $W_2^1(M)$. Moreover, by scaling we can normalize $\{\psi_m\}$ so that

$$\psi_m \in S^1 \equiv \{u \in W_2^1(M) : \int_M |u|^2 dv_g + \int_M |\nabla u|^2 dv_g = 1\}.$$

Hence there exists a subsequence (again denoted by $\{\psi_m\}$) such that $\psi_m \to \phi_\gamma \in W$ weakly in W and strongly in $L_l(M)$, for $2 \le l < 2^*$, and in $L_s(\partial M)$, for $2 \le s < 2^{**}$. Let us show that $\phi_\gamma \neq 0$. Suppose converse $\psi_m \to 0$. Then since $\psi_m \to \phi_\gamma$ in $L_2(M)$, $L_2(\partial M)$ and $\psi_m \in S^1$, we obtain that

$$E(\psi_m) = \int_M |\nabla \psi_m|^2 dv_g + \int_M R(x)|\psi_m|^2 dv_g + \int_{\partial M} h(x)|\psi_m|^2 d\sigma_g \to 1$$

as $m \to \infty$. But this contradicts to the assumption $E(\psi_m) \le 0$. Thus $\phi_\gamma \neq 0$. From here it follows that ϕ_γ is a solution of (60). Since all functionals in (60) are even, we can take $\psi_m \ge 0$ and $\phi_\gamma \ge 0$.

By the Lagrange rule there exist constants ν^1, ν^2, $\nu^1\nu^2 \neq 0$ such that

$$\nu^1 DE(\phi_\gamma)(\xi) = \nu^2 D\mathcal{B}_{\mu_1(\gamma)}(\phi_\gamma)(\xi) \tag{64}$$

for every $\xi \in W^*$. If $\nu^2 = 0$ then

$$DE(\phi_\gamma)(\xi) = 0$$

for every $\xi \in W^*$. Since $\phi_\gamma \ge 0$, it follows that ϕ_γ is the first eigenfunction of problem (5) and $\lambda_g = 0$. But we have supposed that $\lambda_g < 0$. Thus $\nu^2 \neq 0$. If $\nu^1 = 0$ then (64) implies

$$\begin{aligned} D\mathcal{B}_{\mu_1(\gamma)}(\phi_\gamma)(\xi) &= \frac{1}{B^{(+),\gamma}(\phi_\gamma)}(B^{(-),\gamma})'(\phi_\gamma)(\xi) - \frac{B^{(-),\gamma}(\phi_\gamma)}{[B^{(+),\gamma}(\phi_\gamma)]^2}(B^{(+),\gamma})'(\phi_\gamma)(\xi) \\ &= \frac{1}{B^{(+),\gamma}(\phi_\gamma)}[(B^{(-),\gamma})'(\phi_\gamma)(\xi) - \mu_1(\gamma)(B^{(+),\gamma})'(\phi_\gamma)(\xi)] = 0, \end{aligned} \tag{65}$$

for every $\xi \in W^*$.

Consider the case $\mu_1(\gamma) = 0$. Since $\nu^1 = 0$ we derive from (65) that

$$\frac{1}{B^{(+),\gamma}(\phi_\gamma)} D(B^{(-),\gamma})(\phi_\gamma)(\xi) = 0 \text{ for every } \xi \in W^*. \tag{66}$$

This implies that $\phi_\gamma(x) \equiv 0$ on ∂M^-. Suppose $\nu^2 \neq 0$ and $\nu^1 \neq 0$. By (64) and since $0 = \mu_1 = D\mathcal{B}(\phi_\gamma)(\phi_\gamma)/\gamma = 0$ we derive that $B^{(-),\gamma}(\phi_\gamma) = 0$. Thus we again obtain $\phi_\gamma(x) \equiv 0$ on ∂M^-.

Consider now the case $\mu_1(\gamma) > 0$. Suppose that $\nu^1 = 0$. Then from (65) we derive that $\phi_\gamma(x) \equiv 0$ on ∂M. But we have proved above that $\phi_\gamma(x) \neq 0$. The contradiction means that $\nu^1 \neq 0$ and $\nu^2 \neq 0$. Denote $\nu := \nu^2/\nu^1$. It follows from (64) that

$$DE(\phi_\gamma)(\xi) = -\nu \frac{\gamma}{2B^{(+),\gamma}(\phi_\gamma)}[\mu_1(\gamma) DB^{(+),\gamma}(\phi_\gamma)(\xi) - DB^{(-),\gamma}(\phi_\gamma)(\xi)], \tag{67}$$

for every $\xi \in W^*$. Since ϕ_γ is a solution of problem (60) we have $[\mu_1(\gamma)(B^{(+),\gamma})(\phi_\gamma) - (B^{(-),\gamma})(\phi_\gamma)] = 0$. From (67) we derive that $E(\phi_\gamma) = 0$. Thus we have proved (63).
Let us show that $\nu < 0$. Suppose that $\nu > 0$. Let $\varepsilon > 0$ and $\xi_0 \in W^*$ be such that

$$\varepsilon DE(\phi_\gamma)(\xi_0) < 0. \tag{68}$$

Since $DE(\phi_\gamma)(\xi) = \nu D\mathcal{B}_{\mu_1(\gamma)}(\phi_\gamma)(\xi_0)$ we have

$$E(\phi_\gamma + \varepsilon\xi_0) = E(\phi_\gamma) + \varepsilon DE(\phi_\gamma)(\xi_0) + \bar{o}(\varepsilon), \tag{69}$$

$$\mathcal{B}_{\mu_1(\gamma)}(\phi_\gamma + \varepsilon\xi_0) = \mathcal{B}_{\mu_1(\gamma)}(\phi_\gamma) + \varepsilon D\mathcal{B}_{\mu_1(\gamma)}(\phi_\gamma)(\xi_0) + \bar{o}(\varepsilon)$$
$$= \mu_1(\gamma) + \varepsilon\frac{1}{\nu}DE(\phi_\gamma)(\xi_0) + \bar{o}(\varepsilon). \tag{70}$$

Hence since $\nu > 0$ and (63) holds, we obtain that

$$E(\phi_\gamma + \varepsilon\xi_0) < 0, \quad \frac{B^{(-),\gamma}(\phi_\gamma + \varepsilon\xi_0)}{B^{(+),\gamma}(\phi_\gamma + \varepsilon\xi_0)} < \mu_1(\gamma).$$

This contradicts to the definition of $\mu_1(\gamma)$ (see (60)). Thus $\nu < 0$ and if we put $u_\gamma = t_\gamma\phi_\gamma$ with

$$t_\gamma = [-\nu\frac{\gamma}{2B^{(+),\gamma}(\phi_\gamma)}]^{1/(\gamma-2)},$$

we obtain a critical point of $I_{\mu_1(\gamma)}(u)$. □

6. Proofs of Theorems 1.1, 1.2

In this section we shall finish the proof of the remaining statements of Theorems 1.1, 1.2.

6.1. *Proof of statement 1) of Theorem 1.1*

Let us prove

Lemma 6.1. $\hat{I}^2_{\mu_1} = 0$ *and there exists a nonnegative solution* $\phi_{\mu_1} \in W$ *of variational problem* (7). *Furthermore*

1) *If* $\mu_1 = 0$, *then* $\phi_{\mu_1}(x) \equiv 0$ *on* ∂M^-.

2) *Assume that* $\boldsymbol{B}$ *holds and* $\mu_1 > 0$. *Then* ϕ_{μ_1} *is a weak positive solution of boundary value problem* (1) *and*

$$I_{\mu_1}(\phi_{\mu_1}) = 0, \quad B_{\mu_1}(\phi_{\mu_1}) = E(\phi_{\mu_1}) = 0. \tag{71}$$

Proof. First prove 2). Suppose $\mu_1 > 0$. By Proposition 5.1 there exist a solution $\phi_\gamma \in W$ of problem (60) which satisfies the condition: $E(\phi_\gamma) = 0$, $\phi_\gamma(x) \neq 0$, $||\phi_\gamma|| = 1$. Consider $u_\gamma = t_\gamma \phi_\gamma$ where $t_\gamma = ||u_\gamma||$. By Proposition 5.1 2) and since $\mu_1(\gamma) \to \mu_1$ we conclude that

$$I_{\mu_1}(t_\gamma \phi_\gamma) = \frac{t_\gamma^2}{2} E(\phi_\gamma) - \frac{t_\gamma^{2^{**}}}{2^{**}} B_{\mu_1}(\phi_\gamma) \to 0, \tag{72}$$

$$||t_\gamma DE(\phi_\gamma) - t_\gamma^{2^{**}-1} DB_{\mu_1}(\phi_\gamma)|| \to 0 \tag{73}$$

as $\gamma \to 2^{**}$. Thus $u_\gamma \in W$ is a Palais-Smale sequence at the level $\beta = 0$ of I_{μ_1}. By Lemma 4.1 the functional I_{μ_1} satisfies the (P.-S.) condition at the level $\beta = 0$. Therefore there exists a subsequence such that $u_{\gamma_i} \to u_{\mu_1} = t_{\mu_1}\phi_{\mu_1} \in W$ strongly in W as $\gamma_i \to 2^{**}$. Since $u_\gamma = t_\gamma \phi_\gamma$ is a critical point of $I_{\mu_1(\gamma)}(u_\gamma)$ we get in the limit that

$$DI_{\mu_1}(u_{\mu_1})(\psi) = 0, \tag{74}$$

for all $\psi \in C^\infty(M)$. Observe that $\phi_{\mu_1} \neq 0$, since $E(\phi_\gamma) = 0$, $||\phi_\gamma|| = 1$ and $\phi_{\gamma_i} \to \phi_{\mu_1}$ as $\gamma_i \to 2^{**}$.

Let us show that $t_{\mu_1} \neq 0$. Suppose a contrary that $t_{\gamma_i} \to 0$ as $\gamma_i \to 2^{**}$. Then since $u_{\gamma_i} = t_{\gamma_i}\phi_{\gamma_i}$ is a critical point of $I_{\mu_1(\gamma_i)}(u_{\gamma_i})$, we get passing to the limit $\gamma_i \to 2^{**}$ the following identity :

$$\int_M (\nabla_g \phi_{\mu_1}, \nabla_g \psi) dv_g + \int_M R(x)\phi_{\mu_1}\psi dv_g + \int_{\partial M} h(x)\phi_{\mu_1}\psi d\sigma_g = 0 \tag{75}$$

for all $\psi \in C^\infty(M)$. Observe that since $\phi_{\gamma_i} \geq 0$ we have $\phi_{\mu_1} \geq 0$. Furthermore, we may assume that $\phi_{\mu_1} > 0$ on M. Indeed, applying the regularity results from Ref. 4 we derive that $\phi_{\mu_1} \in C^{1,\alpha}(M)$ for some $\alpha \in (0,1)$ and by the Harnack inequality[12] we deduce that $\phi_{\mu_1} > 0$. But then ϕ_{μ_1} is the first eigenfunction of problem (5) and the corresponding eigenvalue is zero, $\lambda_g = 0$. But this contradicts to our assumption that $\lambda_g < 0$. Hence $t_{\mu_1} \neq 0$ and since $\phi_{\mu_1} \neq 0$, we can conclude that $u_\gamma \neq 0$.

From (63) we get passing to the limit that

$$I_{\mu_1}(u) = 0, \quad B_{\mu_1}(u) = E(u) = 0. \tag{76}$$

Thus we have shown that $C^+(\mu_1) = 0$ and have proved the existence of the solution $u_{\mu_1} \in W \setminus \{0\}$.

Prove now 1). Suppose $\mu_1 = 0$. Then it is possible that $\mu_1(\gamma) = 0$ for all γ near 2^{**}. By Proposition 5.1 we know that there exist a solution $\phi_\gamma \in W$ of problem (60) which satisfies the condition: $E(\phi_\gamma) \leq 0$, $\phi_\gamma(x) \neq 0$, $\phi_\gamma(x) \equiv 0$ on ∂M^-. We may assume that $\phi_\gamma \in S^1$. Then there exists a subsequence (again denoted as ϕ_γ) such that it converges $\phi_\gamma \to \phi_{\mu_1} \in W$ as $\gamma \to 2^{**}$

weakly in W and strongly in $L_2(\partial M)$. If $\phi_{\mu_1} = 0$ then we get (since $\phi_\gamma \in S^1$) that $E(u_\gamma) \to 1$ as $\gamma \to 2^{**}$. But this contradicts to the fact that $E(\phi_\gamma) \leq 0$ for $\gamma < 2^{**}$. Hence $\phi_{\mu_1} \neq 0$, $\phi_{\mu_1} \geq 0$ on ∂M, and $\phi_{\mu_1}(x) \equiv 0$ on ∂M^-. Assume now that $\mu_1(\gamma) > 0$ and $\mu_1(\gamma) \to \mu_1 = 0$ as $\gamma \to 2^{**}$. We may assume that $\phi_\gamma \in S^1$. Then arguing as above we see that $\phi_\gamma \to \phi_{\mu_1} \in W$ weakly in W as $\gamma \to 2^{**}$, where $\phi_{\mu_1} \not\equiv 0$, $\phi_{\mu_1} \geq 0$. Consider $u_\gamma = t_\gamma \phi_\gamma$ where $t_\gamma = ||u_\gamma||$. Suppose that $t_\gamma < \infty$. Then since $\mu_1 = 0$, by passing to the limit in $I'_{\mu_1(\gamma)}(u_\gamma) = 0$ we get that

$$DE(u_{\mu_1}) = -DB^{(-),\gamma}(u_{\mu_1}) \text{ and } E(u_{\mu_1}) = -(B^{(-),\gamma})(u_{\mu_1}). \tag{77}$$

On the other hand, since E, $B^{(-)}$ are weakly lower semi-continuous on W we have $E(u_{\mu_1}) \leq liminf_{m\to\infty} E(u_\gamma) = 0$ and $B^{(-)}(u_{\mu_1}) \leq liminf_{m\to\infty} B^{(-),\gamma}(u_\gamma) = 0$. Hence by (77) we get that $B^{(-)}(u_{\mu_1}) = 0$ and therefore $\phi_{\mu_1}(x) \equiv 0$ on ∂M^-.

Suppose that $t_m \to \infty$. From equalities $I'_{\mu_1(\gamma)}(u_\gamma) = 0$ we have

$$\frac{1}{t_\gamma^{\gamma-2}} DE(\phi_\gamma)(\xi) = DB_{\mu_1}(\phi_\gamma)(\xi) + o(1), \ \forall\, \xi \in W^*.$$

If $t_\gamma \to \infty$ then we have $DB(\phi_{\mu_1})(\xi) = 0$ for all $\xi \in W^*$. This is possible only if *supp* $\phi_{\mu_1} \subset \partial M^0 := \{x \in \partial M : H(x) = 0\}$. Thus we get again that $\phi_{\mu_1}(x) \equiv 0$ on ∂M^-.

The proof is complete. □

Now let us conclude the proof of assertion 2)

Suppose $\mu_1 = 0$. Then by Lemma 6.1 there exists a function $u_{\mu_1} \in W$ such that $\mathrm{supp}(u_{\mu_1}) \cap \partial M \subseteq \partial M^+$ and $E(u_{\mu_1}) \leq 0$.

Introduce the following space

$$C^\infty_{\partial M^+}(M) = \{\phi \in C^\infty(M) : \mathrm{supp}(\phi) \cap \partial M \subseteq \partial M^+\}. \tag{78}$$

Consider the following minimization problem:

$$\lambda_g(\partial M^+) = \inf_{\phi \in C^\infty_{\partial M^+}(M)} \frac{E(\phi)}{\int_M |\phi|^2 d\sigma_g}.$$

Lemma 6.2. *There exists a positive solution $\psi(\partial M^+) \in W$ of this problem. Moreover*

$$\begin{cases} -\Delta_g \psi(\partial M^+) + R(x)\psi(\partial M^+) = 0 \quad in \quad M, \\ \frac{\partial \psi(\partial M^+)}{\partial n} + h\psi(\partial M^+) = \lambda_g(\partial M^+)\psi(\partial M^+) \quad on \ \ \partial M^+, \\ \psi(\partial M^+) = 0 \quad on \quad \partial M^-. \end{cases} \tag{79}$$

Proof. is standard.

Remark that $\lambda_g(\partial M^+) \leq 0$ or, the same, that $E(\psi(\partial M^+)) \leq 0$. Indeed, if this is not true, then $E(\psi) > 0$ for any $\psi \in C^\infty_{\partial M^+}(M)$. But this is impossible since the function $u_{\mu_1} \in C^\infty_{\partial M^+}(M)$ satisfies the condition $E(u_{\mu_1}) \leq 0$.

Assume $\mu > 0$. Let us suppose that there exists a positive solution $u_\mu \in W$ of (1). Multiplying (1) by $\psi(\partial M^+)$ and (79) by u_μ, we have, after integration, the following

$$DE(u_\mu)(\psi(\partial M^+)) = DB_\mu(u_\mu)(\psi(\partial M^+)),$$
$$DE(\psi(\partial M^+))(u_\mu) = \lambda_g(\partial M^+)\int_{\partial M^+} u_\mu \psi(\partial M^+)d\sigma.$$

Since

$$DE(u_\mu)(\psi(\partial M^+)) = DE(\psi(\partial M^+))(u_\mu),$$

we obtain

$$\lambda_g(\partial M^+)\int_{\partial M^+} u_\mu \psi(\partial M^+)d\sigma = B_\mu(u_\mu)(\psi(\partial M^+)). \tag{80}$$

The left hand side of this equality is negative, whereas the right hand side is positive. Thus we get a contradiction. □

6.2. *End of the Proofs of Propositions 2.2) in Theorem 1.1 and 2) in Theorem 1.2*

Let $\varepsilon > 0$ be a constant from the statement of Lemma 4.2 and $\mu \in]\mu_1 - \varepsilon, \mu_1[$, $\lambda \in [0, \varepsilon[$.

If φ_m is a minimizing sequence for problem (22), then by Ekeland's principle it follows that $DJ^2_{\lambda,\mu}(\varphi_m) \to 0$ and $J^2_{\lambda,\mu}(\varphi_m) \to \hat{I}^2_{\lambda,\mu}$. Hence φ_m is a (P.-S.) sequence of $J^2_{\lambda,\mu}$ at the level $\hat{I}^2_{\lambda,\mu}$. As above in the proof of assertion 3.1) we may assume that $B_\mu(\varphi_m) = 1$, i.e. $\varphi_m \in \mathcal{B}^+$. From Lemma 4.2 we know that

$$\hat{I}^2_{\lambda,\mu} + |\hat{I}^1_{\lambda,\mu}| < \hat{I}^\infty_{\lambda,\mu}.$$

Therefore by Lemma 4.1 the function $J^2_{\lambda,\mu}$ satisfies the (P.-S.) condition at the level $\hat{I}^2_{\lambda,\mu}$. Thus there exists a subsequence of φ_{m_i} strongly converging in W as $m_i \to \infty$ to a point $\varphi_\mu \in \mathcal{B}^+$. Finally, continuing as above in the proof of existence of $u^1_{\lambda,\mu}$ in Lemma 3.1, we conclude that $u^2_{\lambda,\mu} = t^2_{\lambda,\mu}\varphi_{\lambda,\mu}$ is a weak positive solution of problem (1) such that $I_\mu(u^2_\mu) > 0$.

The case $\lambda = 0$, $\mu = \mu_1$ has been considered in Lemma 6.1, 2).

References

1. Aubin Th., *J. Math. Pures Appl.* **55**, 269(1976).
2. Brezis, H., Nirenberg L., *Comm. Pure Appl. Math.* **36**, 437 (1983).
3. Beckner W.,*Ann. of Math.* **138**, 213 (1993).
4. Cherrier P., *J. Funct. Anal.* **57**, 154 (1984).
5. Coleman, S., Glazer, V., Martin, A., *Comm. Math. Phys.* **58**, 211 (1978).
6. Escobar, J., *J. Diff. Geom.* **35**, 21 (1992).
7. Escobar, J., *Ann. Math.* **136**, 1 (1992).
8. Escobar, J., *Calculus Var. and Partial Differential Equations* **4**, 559 (1996).
9. Il'yasov Ya. Sh., *Izv. Russ. Ac. Nauk, Ser. Mat.* **66**, 19 (2002).
10. Lions, P.L., *Revista Mat. Iberoamer.* **1**, 145 (1985); **2**, 45 (1985).
11. Pohozaev, S.I., *Doklady Acad. Sci. USSR* **247**, 1327 (1979).
12. Trudinger, N.S., *Comm. Pure Appl. Math.* **20**, 721 (1967).
13. Trudinger, N.S., *J. Math. Mech.* **17**, 473 (1967).

Harmonic, Wavelet and p-Adic Analysis
Eds. N. M. Chuong *et al.* (pp. 141–155)

§7. ON CALCULATION OF THE BIFURCATIONS BY THE FIBERING APPROACH

YAVDAT IL'YASOV*

Bashkir State University, Ufa, Frunze 32, Russia
Ilyasov@ic.bashedu.tu

In this contribution we discuss the problems of the nonlocal analysis of bifurcations for equation of variational form. This includes the calculation of the bifurcation values λ_i, the construction of the branches of solutions on $(\lambda_i, \lambda_{i+1})$, and the study of their asymptotic behavior at the bifurcations $\lambda \to \lambda_i$. We present a survey of results where these problems are solved using the method basing on fiber spectral analysis.

1. Introduction

Consider the following families of equations of variational form

$$F_u(u, \lambda) = 0$$

containing a parameter $\lambda \in \mathbb{R}$, where the solution u is being sought on Banach space W.

We focus on the following programm of investigation

Nonlocal Analysis of Bifurcations (NAB):

(i) Existence and calculation of the bifurcation values λ_i.
(ii) Existence and construction of the branches of solutions $\{u_\lambda\}$ on $\lambda \in (\lambda_i, \lambda_{i+1})$.
(iii) Asymptotic behavior of the branch of solutions $\{u_\lambda\}$ at the bifurcation values, i.e. as $\lambda \to \lambda_i$.

It seems today the general theory of such problems exists only in linear cases where the calculation of the bifurcation values λ_i is a subject of Spectral Theory. In the linear cases, there are two well-known variational

*The author was supported in part by grants INTAS 03-51-5007, RFBR 05-01-00370, 05-01-00515

principles for the bifurcation values (solving the problem (i)) λ_i, *Poincarè's and Courant-Weyl's principles.* However in the nonlinear cases so general theory is far beyond from completeness. Global and local bifurcation methods (see Ref. 22) are helpful in solving the problems (ii), (iii) but often can not be applied to any problem directly. As far as we know only so-called Nonlinear Spectral Theory (see Ref. 3 and references there in) concerns a problem (i) in general.

To illustrate the problem let us consider the following class of boundary value problems with indefinite nonlinearities

$$\begin{cases} -\Delta u = \lambda u + f(x)|u|^{\gamma-2}u \text{ in } \Omega, \\ u = 0 \text{ on } \partial\Omega, \end{cases} \tag{1}$$

where Ω is a bounded domain in $\mathbb{R}^N$ with smooth boundary; $\lambda \in \mathbb{R}$; $2 < \gamma < 2^*$, $2^* = 2N/(N-2)$ if $N > 2$, $2^* = +\infty$, if $N \leq 2$; $f \in L^\infty(\Omega)$.

Remark that there are two opposite cases in (1): $f(x) \equiv 1$ and $f(x) \equiv (-1)$ in Ω. It is known that in both cases the set of bifurcation values $\{\lambda_i\}$ coincides with the discreet spectrum of linear Dirichlet boundary value problem

$$\begin{cases} -\Delta\phi_i = \lambda_i\phi_i \text{ in } \Omega \\ \phi_i = 0 \text{ on } \partial\Omega. \end{cases} \tag{2}$$

However the structure of the branch of solutions u_λ in these cases are different, i.e., in the first case the branches of solutions bifurcate from $\lambda = \lambda_i$ to $\lambda = -\infty$ and in the second one from $\lambda = \lambda_i$ to $\lambda = +\infty$.

It can be stated the following problem:

What is the set of bifurcation points in the mixed case, when a sign of nonlinearity is indefinite, i.e. $f(x)$ may change a sign on Ω?

Some answers to the problem can be found in the papers devoted to the problem on the existence of multiple solutions. For instance from the works by Berestycki, Capuzzo–Dolcetta & Nirenberg, L.,[4] Drábek & Pohozaev,[7] Ouyang,[16] M.del Pino[6] it follows that the problem (23) possess a bifurcations point $\Lambda^* < +\infty$ such that for $\lambda_1 < \lambda < \Lambda^*$ there exist two positive solutions and there are no positive solutions as $\Lambda^* < \lambda$.

It is remarkable that by Ouyang[16] it has been found *new type of characteristic point:* corresponding to the problem

$$\lambda^* = \inf\left\{\frac{\int|\nabla u|^2dx}{\int|u|^2dx} : \int f(x)|u|^\gamma dx \geq 0,\ u \in W_0^{1,2}\right\}. \tag{3}$$

Furthermore, by M. del Pino,[6] in particular cases, it has been shown that λ^* actually is a bifurcation point.

These observations and some other results on the multiplicity of solutions (see Ref. 1) are important from the following point of view. It follows that we deal with the following phenomenon

(NB.) Nonlinear type of bifurcations values: *There exist boundary value problems that possesses a new type of bifurcation values which are not contained into the discrete spectrum of a corresponding linear boundary value problem.*

Based on this observation one may state the following conjecture on the existence of New Variational Principles:

(NVP.) New Variational Principles: *For the equations of variational form the corresponding set of bifurcation values are expressed in terms of variational principles.*

The aim of this contribution is to show that it can be achieved progress in the solving of the programm (NAB). Below we will discuss an approach to the problems based on the fibering method[17–20] which we call *Fiber Spectral Analysis* (SAF).[9,10,12,14] In general, this method allows to solve all of the items (1)-(3) of the programm (NAB) (see Refs. 9,10). According to this approach the conception of the bifurcation value is treated in a more wide sense which we call *characteristic value by fibering*.[9,10] In order to find characteristic values it is natural to search the critical points $(u_{\lambda_i}, \lambda_i)$ $(F_u(u_{\lambda_i}, \lambda_i) = 0)$ such that the operator

$$F_{uu}(u_{\lambda_i}, \lambda_i) \tag{4}$$

is not regular in a certain sense. It can be say that the fiber spectral analysis is a solving for values λ_i where the operator (4) is not regular in *the sense of a fibering approach.* The main advantage of this approach is that it is constructive and very simple way to find a set of characteristic points corresponding to the considering equation. This set contains a prior bifurcation values which in some cases correspond to the bifurcations. It is remarkable that characteristic values determined by this method are expressed in terms of variational principles which include the well-known Poincarè's and Courant-Weyl's principles as special cases. Thus it is natural to call this set of characteristic values obtained by (SAF) also as a *fiber spectrum.*

Remark 1.1. We call the programm of analysis of bifurcations (i)-(iii) as nonlocal, since by this programm, in bvcontrast to the local methods (cf. Ref. 22), the study of the branch of solutions begins outside of the bifurcation points.

Remark 1.2. Another approach to the problems (i)-(iii) (NAB), so-called a *dual method of the calculation of bifurcations* is developed by the author in Refs. 8,15.

2. Fiber Spectral Analysis

The statement and the proof of the method Fiber Spectral Analysis is given in Refs.9,10,12. Here we present only an idea the method by application of its to the boundary valuer problem (1).

The problem (1) has a variational form with the Euler functional

$$I_\lambda(u) = \frac{1}{2}H_\lambda(u) - \frac{1}{\gamma}F(u), \ u \in W,$$

where $W = W_0^{1,2}(\Omega)$ is a Sobolev space and $F(u) = \int f(x)|u|^\gamma dx$, $H_\lambda(u) = \int |\nabla u|^2 dx - \lambda \int |u|^2 dx$.

First, with respect to the fibering method[17,18] we consider on the fiber space $\mathbb{R}^+ \times S^1$ with $S^1 = \{w \in W : ||w|| = 1\}$ the fiber functionals

$$\tilde{I}_\lambda(t,v) = I_\lambda(tv) = \frac{t^2}{2}H_\lambda(v) - \frac{t^\gamma}{\gamma}F(v), \ (t,v) \in \mathbb{R}^+ \times S^1, \tag{5}$$

and

$$Q_\lambda(t,v) = \frac{\partial}{\partial t}\tilde{I}_\lambda(t,v) = t(H_\lambda(v) - t^{\gamma-1}F(v)), \tag{6}$$

$$L_\lambda(t,v) = \frac{\partial^2}{\partial t^2}\tilde{I}_\lambda(t,v) = (H_\lambda(v) - (\gamma-2)t^{\gamma-2}F(v)). \tag{7}$$

Then we extract in $\mathbb{R}^+ \times S^1$ the following submanifolds

$$\Sigma_\lambda^1 = \{(t,v) \in \mathbb{R}^+ \times S^1 | Q_\lambda(t,v) = 0, \quad L_\lambda(t,v) > 0\}, \tag{8}$$

$$\Sigma_\lambda^2 = \{(t,v) \in \mathbb{R}^+ \times S^1 | Q_\lambda(t,v) = 0, \quad L_\lambda(t,v) < 0\}. \tag{9}$$

The following result holds[9,10,12]

Theorem 2.1. *Assume that $I_\lambda(u) \in C^1(W \setminus \{0\})$ and $Q_\lambda(t,v) \in C^1(\mathbb{R}^+ \times S^1)$. Let $j = 1,2$.*

If $(t_0^j, v_0^j) \in \Sigma_\lambda^j$ is a critical point of the restricted functional $\tilde{J}_\lambda^j := \tilde{I}_\lambda|_{\Sigma^j}$ on the submanifold Σ^j, then

$$u_0^j = t_0^j v_0^j \in W \setminus \{0\} \tag{10}$$

is a critical point of I_λ on $W \setminus \{0\}$.

Remark 2.1. It is important to note, that in comparison with the usual constraint minimization method (cf. 21), by the assumptions of Theorem 2.1 the point $(t_0^j, v_0^j) \in \Sigma_\lambda^j$ is not necessary extremal one, i.e. it may be not a local minimum or maximum of $\tilde{J}_\lambda^j$ on Σ_λ^j. This property for example allow us to apply the Lusternik-Schnirelman theory over constraints Σ_λ^j (see below and Ref. 14).

Remark 2.2. Let us emphasize, that the assertion of the theorem holds under assumption that the point (t_0^j, v_0^j) should be internal in the set Σ_λ^j, i.e. (t_0^j, v_0^j) does not belong to $\partial\Sigma_\lambda^j$ Refs. 9,10,12.

We call the following variational problems defined by

$$\hat{I}_\lambda^j = inf\{\tilde{I}_\lambda(t,v)|\ (t,v) \in \Sigma_\lambda^j\}, \quad j = 1,2, \tag{11}$$

the *ground minimization problems* with respect to the fibering scheme. It can be prove[9,10,12] that the solutions of these problems correspond to ground states of I_λ on $W \setminus \{0\}$ (modulo Morse index).

Let us now explain the idea of the Fiber Spectral Analysis. By Theorem 2.1 we should avoid those values of λ where the solution $(t_\lambda^j, v_\lambda^j)$ of the problem (11) could be belongs to $\partial\Sigma_\lambda^j$ (see Remark 2.2). Thus it is natural to consider as the characteristic values the following ones

$$\lambda_{inf}^j(\lambda_{sup}^j) = \inf(\sup)\left\{\lambda \mid \hat{I}_\lambda^j = inf\{\tilde{I}_\lambda(t,v)|\ (t,v) \in \partial\Sigma_\lambda^j\}\right\}. \tag{12}$$

Observe by (8), (9) the boundaries $\partial\Sigma_\lambda^j$ are described by the set of solutions $(t,v) \in \overline{\mathbb{R}}^+ \times S^1$, $\overline{\mathbb{R}}^+ = 0 \cup \mathbb{R}^+ \cup +\infty$ of the following system

$$\begin{cases} \frac{\partial}{\partial t}\tilde{I}_\lambda(t,v) = 0, \\ \frac{\partial^2}{\partial t^2}\tilde{I}_\lambda(t,v) = 0. \end{cases} \tag{13}$$

Hence the characteristic values (12) correspond to the limit of the set $\{\lambda, t_\lambda^j, v_\lambda^j\}$ where the map $D^2\tilde{I}_\lambda(t,v)$ loses a regularity (modulo fibering).

In simple cases to find characteristic values, it is sufficient to analyze only the system (13). Let us consider this system in the case of the problem (1)

$$\begin{cases} H_\lambda(v) - t^{\gamma-2}F(v) = 0, \\ H_\lambda(v) - (\gamma-1)t^{\gamma-2}F(v) = 0. \end{cases}$$

It is easy to see that the set of solution of this system is a sum of the following subset

$$d_\lambda^0 = \{(t,v) : H_\lambda(v) = 0,\ t = 0\}, \quad d_\lambda^\infty = \{(t,v) : \forall t \in \overline{\mathbb{R}}^+,\ H_\lambda(v) = 0,\ F(v) = 0\}.$$

By our idea of finding the limit points we consider

$$\lambda^0_{inf} = \inf\{\lambda \in \mathbb{R} : d^0_\lambda \neq \emptyset\} = \inf\{\lambda \in \mathbb{R} : H_\lambda(u) = 0, u \in S^1\} =$$
$$\inf\{\lambda \in \mathbb{R} : \int_\Omega |\nabla u|^2 dx - \lambda \int_\Omega |u|^2 dx = 0, u \in S^1\} =$$
$$\inf\{\frac{\int_\Omega |\nabla u|^2 dx}{\int_\Omega |u|^2 dx} : u \in W^{1,2}_0 \setminus \{0\}\}.$$

Hence we obtain the well-known Poincaré's and Courant-Weyl's principle for the first eigenvalue $\lambda_1 = \lambda^0_{inf}$ of the problem (2), i.e. the linear characteristic point.

Now consider the second set d^∞_λ

$$\lambda^\infty_{inf} = \inf\{\lambda \in \mathbb{R} : d^\infty_\lambda \neq \emptyset\} = \inf\{\lambda \in \mathbb{R} : H_\lambda(u) = 0, F(u) = 0, \ u \in S^1\} =$$
$$\inf\{\lambda \in \mathbb{R} : \int_\Omega |\nabla u|^2 dx - \lambda \int_\Omega |u|^2 dx = 0, F(u) = 0, \ u \in S^1\} =$$
$$\inf\{\frac{\int_\Omega |\nabla u|^2 dx}{\int_\Omega |u|^2 dx} : F(u) = 0, \ u \in S^1\}.$$

Hence we obtain the Ouyang's characteristic point (3) $\lambda^* = \lambda^\infty_{inf}$, i.e. the nonlinear characteristic point.

Remark 2.3. Applying more detail approach (see (12)) we get[9,10]

$$\lambda^* = \inf\{\frac{\int_\Omega |\nabla u|^2 dx}{\int_\Omega |u|^2 dx} : F(u) \geq 0, \ u \in W^{1,2}_0\}.$$

Hence applying the general Theorem 4.1. from Ref. 12 it can be proved the results of Ouyang[16]

Theorem 2.2. *Let* $2 < \gamma < 2^*$, $f \in L^\infty(\Omega)$.
1) Assume $f^+ \neq 0$. *Then* $\lambda^*_1 < \infty$ *and for any* $\lambda \in (-\infty, \lambda^*_1)$ *there exists a positive solution* u^1_λ.

2) Assume $F(\phi_1) < 0$. *Then* $\lambda_1 < \lambda^*_1$ *and for any* $\lambda \in (\lambda_1, \lambda^*_1)$ *there exists a positive solution* u^2_λ.

Remark 2.4. Theorem 2.2 gives an answer to the problem (i) and a partial answer to the problem (ii) of (NAB). To solve the problems (i)-(iii) of (NAB) on the whole it is not sufficient to use only the trivial fiber space $\mathbb{R}^+ \times S^1$. In the paper Ref. 9 it is shown the solution of the problem (NAB) in the large using the fiber bundel over the projective space $P(W)$.

3. The Problem with Inhomogeneous Indefinite Nonlinearities

In the papers Refs.12,14 it is studied using the fiber spectral analysis the classes of inhomogeneous boundary value problems with indefinite nonlinearities.

Let us show for instance the results that has been obtained in Ref. 12. It is considered the following class of inhomogeneous Neumann boundary value problems with indefinite nonlinearities

$$\begin{cases} -\Delta_p u - \lambda |u|^{p-2}u = D(x)|u|^{q-2}u + K(x)|u|^{\gamma-2}u \ \text{ in } \ \Omega, \\ |\nabla u|^{p-2}\frac{\partial u}{\partial n} + |u|^{p-2}u = 0 \ \text{ on } \ \partial\Omega, \end{cases} \tag{14}$$

where Ω is a bounded domain in Real^n, $n \geq 2$, with smooth boundary $\partial\Omega$. The functions K and D may change the sign, i.e. nonlinearities are indefinite, Δ_p is a p-Laplacian.

In comparison with homogeneous case (1), the geometry of variational functional corresponding to the problem (14) and the dependence of its from the parameter λ is more complicated. For example, it is not clean in which a prior interval $(\lambda_j, \lambda_{j+1}) \subseteq$ Real there exists a solution u_λ. However we are able to apply here the fiber spectral analysis.

Let $W = W_0^{1,p}(\Omega)$. Denote

$$F(u) = \int_\Omega K(x)|u|^\gamma dx, \quad B(u) = \int_\Omega D(x)|u|^q dx,$$

$$H(u) = \int_\Omega |\nabla u|^p dx + \int_{\partial\Omega} |u|^p dx, \ T(u) = \int_\Omega |u|^p dx.$$

Applying the nonlinear fiber spectral analysis we find the following characteristic values:

$$\lambda_1 = \inf\left\{\frac{H(u)}{T(u)} \mid u \in W\right\}, \tag{15}$$

$$\lambda_D^* = \inf\left\{\frac{H(u)}{T(u)} \mid B(u) \geq 0, \ u \in W\right\}, \tag{16}$$

$$\lambda_{D/K}^* = \inf\left\{\frac{(H(u))|F(u)|^{\frac{(q-p)}{(\gamma-p)}} + R|B(u)|^{\frac{(\gamma-p)}{(\gamma-q)}}}{T(u)|F(u)|^{\frac{(q-p)}{(\gamma-p)}}} \,|B(u) \tag{17}$$

$$\leq 0, \ F(u) > 0\right\},$$

where

$$R = \left(\frac{q-p}{\gamma-p}\right)^{\frac{(q-p)}{(\gamma-p)}}\left(\frac{\gamma-q}{\gamma-2}\right).$$

Observe that $\lambda_1 \geq 0$ is the first eigenvalue of the problem

$$-\Delta_p \phi_1 = \lambda_1 |\phi_1|^{p-2}\phi_1 \text{ in } \Omega,$$
$$|\nabla u|^{p-2}\frac{\partial \phi_1}{\partial n} + |\phi_1|^{p-2}\phi_1 = 0 \text{ on } \partial\Omega,$$

and the variational formula (15) coincides (in case $p = 2$) with Poincare's and Courant-Weyl's principles. Moreover it can be proved the existence of solutions of the variational problems 1)-4) and to show that these solutions correspond to the solution of the problem (14).

We are able to prove the following result on the existence of positive solution.

Theorem 3.1. *Suppose that* $K(\cdot), D(\cdot) \in L^\infty(\Omega)$, $p < q < \gamma \leq p^*$. *Assume that* $B(\phi_1) < 0$. *Then* $\lambda_1 < min\{\lambda^*_{D/K}, \lambda^*_{D,K}\}$ *and for every* $\lambda \in (\lambda_1, min\{\lambda^*_{D/K}, \lambda^*_{D,K}\})$ *there exists a weak positive solution* $u^1_\lambda \in W^1_p(\Omega)$ *of (14).*

Denote by $\phi_{\lambda,D} \in W^1_p(\Omega)$, $\lambda < \lambda^*_D$, the positive solution of the following minimization problem

$$\min\{H(u) - \lambda T(u) \mid B(u) \geq 0, \ u \in W\}. \tag{18}$$

The next main result on the existence the second positive solution for (14) is the following.

Theorem 3.2. *Suppose* $K(\cdot), D(\cdot) \in L^\infty(\Omega)$, $p < q < \gamma \leq p^*$. *Assume that the set* $\{x \in \Omega \mid K(x) > 0\}$ *is not empty and* $F(\phi_{\lambda,D}) \geq 0$.
Let $\lambda < \min\{\lambda^*_D, \lambda^*_{D/K}\}$. *Then there exists a positive solution* $u^2_\lambda \in W^1_p(\Omega)$ *of (14).*

The main multiplicity result on the positive solutions is the following

Theorem 3.3. *Suppose* $K(\cdot), D(\cdot) \in L^\infty(\Omega)$, $p < q < \gamma \leq p^*$. *Assume* $B(\phi_1) < 0$, *the set* $\{x \in \Omega \mid K(x) > 0\}$ *is not empty and* $F(\phi_{\lambda,D}) \geq 0$, *then* $\lambda_1 < \min\{\lambda^*_D, \lambda^*_{D,K}, \lambda^*_{D/K}\}$ *and for every* $\lambda \in (\lambda_1, \min\{\lambda^*_D, \lambda^*_{D,K}, \lambda^*_{D/K}\})$ *there exist at least two positive solutions* $u^1_\lambda, u^2_\lambda \in W^1_p(\Omega)$ *of (14).*

4. A Problem with Concave-Convex Nonlinearity

In Ref. 14 it has been considered the following generalized Ambrosetti-Brezis-Cerami problem[1] with concave-convex nonlinearity

$$\begin{cases} -\Delta_p u = \lambda |u|^{q-2}u + f(x)|u|^{\gamma-2}u, \ x \in \Omega, \\ \\ u|_{\partial\Omega} = 0, \end{cases} \tag{19}$$

where Ω is a bounded domain in $\mathbb{R}^N$, $N \geq 1$, with smooth boundary $\partial\Omega$, Δ_p is the p-Laplacian and

$$1 < q < p < \gamma \leq p^*, \text{ where } p^* = \begin{cases} \frac{pN}{N-p} & \text{if } p < N, \\ +\infty & \text{if } p \geq N. \end{cases} \tag{20}$$

We suppose that $f(x) \geq 0$ on Ω and

$$f \in L_r(\Omega), \text{ where } \begin{cases} r > p^*/(p^* - \gamma) & \text{if } p < N \text{ and } \gamma < p^*, \\ r = +\infty & \text{if } p < N \text{ and } \gamma = p^*, \\ 1 < r & \text{if } p \geq N. \end{cases} \tag{21}$$

The problem (19) has the variational form with the Euler functional $I_\lambda(u)$, defined on Sobolev space $W = W_0^{1,p}(\Omega)$ by

$$I_\lambda(u) = \frac{1}{p}H(u) - \frac{\lambda}{q}G(u) - \frac{1}{\gamma}F(u), \tag{22}$$

where

$$H(u) = \int_\Omega |\nabla u|^p dx,\ G(u) = \int_\Omega |u|^q dx \text{ and } F(u) = \int_\Omega f(x)|u|^\gamma dx. \tag{23}$$

Applying the fiber spectral analysis the following variational principle

$$\Lambda^* = \frac{\gamma - p}{p - q}\left(\frac{p-q}{\gamma - q}\right)^{\frac{\gamma-q}{\gamma-p}} \inf_{v \in W\setminus\{0\}} \left(\frac{H^{\frac{\gamma-q}{\gamma-p}}(v)}{G(v)\, F^{\frac{p-q}{\gamma-p}}(v)} \right), \tag{24}$$

is introduced. Remark that it can be prove that there exists a solutions of the variational problem (24) and it corresponds to the solution of the problem (19).

With respect to the fiber spectral analysis it is introduced the functional $\mathcal{L}_\lambda$ defined on W by

$$\mathcal{L}_\lambda(u) = p\,H(u) - \lambda\, q\, G(u) - \gamma\, F(u).$$

This functional allows us to separate multiple critical points of the Euler functional I_λ.

The first example of application of the fiber spectral analysis is the following

Theorem 4.1. *Let $1 < q < p < \gamma < p^*$, $f(x) \geq 0$ on Ω and suppose (21) holds. Then*

(i) for every $\lambda \in (-\infty, \Lambda^*)$ there exists a positive solution $u_\lambda^1 \in C^{1,\alpha}(\Omega)$ for some $\alpha \in (0,1)$ of (19). Moreover

$$u_\lambda^1 \in K_\lambda^1 := \{u \in W \,:\, I'_\lambda(u) = 0,\ \mathcal{L}_\lambda(u) < 0\}.$$

(ii) for every $\lambda \in (0, \Lambda^*)$ there exists a second positive solution $u^2_\lambda \in C^{1,\alpha}(\Omega)$ of (19). Moreover

$$u^2_\lambda \in K^2_\lambda := \{u \in W \, : \, I'_\lambda(u) = 0, \; \mathcal{L}_\lambda(u) > 0\}.$$

To prove the existence of solution in the critical case of the exponent $\gamma = p^*$, we introduce (also applying an idea of the fiber spectral analysis) characteristic values $\lambda^*_{ps}, \lambda^*_0$. They allow to find an interval where I_λ satisfies to Palais-Smale (P.-S.) condition

$$\lambda^*_{ps} = \frac{p}{N(p-q)} S^{\frac{N(p^*-q)}{p^*}} \inf_{u \in W \setminus 0} \frac{[\int f|u|^{p^*} dx]^{\frac{q}{p^*}}}{\int |u|^q dx}, \tag{25}$$

and

$$\lambda^*_0 := \frac{q}{\gamma} \left(\frac{\gamma}{p}\right)^{\frac{\gamma-q}{\gamma-p}} \Lambda^*. \tag{26}$$

Here S is the best Sobolev constant. It can be shown that $\lambda^*_0 < \Lambda^*$ and $\lambda^*_{ps} > 0$ as $1 < q < p < \gamma \leq p^*$.

Theorem 4.2. *Let $1 < q < p < \gamma = p^*$, $f \in C(\overline{\Omega})$, $f(x) > 0$ on $\overline{\Omega}$.*

(i) Then for every $\lambda \in (0, min\{\Lambda^*, \lambda^*_{ps}\})$ there exists a positive solution $u^1_\lambda \in C^{1,\alpha}(\Omega)$ for some $\alpha \in (0,1)$ which belongs to K^1_λ.

(ii) Suppose that $\lambda^*_0 < min\{\Lambda^*, \lambda^*_{ps}\}$. Then for every $\lambda \in [\lambda^*_0, min\{\Lambda^*, \lambda^*_{ps}\})$ there exists a second positive solution $u^2_\lambda \in C^{1,\alpha}(\Omega)$ for some $\alpha \in (0,1)$ which belongs to K^2_λ.

The next theorem on the existence of two disjoint sets of solutions is the sonsequence of the general Theorem 4.2 from Ref. 14

Theorem 4.3.

(I) Let $1 < q < p < \gamma < p^*$, $f \in C(\overline{\Omega})$, $f(x) > 0$ on $\overline{\Omega}$. Then

(i) for every $\lambda \in (0, \Lambda^*)$ there exists an infinite set $(u^{1,n}_\lambda)$ of solutions of (19) such that $u^{1,n}_\lambda \in K^1_\lambda$, $I_\lambda(u^{1,n}_\lambda) < 0$ and $I_\lambda(u^{1,n}_\lambda) \to 0$ as $n \to \infty$;

(ii) for every $\lambda \in (-\infty, \Lambda^*)$ there exists a second infinite set $(u^{2,n}_\lambda)$ of solutions of (19) such that $u^{2,n}_\lambda \in K^2_\lambda$ and $I_\lambda(u^{2,n}_\lambda) \to +\infty$ as $n \to \infty$.

(II) In the critical case let $1 < q < p < \gamma = p^*$, $f \in C(\overline{\Omega})$, $f(x) > 0$ on $\overline{\Omega}$. Then for every $\lambda \in (0, min\{\Lambda^*, \lambda^*_{ps}\})$ there exists at least one infinite set $u^{1,n}_\lambda$ of solutions to (19) such that $u^{1,n}_\lambda \in K^1_\lambda$, $I_\lambda(u^{1,n}_\lambda) < 0$ and $I_\lambda(u^{1,n}_\lambda) \to 0$ as $n \to \infty$.

5. The Equations with p&q- Laplacian

The paper[5] is devoted to the study of the following equations with p&q-Laplacian

$$\begin{cases} -\Delta_p u - \Delta_q u + q(x)|u|^{p-2} + w(x)|u|^{q-2}u = \lambda f(x)|u|^{\gamma-2}u \text{ in } \Omega, \\ u = 0 \quad \text{on} \quad \partial\Omega, \end{cases} \tag{27}$$

here Ω is a smooth, bounded domain in Real^n, $(n \geq 1)$, $\lambda \geq 0$, Δ_s, $(s = p, q)$ denote the s-Laplacian defined by $\Delta_s = \operatorname{div}(|\nabla u|^{s-2}\nabla u)$ for $s \in (1, \infty)$,

$$1 < p < \gamma < q \quad \text{and } \gamma < p^*, \text{ where } p^* = \begin{cases} \frac{pn}{n-p} & \text{if } p < n, \\ +\infty & \text{if } p \geq n. \end{cases} \tag{28}$$

We assume that $f,\ q,\ w\ \in L^r(\Omega)$ with $r > \frac{p^*}{p^*-\gamma}$ if $p < n$, and $r > 1$ if $p \geq n$.

A major difficulty associated with (27) is the absence of a priori information on the parameter λ for which the problem (27) may has or no solution. In Ref. 5 the main idea to overcome this difficulty lies on the fiber spectral analysis. Based on this idea, it is introduced constructively a well-defined variational principle, that is bounded, 0-homogeneous and weakly lower semi-continuous below. Furthermore, the critical points of its correspond to problem (27) on a discrete subset of the spectral values λ.

Introduce the following functionals

$$H_p(v) = \int_\Omega (|\nabla v|^p + q(x)|v|^p)dx, \quad H_q(v) = \int_\Omega (|\nabla v|^q + w(x)|v|^q)dx,$$

$$\text{and } F_\gamma(v) = \int_\Omega f(x)|v|^\gamma dx, \quad v \in W_0^{1,q}(\Omega), \tag{29}$$

where $W_0^{1,q}(\Omega)$ is the Sobolev space. Under the assumption (28) the functionals (29) are well-defined on the Sobolev space $W_0^{1,q}$ and belong to the class $C^1(W_0^{1,q})$.

The problem (27) has a variational form with the following Euler functionnal on $W_0^{1,q}$

$$I_\lambda(u) = \frac{1}{p}H_p(u) + \frac{1}{q}H_q(u) - \frac{\lambda}{\gamma}F_\gamma(u). \tag{30}$$

Introduce the following assumption

A. $H_p(u) > 0$, $H_q(u) > 0$ for all $u \in W_0^{1,q}$.

Following the strategy of the fiber spectral analysis we introduce a characteristic value Λ^* by the following variational principle

$$\Lambda^* = \inf\{\ \lambda(v)\ :\ F_\gamma(v) > 0,\ v \in W_0^{1,q} \setminus \{0\}\}, \tag{31}$$

where $\Lambda^* = +\infty$ in case $\Omega^+ := \{x \in \Omega : \ f(x) > 0\} = \emptyset$ and

$$\lambda(v) = \frac{C_{p,q,\gamma} H_p(v)^{\frac{q-\gamma}{q-p}} H_q(v)^{\frac{\gamma-p}{q-p}}}{F_\gamma(v)}, \quad C_{p,q,\gamma} = \frac{q-p}{(\gamma-p)^{\frac{\gamma-p}{q-p}}(q-\gamma)^{\frac{q-\gamma}{q-p}}}. \tag{32}$$

The following theorem plays a decisive role in Ref. 5

Theorem 5.1. *Suppose that (28) and* **A.** *hold. Assume that* $\Omega^+ \neq \emptyset$*, then 1)* $0 < \Lambda^* < +\infty$*; 2) there exists non-negative solution* $v^* \in W \setminus \{0\}$ *of the variational problems (31), i.e.* $\Lambda^* = \lambda(v^*)$ *and* $F_\gamma(v^*) > 0$.
Furthermore, there exists a constant $t^* > 0$ *such that the function* $u_{\lambda^*} = t^* \cdot v^*$ *is a weak nontrivial solution of the problem (27) with*

$$\lambda = \lambda^* = \frac{\gamma}{p^\alpha q^{1-\alpha}} \Lambda^*, \quad \alpha = \frac{q-\gamma}{q-p}. \tag{33}$$

Moreover, the strong inequality $\Lambda^* < \lambda^*$ *holds.*

The main result on the existence and non-existence of non-negative solutions for (27) obtained in Ref. 5 is the following

Theorem 5.2. *Suppose that (28) and* **A.** *hold.*

(i) Then for every $\lambda \in [0, \Lambda^*[$ *the problem (27) has no non-trivial solution.*
(ii) Assume that $\Omega^+ \neq \emptyset$*. Then for every* $\lambda \geq \lambda^*$ *there exists a non-negative weak solution* $u_\lambda \in W_0^{1,q}(\Omega) \setminus \{0\}$ *of the problem (27). Moreover,* $I_{\lambda^*}(u_{\lambda^*}) = 0$ *and* $I_\lambda(u_\lambda) > 0$ *for every* $\lambda > \lambda^*$.

Remark 5.1. It is interesting to see that in the case of a single Laplacian in (27), the behavior with respect to λ is different. For instance, when equation (27) contains only the p-Laplacian, $\Omega^+ \neq \emptyset$ and **A.** holds , it is well known (see for example Ref. 9) that for all $\lambda > 0$ (27) possesses a positive solution.

6. Solutions of Minimal Period for a Hamiltonian System with Potential Indefinite in Sign

Another kind of application of the fibe spectral analysis is given in Ref. 13. This paper is concerned with the existence of periodic solutions for the following second-order Hamiltonian systems:

$$\begin{cases} -\ddot{x} = Bx + f(t)|x|^{\gamma-2}x, \quad x = (x_1, ..., x_n) \in \text{Real}^n, \\ \\ x(0) = x(T), \quad \dot{x}(0) = \dot{x}(T), \end{cases} \tag{34}$$

here B is a positive definite, symmetric matrix with eigenvalues $0 < \omega_1^2 \leq ... \leq \omega_n^2$; $2 < \gamma < \infty$; $f(t) = diag(f^1(t), ..., f^n(t))$ is a continuous T-periodic matrix-valued function. $f^i(t)$ may change sign, i.e., the problem is with the potential indefinite in sign.

In Refs. 2,23 it has been stated the following question:

(P) *May the Hamiltonian systems with potential indefinite in sign has solution as* $T > \frac{2\pi}{\omega_n}$*?*

The main goal of the note[13] is to give an answer for this question. The answer in Ref. 13 to this question is positive.

Let us state the main results obtained in Ref. 13. The solutions of the problem is sought in the following closed subspace

$$E(0,T) = \{x \in H(0,T) : \int_0^T x dt = 0, \; x(0) = x(T)\}$$

where $H(0,T) = H_2^1(0,T)$ is the usual Sobolev space.

The problem (34) is the Euler-Lagrange equation of the functional

$$I_T(x) = \frac{1}{2}\int_0^T |\dot{x}|^2 dt - \frac{1}{2}\int_0^T (Bx,x)dt - \frac{1}{\gamma}\int_0^T |x|^{\gamma-2}(f(s)x,x)ds \quad (35)$$

on the subspace $E(0,T)$.

Applying the fiber spectral analysis it is introduced the following Ouyang's characteristic point[16]

$$T^* = \inf\left\{ \frac{\int_0^1 |\dot{y}|^2 ds}{\int_0^1 (By,y)ds} : \int_0^1 |y|^{\gamma-2}(g(s)y,y)ds \geq 0, \; y \in E(0,1) \right\} \quad (36)$$

where $g(t) = f(tT)$ and $T^* = +\infty$ in the case when the set $\{y \in E : \int_0^1 |y|^{\gamma-2}(gy,y)ds \geq 0\}$ is empty.

Let $\psi_n \in \text{Real}^n$ be a unit eigenvector of B associated to the eigenvalue ω_n^2. Then $T_1^2 = (\frac{2\pi}{\omega_n})^2$ is the simple first eigenvalue of the following boundary value problem

$$-\ddot{\phi}_1 = T_1^2 B\phi_1, \quad (37)$$
$$\phi_1 \in E(0,1),$$

where $\phi_1(t) = \psi_n \sin(2\pi t)$. Let us denote $\phi_{1,T}(t) = \psi_n \sin(\frac{2\pi t}{T})$. It is introduced the following hypothesis

$$F(\phi_{1,T}) := \int_0^T |\phi_{1,T}|^{\gamma-2}(f(t)\phi_{1,T},\phi_{1,T})dt < 0. \quad (38)$$

The answer to the problem **P.** is given by the following result

Theorem 6.1. *Suppose that* $2 < \gamma < \infty$ *and the hypothesis (38) holds. Then 1)* $\frac{2\pi}{\omega_n} < T^*$*; 2) for every* $T \in (\frac{2\pi}{\omega_n}, T^*)$ *there exists a classical solution* $x_T^1 \in E(0,T)$ *of (27) with minimal period* T*. Moreover* $I_T(x_T^1) < 0$*.*

In the next theorem the existence of second classical solution with minimal period is proved

Theorem 6.2. *Suppose that* $2 < \gamma < \infty$. *Assume* $T \in (0, T^*)$ *and* $f^i(s_0) > 0$ *for some* $i = 1, 2, ..., n$ *and* $s_0 \in (0, T)$. *Then there exists a classical solution* $x_T^2 \in E(0, T)$ *of (27) with minimal period* T. *Moreover* $I_T(x_T^2) > 0$.

The main result on the existence of infinitely many T-periodic solutions is the following

Theorem 6.3. *Suppose that* $2 < \gamma < \infty$. *Assume* $T \in (0, T^*)$ *and* $f^i(s_0) > 0$ *for some* $i = 1, 2, ..., n$ *and* $s_0 \in (0, T)$. *Then there exists an infinite set* $(x_T^{2,m}) \subset E(0, T)$ *of classical* T*-periodic solutions of (27) such that* $I_T(x_T^{2,m}) > 0$ *and* $I_T(x_T^{2,m}) \to +\infty$ *as* $m \to \infty$.

Another the set of T-periodic solutions is given in the next theorem. Let us denote by $T_1^2 < T_2^2 \leq ... \leq T_m^2 \leq ...$, $m = 1, 2, ...$ the eigenvalues of the linear problem (37).

Theorem 6.4. *Let* $2 < \gamma < \infty$ *and the hypothesis (38) holds. Assume that* $T \in (\frac{2\pi}{\omega_n}, T^*)$ *and* $\frac{2\pi}{\omega_n} \equiv T_1 \leq T_{N(T)} < T$ *for some integer* $N(T) \geq 1$. *Then the problem (27) possesses at least* $N(T)$ *classical* T*-periodic solutions* $(x_T^{1,m}) \subset E(0, T)$, $m = 1, 2, ..., N$. *Furthermore,* $I_T(x_T^{1,m}) < 0$, $m = 1, 2, ..., N$.

Observe that $T^* = +\infty$ if $\{y \in E(0, 1) : \int_0^1 |y|^{\gamma-2}(g(s)y, y)ds \geq 0\} = \emptyset$ and $T^* \to +\infty$ as $mes\{x \in (0, T) : f^i(s) \geq 0, \ i = 1, 2, ..., n\} \to 0$.

To obtain solution with minimal period it is important to find suitable constrained minimization problem. To this aim we follow the arguments[9] which allow us to use ground constrained minimization problem (11). Another advantages of using the arguments of fiber spectral analysis is the following. By this way the points $\frac{2\pi}{\omega_n}, T^*$ are introduced constructively that bring some light on the nature of these constants. In other words the answer for the question **(P)** in Ref. 13 contains not only in the statements of the main Theorems 6.1-6.4.

The proofs of the existence of solutions in Theorems 6.1-6.4 base on Lyusternik-Schnirelman theory in the framework of fibering approach. The main difficulty here is that the fibering constrains Σ_T^j, $j = 1, 2$ (see (8), (9)) generally is not necessary to be a *complete* manifolds. In Ref. 13 the main idea to overcome this difficulty again lies on the fiber spectral analysis. Another problem is to prove that the restricted fiber functionals $\tilde{J}_T^j :=$

$\tilde{I}_T|_{\Sigma_T^j}$ satisfy to the Palais-Smale (P-S) condition on the submanifolds Σ_T^j. The fiber spectral analysis gives an elegant treatment of this problem.

References

1. Ambrosetti, A., Azorero J.G., Peral I., *Rend. Mat. Appl.* **20**, 167 (2000).
2. Antonacci, F., *Nonlinear Anal.* **29**, 1357 (1997).
3. Appell J., De Pascale E., Vignoli A., *Gruyter Series in Nonlinear Analysis and Applications 10.* (Berlin: de Gruyter. 2004).
4. Berestycki, H., Capuzzo–Dolcetta, I., Nirenberg, L., *NoDEA* **2**, 553(1995).
5. Cherfils L.& Il'yasov Y., *Commun. Pure and Appl. Anal.* **4**, 9 (2005).
6. Del Pino, M., *Nonlinear Anal.* **22**, 1423 (1994).
7. Drábek, P.& Pohozaev, S.I.,: *Proc. Roy. Soc. Edinb. Sect. A* **127**, 703 (1997).
8. Il'yasov, Y., *C. R. Acad. Sci., Paris* **332**, 533 (2001).
9. Il'yasov Ya. Sh., *Izv. Russ. Ac.N. Ser. Mat.* **66**, 19 (2002).
10. Il'yasov, Ya. Sh., *The fibering method,* in *Nonlinear analysis and nonlinear differential equations*, eds. Trenogin B.A., Fillipov B.A, (Moscow: Fizmatlit, 2003), pp. 464.
11. Il'yasov Y.& Runst T. *Top. Meth. in Nonl. Anal.* **24**, 41 (2004).
12. Ilyasov Y.& Runst T., *Calculus Var. & Part. Diff. Eq.,* **22,** 101 (2005).
13. Il'yasov Y.& Sari N., *Commun. Pure and Appl. Anal.* **4**, 175 (2005).
14. Il'yasov Y., *Nonliner Analysis TMA*, **61**, 211 (2005).
15. Il'yasov Y., *Diff. Eq.* **41**, 548 (2005).
16. Ouyang, T.C., *Indiana Univ. Math. J.* **40**, 1083 (1991).
17. Pohozaev, S.I., *Doklady Acad. Sci. USSR* **247**, 1327 (1979).
18. Pohozaev, S.I., *Proc. Stekl. Ins. Math.* **192**, 157 (1990).
19. Pohozaev, S.I., *Rend. Inst. Math. Univ. Trieste,* **v. XXXI**, 235 (1999).
20. Pohozaev S. & Veron L., *Appl. Anal.*, **74**, 363 (2000).
21. M. Struwe, *Application to Nonlinear Partial Differential Equations and Hamiltonian Systems,* (Springer - Verlag Berlin, Heidelberg, New-York, 1996).
22. E. Ziedler, *Nonliner functional analysis and its applications I-IV*, (Springer, New-York-Heidelberg-Berlin, 1988-1990).
23. Zou, Wenming, Li, Shujie, *J. Diff. Eq.* **V 186**, 141 (2002).

Harmonic, Wavelet and p-Adic Analysis
Eds. N. M. Chuong *et al.* (pp. 157–174)

§8. ON A FREE BOUNDARY TRANSMISSION PROBLEM FOR NONHOMOGENEOUS FLUIDS

BUI AN TON
Department of Mathematics, University of British Columbia,
Vancouver, Canada V6T1Z2
E-mail: bui@math.ubc.ca

The existence of a weak solution of a free boundary transmission problem for two fluids with different densities, arising in the study of water pollution, is established.

1. Introduction

Let G be a bounded open subset of R^3 with a smooth boundary and consider the motion of two fluids of densities $\rho_\pm$ in $G_\pm(t)$ with $G_+(t) \subset \text{Int}(G)$ and $G_-(t) = G\big/\overline{G_+(t)}$.

The free boundary transmission problem is described by the system

$$\begin{aligned} \rho_\pm\{\vec{u}'_\pm + (\vec{u}_\pm \cdot \nabla)\vec{u}_\pm\} - \Delta\vec{u}_\pm + \nabla p_\pm \\ = \vec{f}_\pm(t, \vec{u}_\pm) \text{ in } \bigcup_t (G_\pm(t) \times \{t\}) \\ \nabla \cdot \vec{u}_\pm = 0 \text{ in } \bigcup_t (G_\pm(t) \times \{t\}) \end{aligned} \tag{1}$$

$$\vec{u}_- = 0 \text{ on } \partial G \times (0,T), \vec{u}_\pm(x,0) = \vec{u}^0_\pm \text{ in } G_\pm(0)$$

The conservation of mass is expressed by the initial value problem

$$\begin{aligned} \rho'_\pm + \vec{u}_\pm \cdot \nabla\rho_\pm = 0 \text{ in } \bigcup_t (G_\pm(t) \times \{t\}), \\ \rho_\pm(\cdot, 0) = \rho^0_\pm > 0 \text{ in } G_\pm(0) \end{aligned} \tag{2}$$

On the free boundary $\varGamma_t = \partial G_+(t)$ the transmission conditions are

$$\rho_+ = \rho_- \ ; \ \vec{u}_+ = \vec{u}_- \text{ on } \bigcup_t (\varGamma_t \times \{t\})$$
$$n \cdot (\nabla \vec{u}_+ - \nabla \vec{u}_-) = 0 \text{ on } \bigcup_t (\varGamma_t \times \{t\}) \tag{3}$$

and the free boundary $\varGamma_t$ is described in Lagrangian coordinates by

$$\frac{d}{dt} X(\xi, t) = \vec{u}(X(\xi, t), t); \ X(\xi, 0) = \xi \tag{4}$$

where $\vec{u} = \vec{u}_\pm$ in $G_\pm(t)$ and $X(\xi, t)$ is the position of a fluid particle which at $t = 0$ is at a point $\xi \in G$. The free boundary $\varGamma_t$ is then

$$\varGamma_t = \{X(\xi, t) : \ X(\xi, t) \text{ solution of } (4), \Lambda(\xi) = c = \varGamma_0\}$$

where $\Lambda(\xi) = c$ represents the initial surface of separation of the fluids.

With the sources in (1) being in $L^2(0, T; L^2(G))$, a solution $\vec{u}$ of the Navier Stokes equations is only in $L^2(0, T; J_0^1(G))$ and thus the expression $\vec{u}(X(\xi, t), t)$ may be meaningless. Since $X(\xi, t)$ is the material position of the interface and is inside G, X is in $L^\infty(0, T; L^\infty(G))$ and we are led to consider the set

$$\mathcal{K} = \{\vec{v} : \|\vec{v}\|_{L^2(0,T;H_0^1(G))}, \ \|\vec{v}\|_{L^\infty(0,T;C^\lambda(G))} \le C$$
$$\|\vec{v}'\|_{L^2(0,T;H^{-3}(G))} \le C\}$$

We replace (4) by

$$\frac{d}{dt} X(\xi, t; \vec{u}) = \mathcal{P}\vec{u}(X(\xi, t; \vec{u}), t); \ X(\cdot, 0; \vec{u}) = \xi \ \text{ in } G \tag{5}$$

where $\mathcal{P}$ is the projection of $L^2(0, T; H^{-3}(G))$ onto the closed bounded convex subset $\mathcal{K}$ of the Hilbert space $L^2(0, T; H^{-3}(G))$.

The system (1)-(5) may be considered as a model describing the evolution of the interface between the polluted region, originating from a source $\vec{f}_+$ in $G_+(t)$, and the unpolluted one. It is a hybrid "hyperbolic-parabolic" free boundary nonlinear system.

Initial boundary value problems for hyperbolic-parabolic systems arise in the study of water waves, in the theory of compressible fluids and have been studied by Y. Belov and N. Yalenko,[2] B. Gustafsson and A. Sundstrom,[3] V. Kajikov,[4] J. L. Lions[6] and others. The Euler equations for nonhomogeneous fluids have been investigated by H. Beirao da Viega and A. Valli.[1] Free boundary transmission problem nonlinear hybrid stochastic wave-Navier Stokes system has been studied by the author in Ref. 9.

The notations, the main assumptions of the paper and some preliminary results are given in Sec. 2. The existence of a unique solution of the initial value problem (2) is establihed in Sec. 3. A linearized version of (1)-(5) is studied in Sec. 4 and the main result of the paper which seems new, is proved in Sec. 5. We shall adapt some of the techniques used earlier by the author in the study of a free boundary problem arising from the flow of blood around the heart valves.[8]

2. Notations, assumptions, preliminary results

Let Γ_0 be a smooth closed surface lying entirely inside of G and let $G_+(0)$ be the region bounded by Γ_0, $\partial G_+(0) \cap \partial G = \emptyset$ and $G_-(0) = G/\overline{G_+(0)}$ with

$$G = \overline{G_+(0)} \cup G_-(0),\ \partial G_+(0) = \Gamma_0,\ \partial G_-(0) = \partial G \cup \Gamma_0.$$

Throughout the paper we assume that

$$\vec{u}^0_\pm \in L^2(G_\pm(0)),\ \vec{u}^0_+ = \vec{u}^0_- \ \text{ on } \Gamma_0;\ \vec{u}_0(x) = \vec{u}^0_\pm \ \text{ in } G_\pm(0)$$

Lemma 2.1. *Let $\vec{v} \in L^2(0,T;L^2(G))$, then for any $\xi \in G$, there exists a unique solution X of (5) with*

$$\|X(\cdot,\vec{v})\|_{C(0,T;L^\infty(G))} + \|X'(\cdot,\vec{v})\|_{L^\infty(0,T;L^\infty(G))} \le C(1+ \mid G \mid)$$

The constant C is independent of $\vec{v}$.

Suppose that $\vec{v}_n \to \vec{v}$ weakly in $L^2(0,T;L^2(G))$ then

$$X(\cdot,t;\vec{v}_n) \to X(\cdot,;\vec{v}) \ \text{ in } C(0,T;C(G))$$

and $X(\,.;\vec{v})$ is the unique solution of (5). Moreover

$$\Gamma(\vec{v},t) = \bigcap_{n\ge n_0} \Gamma(\vec{v}_n,t)\,;\ G^\pm(\vec{v}) = \bigcap_{n\ge n_0} G^\pm(\vec{v}_n)$$

Proof. (Ref. 7, Lemmas 2.1,2.2,2.3) Let $\hat{J}$ be the duality mapping of $L^2(0,T;H^{-1}(G))$ into $L^2(0,T;H^1_0(G))$ with gauge function $\Phi(r) = r$. We have

$$\|\hat{J}(v)\|_{L^2(0,T;H^1_0(G))} = \Phi(\|v\|_{L^2(0,T;H^{-1}(G))}) = \|v\|_{L^2(0,T;H^{-1}(G))}$$

$$\int_0^T (\hat{J}(v),v)dt = \|v\|^2_{L^2(0,T;H^{-1}(G))} \quad \forall v \in L^2(0,T;H^{-1}(G)).$$

□

Definition 2.1. Let $\vec{g}$ be a mapping of $L^2(0,T;H^{-1}(G))$ with $D(\vec{g}) = L^2(0,T;L^2(G))$ and values in $L^2(0,T;H^{-1}(G))$. Then $\vec{g}$is said to be accretive with respect to the duality mapping $\hat{J}$ if

$$\int_0^T (\vec{g}(\vec{u}) - \vec{g}(\vec{v}),\ \hat{J}(\vec{u}-\vec{v}))dt \geq 0 \quad \forall \vec{u}, \vec{v} \in D(\vec{g})$$

Assumption 2.1. *Let $\vec{f}_\pm$ be continuous mappings of $L^2(0,T;L^2(G))$ into $L^2(0,T;L^2(G))$. Suppose that*

$$\|\vec{f}_\pm(\cdot,\vec{u})\|_{L^2(0,T;L^2(G))} \leq C\{1 + \|\vec{u}\|_{L^2(0,T;L^2(G))}\} \quad \forall \vec{u} \in L^2(0,T;L^2(G))$$

and that $\lambda I + \vec{f}_\pm$ is accretive in the sense of Definition 2.1 for some $\lambda \geq \lambda_0 > 0$.

The following lemma has been proved in Ref. 7

Lemma 2.2. *Let $\vec{f}$ be as in Assumption 2.1 and suppose that*

$$\vec{v}_n \to \vec{v} \text{ in } L^2(0,T;H^{-1}(G)) \cap (L^2(0,T;L^2(G)))_{weak}$$

then there exists a subsequence such that $\vec{f}(\cdot,\vec{v}_n) \to \vec{f}(\cdot,\vec{v})$ weakly in $L^2(0,T;L^2(G))$.

We denote by $J_0^k(G)$ the closure in the $H^k(G)$-norm of the set of all infinitely smooth soleinoidal vectors with compact support in G.

Remark 2.1.

(1) Lemma 2.2 is still valid if $L^2(G), H^{-1}(G)$ are replaced by $J_0(G)$ and by $J^{-1}(G) = (J_0^1(G))^*$.

(2) If $\vec{f}_\pm$ are continuous functions from $[0,T] \times R^3$ into R^3 and

$$| \vec{f}_\pm(t,\vec{x}) - \vec{f}_\pm(\vec{y}) | \leq C | \vec{x} - \vec{y} | \quad \forall \vec{x}, \vec{y} \in R^3$$

then $\lambda I + \vec{f}_\pm$ is accretive from $L^2(0,T;H^{-1}(G))$ into $L^2(0,T;H^{-1}(G)))$ with respect to the duality mapping $\hat{J}$.

Let ρ_0 be a scalar function representing the initial density of the fluid, be defined by

$$\rho_0(x) = \rho_0^\pm(x) \text{ in } G_0^\pm,\ \rho_0^+(x) = \rho_0^-(x) \text{ on } \Gamma_0.$$

Definition 2.2. Let $\vec{v} \in L^2(0,T;J_0(G))$,let $\rho_0 = \rho_0^\pm$ in $G^\pm(0)$ with $\rho_0^+(x) = \rho_0^-(x)$ on Γ_0 and suppose that

$$0 < a \leq \rho_0(x) \quad \forall x \in G$$

Then $\rho \in L^\infty(0,T;L^\infty(G))$ with $\rho(x,t) > 0$ for all (x,t) in $G \times [0,T]$, is said to be a weak solution of the initial value problem

$$\rho' + \vec{v} \cdot \nabla\rho = 0,\ \rho > 0;\ \ \rho(x,0) = \rho_0 \text{ in } G$$

if

$$-\int_0^T (\rho,\, \varphi')dt - \int_0^T (\rho\vec{v},\, \nabla\varphi)dt = (\rho_0,\, \varphi(\cdot,0))$$

for all φ in $L^2(0,T;H^1(G))$, φ' in $L^2(0,T;L^2(G))$, $\varphi(\cdot,T) = 0$.

The $L^2(G)$ inner product is denoted by $(\cdot,\cdot)$.

Definition 2.3. Let $\{\vec{u}_0,\, \rho_0\}$ be in $J_0(G) \times L^\infty(G)$ with

$$0 < a \le \rho_0(x), \quad \forall x \in G.$$

Then $\{\vec{u},\, \rho\}$ is said to be a weak solution of (1)-(5) if

- $\{\vec{u}, \rho, \rho', (\rho\vec{u})'\} \in \{L^\infty(0,T;L^2(G)) \cap L^2(0,T;J_0^1(G))\}$
$\times L^\infty(0,T;L^\infty(G)) \times L^2(0,T;H^{-2}(G)) \times L^2(0,T;(J_0^1(G) \cap H^2(G))^*)$
- ρ is a solution,in the sense of Definition 2.1, of the initial value problem

$$\rho' + \vec{u} \cdot \nabla\rho = 0, \rho > 0 \text{ in } G \times (0,T); \rho(\cdot,0) = \rho_0 \text{ in } G$$

- Let $X(\cdot,\vec{u})$ be the unique solution of (5) and let $\Gamma(\vec{u})$ be the surface separating G into G_+, G_- with

$$G = \overline{G_+} \cup G_-, \partial G_+ \cap \partial G = \emptyset; \Gamma(\vec{u}) = \partial G_+ \cap \partial G_-$$

- Set $\vec{u}_\pm = \vec{u}\,|_{G_\pm}, \rho_\pm = \rho\,|_{G_\pm}$, then

$$\begin{aligned}
&((\rho_+\vec{u}_+)(t),\phi)_{G_+} + ((\rho_-\vec{u}_-)(t),\phi)_{G_-} + \int_0^t (\nabla\vec{u}_+, \nabla\phi)_{G_+} ds \\
&+ \int_0^t (\nabla\vec{u}_-, \nabla\phi)_{G_-} ds - \int_0^t ((\rho_+\vec{u}_+ \cdot \nabla)\phi, \vec{u}_+)_{G_+} ds \\
&- \int_0^t ((\rho_-\vec{u}_- \cdot \nabla)\phi, \vec{u}_-)_{G_-} ds = (\rho_0^+\vec{u}_0, \phi)_{G_+(0)} + (\rho_0^-\vec{u}_0, \phi)_{G_-(0)} \\
&+ \int_0^t \{(\vec{f}_+(\cdot,\vec{u}_+),\phi)_{G_+} + (\vec{f}_-(\cdot,\vec{u}_-),\phi)_{G_-}\} ds
\end{aligned}$$

for all $\phi \in J_0^1(G) \cap H^2(G)$.

3. The initial value problem (2)

Let $\vec{v}$ be in $L^2(0,T;J_0(G))$ and consider the initial value problem

$$\rho' + \vec{v}\cdot\nabla\rho = 0, \rho(x,t) > 0 \text{ in } G\times(0,T)$$
$$\rho(\cdot,0) = \rho_0(x) \text{ in } G. \tag{6}$$

The main result of the section is the following theorem.

Theorem 3.1. *Let $\vec{v}\in L^2(0,T;J_0(G))$ and let $\rho_0 = \rho_0^{\pm}$ in $G_0^{\pm}$ be in $L^\infty(G)$ with*

$$\rho_0^+ = \rho_0^- \text{ on } \Gamma_0; 0 < a \leq \rho_0(x) \quad \forall x\in G$$

Then there exists a unique solution ρ of (6) with $a \leq \rho(x,t)$ in $G\times[0,T]$. Furthermore

$$\|\rho'\|_{L^2(0,T;H^{-1}(G))} + \|\rho\|_{L^\infty(0,T;L^\infty(G))} \leq C\|\rho_0\|_{L^\infty(G)}\{1+\|\vec{v}\|_{L^2(0,T;J_0(G))}\}$$

The constant C is independent of $\vec{v}$.

Proof. Let $\vec{v}_n$ be in $C(0,T;C_0^1(G))$ with $\nabla\cdot\vec{v}_n = 0$ and

$$\vec{v}_n \to \vec{v} \text{ in } L^2(0,T;J_0(G)).$$

The existence of a unique solution $\rho_n\in C^1(0,T;W^{1,\infty}(G))$ of the initial value problem (6) with $\vec{v}_n$ instead of $\vec{v}$ is well known. We shall now establish the estimates of the theorem.

(1) We have

$$\rho_n' + \vec{v}_n\cdot\nabla\rho_n = 0 \text{ in } G\times(0,T); \rho_n(x,0) = \rho_0 \text{ in } G.$$

Let s be a large positive integer, then ρ_n^{s-1} is in $L^\infty(0,T;W^{1,\infty}(G))$. A simple integration by parts gives

$$(\vec{v}_n\cdot\nabla(\rho_n),\rho_n^{s-1}) = -(s-1)(\vec{v}_n\cdot\nabla\rho_n,\rho_n^{s-1})(\vec{v}_n\cdot\nabla(\rho_n),\rho_n^{s-1})$$
$$= -(s-1)(\vec{v}_n\cdot\nabla\rho_n,\rho_n^{s-1})$$

and thus,

$$s^{-1}\|\rho_n(\cdot,t)\|_{L^s(G)}^{s-1}\frac{d}{dt}\|\rho_n(\cdot,t)\|_{L^s(G)} = 0$$

It follows that

$$\|\rho_n(\cdot,t)\|_{L^s(G)} = \|\rho_0\|_{L^s(G)}$$

Since ρ_n and ρ_0 are in $L^\infty(G)$, we obtain by letting $s\to\infty$

$$\|\rho_n\|_{L^\infty(0,T;L^\infty(G))} \leq \|\rho_0\|_{L^\infty(G)} \tag{7}$$

(2) It is clear that

$$\begin{aligned}\|\rho_n'\|_{L^2(0,T;H^{-1}(G))} &\leq C\|\vec{v}_n\|_{L^2(0,T;J_0(G))}\|\rho_n\|_{L^\infty(0,T;L^\infty(G))}\\ &\leq C\|\vec{v}\|_{L^2(0,T;J_0(G))}\|\rho_0\|_{L^\infty(G)} \end{aligned} \tag{8}$$

(3) We now show that $\rho_n \geq a > 0$ for all x, t in $G \times (0,T)$. Consider the one-parameter family of transformations

$$x = \xi + \int_0^t \vec{v}_n(\xi, s)ds = X_n(\xi, t)$$

of G onto G_t. Then

$$\frac{d}{dt}\rho_n + \vec{v}_n \cdot \nabla \rho_n = \frac{D}{Dt}\tilde{\rho}_n(\xi, t) = 0$$

and thus,

$$\rho_n(x,t) = \tilde{\rho}_n(X^{-1}(x,t),t) = \rho(\xi) \geq a > 0 \tag{9}$$

for $t \in [0, T_n]$ with

$$(T_n + T_n^\alpha)\|\vec{v}_n\|_{C^{2+\lambda,(1+\lambda)/2}(Q_T)} \leq \delta, 0 < \delta < 1/8$$

and $Q_T = G \times (0,T)$. By continuation, we get (3.4) for all of [0,T].

(4) Let $n \to \infty$ and we obtain by taking subsequences

$$\{\rho_n, \rho_n'\} \to \{\rho, \rho'\} \text{ in } (L^\infty(0,T; L^\infty(G)))_{weak^*} \times (L^2(0,T; H^{-1}(G)))_{weak}$$

with $a \leq \rho(x,t)$ for almost all $(x,t) \in G \times (0,T)$. The estimates of the theorem are immediate consequences of (7)-(9) and it is trivial to check that ρ is the unique solution of (6) □

4. A linear transmission problem in non cylindrical domains

Let $\vec{v}$ be in $L^2(0,T; J_0(G))$, let $\rho, X(\xi, t; \vec{v})$ be the unique solution of (3), (5) respectively. For simplicity of notations, we shall write $G_\pm(t), \Gamma(t)$ for $G_\pm(\vec{v}), \Gamma(\vec{v})$ when there is no confusion possible and the restrictions of ρ to $G_\pm(t)$ are denoted by $\rho_\pm$. We consider the transmission problem

$$\begin{aligned}\rho_\pm\{\vec{u}_\pm' + (\vec{v}\cdot\nabla)\vec{u}_\pm\} - \Delta\vec{u}_\pm + \nabla p_\pm &\\ = \vec{f}_\pm(t, \vec{v}) \text{ in } \bigcup_t (G_\pm(t) \times \{t\})&\end{aligned} \tag{10}$$

$$\nabla \cdot \vec{u}_\pm = 0 \text{ in } \bigcup_t (G_\pm(t) \times \{t\})$$

$$\vec{u}_\pm(\cdot, 0) = \vec{u}_0 \text{ in } G \;\; ; \;\; \vec{u}_- = 0 \text{ on } \partial G \times (0,T)$$

with

$$\vec{u}_- = \vec{u}_+, n \cdot \nabla(\vec{u}_+ - \vec{u}_-) = 0 \text{ on } \bigcup_t (\Gamma(t) \times \{t\}) \tag{11}$$

Set

$$\vec{u} = \vec{u}_+ \text{ in } G_+(\vec{v}), \vec{u} = \vec{u}_- \text{ in } G_-(\vec{v})$$

We now state the main result of the section.

Theorem 4.1. *Let $\{\vec{v}, \rho\}$ be in $L^2(0,T; J_0(G)) \times L^\infty(0,T; L^\infty(G))$ with $\rho \geq a > 0$ a.e. in $G \times (0,T)$. Suppose that Assumption 2.1 is satisfied, then there exists a unique weak solution $\vec{u} = \{\vec{u}_+, \vec{u}_-\}$ of (10)-(11) with*

$$\|\vec{u}\|_{L^\infty(0,t;L^2(G(\vec{v})))} + \|\vec{u}\|_{L^2(0,t;J^1(G(\vec{v})))} \leq C\|\vec{u}_0\|_{L^2(G)}\{1 + \|\vec{v}\|_{L^2(0,t;J_0(G))}\}$$

Moreover

$$\|(\rho\vec{u})'\|_{L^2(0,T;(J_0^1(G)\cap H^2(G))^*)} \leq C\{1 + \|\vec{v}\|_{L^2(0,T;J_0(G))}\}$$

The constant C is independent of $\vec{v}, \rho$.

We shall use the penalty method and introduce a transmission problem for cylindrical domains. Let G^+, be a bounded open subset of R^2 with

$$\bigcup_t (G_+(\vec{v}) \times \{t\}) \subset G^+ \times (0,T)$$

and let M_1 be a $L^\infty(0,T; L^\infty(G))$-function with

$$M_1 = 0 \text{ on } \bigcup (G_+(\vec{v}) \times \{t\}); M_1 = 1 \text{ otherwise}$$

Let $M_2, \mathcal{M}$ be $L^\infty(0,T; L^\infty(G))$-function with

$$M_2 = 0 \text{ on } \Gamma = \bigcup_t \Gamma(\vec{v}); M_2 = 1 \text{ otherwise}$$

and

$$\mathcal{M} = 0 \text{ on } \bigcup_t (G_-(\vec{v}) \times \{t\}); \mathcal{M} = 1 \text{ otherwise}$$

Set

$$\tilde{\vec{u}} = \begin{cases} \vec{u}_1(x,t) \text{ if } x \in G^+ \\ \vec{u}_2(x,t) \text{ if } x \in G^- \end{cases}$$

Consider the transmission problem

$$\begin{aligned}(\rho\tilde{\vec{u}}', \vec{\phi}) &+ (\nabla\tilde{\vec{u}}, \nabla\vec{\phi}) + (\rho(\vec{v}\cdot\nabla\tilde{\vec{u}}), \vec{\phi}) \\ &+ \varepsilon^{-1}(M_1\tilde{\vec{u}}, \vec{\phi}) + \varepsilon^{-1}(M_2\rho \mid \vec{u}_1 - \vec{u}_2 \mid \vec{u}_1, \vec{\phi}) \\ &+ \varepsilon^{-1}(\mathcal{M}\vec{u}_2, \vec{\phi}) = (\vec{f}(\cdot,\vec{v}), \vec{\phi})\end{aligned} \tag{12}$$

with

$$\begin{aligned} \vec{u}_1 = \vec{u}_2, \quad n \cdot (\nabla \vec{u}_1 - \nabla \vec{u}_2) = 0 \text{ on } \partial G^+ \times (0,T) \\ \vec{u}_2 = 0 \text{ on } \partial G \times (0,T) \\ \{\vec{u}_1, \vec{u}_2\} \mid_{t=0} = \{\vec{u}_{1,0}, \vec{u}_{2,0}\}, \quad \forall \vec{\phi} \in L^2(0,T; J_0^1(G)) \end{aligned} \tag{13}$$

Theorem 4.2. *Suppose all the hypotheses of Theorem 4.1 are satisfied and suppose further that* $\vec{v} \in L^\infty(0,T; J_0^3(G))$. *Then there exists a weak solution* $\vec{u}^\varepsilon = \{\vec{u}_1^\varepsilon, \vec{u}_2^\varepsilon\}$ *of (12)-(13) with*

$$\|\vec{u}^\varepsilon\|_{L^\infty(0,t;J_0(G))} + \|\vec{u}^\varepsilon\|_{L^2(0,t;J^1(G))} \le C\|\vec{u}_0\|_{L^2(G)}\{1 + \|\vec{v}\|_{L^2(0,t;J_0(G))}\}$$

with

$$\begin{aligned} \|M_1\vec{u}_1^\varepsilon\|^2_{L^2(0,T;L^2(G^+))} + \|M_2\{\rho \mid \vec{u}_1^\varepsilon - \vec{u}_2^\varepsilon \mid\}^{1/2}\vec{u}_1^\varepsilon\|^2_{L^2(0,T;L^2(G^+))} \\ + \|\mathcal{M}\vec{u}_2^\varepsilon\|^2_{L^2(0,T;L^2(G^-))} \le C\varepsilon \end{aligned}$$

Furthermore

$$\|(\rho\vec{u}^\varepsilon)'\|_{L^2(0,T;J_0^1(G)\cap H^2(G))^*)} \le C\{1 + \|\vec{v}\|_{L^2(0,T;J_0(G))}\}$$

The generic constant C *is independent of* $\varepsilon, \vec{v}$.

Let V be the Hilbert space

$$V = \left\{\{\vec{h}, \vec{k}\} : \{\vec{h}, \vec{k}\} \in J^1(G^+) \times J^1(G^-), \vec{k} = 0 \text{ on } \partial G, \vec{h} = \vec{k} \text{ on } \Gamma\right\}$$

with the obvious norm and inner product. Let $\vec{\phi}_j = \{\vec{\varphi}_j, \vec{\psi}_j\}$ be a basis of V and set

$$\vec{u}_n = \{\vec{u}_{1,n}, \vec{u}_{2,n}\} = \sum_{j=1}^n \left\{a_{j,n}\vec{\varphi}_j, a_{j,n}\vec{\psi}_j\right\}$$

with $a_{j,n}(0) = \alpha_{j,n}$ and

$$\sum_{j=1}^n \{\alpha_{j,n}\vec{\varphi}_j, \alpha_{j,n}\vec{\psi}_j\} \to \{\vec{u}_{1,0}, \vec{u}_{2,0}\} \text{ in } L^2(G^+) \times L^2(G^-).$$

Let $\vec{\phi}_j$ be the eigenfunctions of

$$-\Delta\vec{\phi}_j + \mathrm{grad} p_j = \lambda_j\vec{\phi}_j \text{ in } G; \vec{\phi}_j = 0 \text{ on } \partial G.$$

We shall take $\vec{\varphi}_j = \vec{\phi}_j \mid_{G+}; \vec{\psi}_j = \vec{\phi}_j \mid_{G^-}$.

Lemma 4.1. *Suppose all the hypotheses of Theorem 4.2 are satisfied. There exists an approximate solution* $\vec{u}_n = \{\vec{u}_{1,n}, \vec{u}_{2,n}\}$ *of (12)-(13) with*

$$\|\vec{u}_n\|_{L^\infty(0,t;J_0(G))} + \|\vec{u}_n\|_{L^2(0,t;J^1(G))} \le C\|\vec{u}_0\|_{L^2(G)}\{1 + \|\vec{v}\|_{L^2(0,t;J_0(G))}\}$$

Moreover

$$\|M_1\vec{u}_{1,n}\|^2_{L^2(0,T;L^2(G^+))} + \|M_2(|\,\vec{u}_{1,n} - \vec{u}_{2,n}\,|^{1/2}\,\vec{u}_{1,n})\|^2_{L^2(0,T;L^2(G^+))} + \|\mathcal{M}\vec{u}_{2,n}\|^2_{L^2(0,T;L^2(G^-))} \le C\varepsilon$$

Furthermore

$$\|(\rho\vec{u}_n)'\|_{L^2(0,T;(J^1_0(G)\cap H^2(G))^*)} \le C(1 + \|\vec{v}\|_{L^2(0,T;J_0(G))})$$

The constant C is independent of $\varepsilon, n, \vec{v}$.

Proof. The existence of a local (in time) solution of the system is obtained by well known arguments. We now show that it is a global solution and establish the estimates of the lemma.

- We have

$$\frac{1}{2}\int_G \rho\frac{d}{dt}\,|\,\vec{u}_n(\cdot,t)\,|^2\,dx + \|\nabla\vec{u}_n\|^2_{L^2(G)} + (\rho\{\vec{v}\cdot\nabla\vec{u}_n\},\vec{u}_n)$$
$$+\varepsilon^{-1}\Big\{\|M_1\vec{u}_{1,n}\|^2_{L^2(G^+)} + \|\mathcal{M}\vec{u}_{2,n}\|^2_{L^2(G^-)} + \|M_2(\rho\,|\,\vec{u}_{1,n} - \vec{u}_{2,n})\,|)^{1/2}\vec{u}_{1,n}\|^2_{L^2(G^+)}\Big\}$$
$$= (\vec{f}(t,\vec{v}),\vec{u}_n) \tag{14}$$

From Theorem 3.1 we get

$$(\rho',\varphi) - (\vec{v}\cdot\nabla\varphi,\rho) = 0 \quad \forall\varphi\in L^2(0,T;W^{1,1}(G))$$

as $\vec{v}$ is in $L^\infty(0,T;J^2_0(G))$ and G is a bounded subset of R^2. Since $\vec{u}_n\in L^\infty(0,T;L^2(G))$ and $\nabla\vec{u}_n\in L^2(0,T;L^2(G))$ and with G being a bounded open subset of R^3, we deduce that

$$\vec{u}^2_{j,n}\in L^2(0,T;W^{1,1}(G)).$$

With $\vec{u}_n$ being a solution of 12)-(13), we have

$$\vec{u}_{1,n} = \vec{u}_{2,n} \text{ on } \partial G^+\times(0,T)$$

and hence $\vec{u}^2_n\in L^2(0,T;W^{1,1}(G))$. Applying Theorem 3.1 and we get

$$(\vec{u}^2_n,\rho') - 2(\rho(\vec{v}\cdot\nabla)\vec{u}_n,\vec{u}_n) = 0. \tag{15}$$

Since $\rho>0$, we obtain by adding (14)-(15)

$$\frac{d}{dt}\|\sqrt{\rho}\vec{u}_n(\cdot,t)\|^2_{L^2(G)} + \|\nabla\vec{u}_n\|^2_{L^2(G)} + \varepsilon^{-1}\|M_1\vec{u}_{1,n}\|^2_{L^2(G^+)}$$
$$+\,\varepsilon^{-1}\{M_2(\rho\,|\,\vec{u}_{1,n} - \vec{u}_{2,n}\,|)^{1/2}\vec{u}_{1,n}\|^2_{L^2(G^+)} \tag{16}$$
$$+\,\|\mathcal{M}\vec{u}_{2,n}\|^2_{L^2(G^-)}\}$$
$$\le Ca^{-1/2}\{1 + \|\vec{v}\|_{L^2(G)} + \|\rho\|_{L^2(G)}\}\|\sqrt{\rho}\vec{u}_n\|_{L^2(G)}$$

Since $\vec{u}_{1,} = \vec{u}_{2,n}$ on $\partial G^+ \times (0,T)$ we obtain by integrating from 0 to t

$$\begin{aligned}
&\|\sqrt{\rho}\vec{u}_n\|^2_{L^\infty(0,t;L^2(G))} + \|\vec{u}_n\|^2_{L^2(0,t;J^1(G))} \\
&+\varepsilon^{-1}\{\|M_1\vec{u}_{1,n}\|^2_{L^2(0,t;L^2(G^+))} + \|\mathcal{M}\vec{u}_{2,n}\|^2_{L^2(0,t;L^2(G^-))} \\
&+\|M_2(\rho \mid \vec{u}_{1,n} - \vec{u}_{2,n} \mid)^{1/2}\vec{u}_{1,n}\|_{L^2(0,t;L^2(G^+))}\} \qquad (17)\\
\le\; & C\{\|\rho_0\|_{L^\infty(G)} + \|\vec{v}\|^2_{L^2(0,t;J_0(G))})\}
\end{aligned}$$

- We now establish the estimates on the time-derivative of $\vec{u}_n$. We have

$$\begin{aligned}
&\frac{d}{dt}(\rho\vec{u}_n) - \Delta\vec{u}_n + (\rho\vec{v}\cdot\nabla)\vec{u}_n + (\vec{v}\cdot\nabla\rho)\vec{u}_n + \text{grad}p \\
&\quad + \varepsilon^{-1}\{M_1\vec{u}_{1,n} + M_2(\rho \mid \vec{u}_{1,n} - \vec{u}_{2,n} \mid)\vec{u}_{1,n} + \mathcal{M}\vec{u}_{2,n}\} = \vec{f}(t,\vec{v})
\end{aligned}$$

It follows that

$$\begin{aligned}
&\|(\rho\vec{u}_n)'\|_{L^2(0,T;(J^1(G)\cap H^2(G))^*)} \\
&\le C\|\vec{u}_n\|_{L^2(0,T;J^1(G))}\{1 + \|\rho\|_{L^\infty(0,T;L^\infty(G))}\|\vec{v}\|_{L^2(0,T;J_0(G))}\} \\
&+ C\varepsilon^{-1}\left\{\|M_1\vec{u}_{1,n}\|_{L^2(0,T;L^2(G^+))}\right. \\
&+ \|(\rho \mid \vec{u}_{1,n} - \vec{u}_{2,n} \mid)^{1/2}\vec{u}_{1,n}\|_{L^2(0,T;L^2(G^-))} \qquad (18)\\
&+ \left.\|\mathcal{M}\vec{u}_{2,n}\|_{L^2(0,T;L^2(G^-))}\right\}
\end{aligned}$$

Thus,

$$\|(\rho\vec{u}_n)'\|_{L^2(0,T;(J_0^1(G)\cap H^2(G))^*)} \le C\{1 + \|\vec{v}\|_{L^2(0,T;J_0(G))}\}$$

We have used the estimates (16) in the above calculations and the lemma is proved □

Proof of Theorem 4.2. Let $\vec{u}_n^\varepsilon$ be as in Lemma 4.1, then we obtain by taking subsequences if necessary

$$\{\vec{u}_n^\varepsilon, \rho\vec{u}_n^\varepsilon\} \to \{\vec{u}^\varepsilon, \rho\vec{u}^\varepsilon\}$$

in

$$(L^2(0,T;J^1(G)))_{weak} \cap (L^\infty(0,T;L^2(G)))_{weak^*} \times (L^\infty(0,T;L^2(G)))_{weak^*}$$

The estimates of the theorem are an immediate consequence of those of the lemma and it is trivial to check that $\vec{u}^\varepsilon$ is a solution of (12)-(13) □

We shall now remove the extra hypothesis on $\vec{v}$.

Lemma 4.2. *Suppose all the hypotheses of Theorem 4.1 are satisfied. Then there exists a weak solution* $\vec{u}^\varepsilon = \{\vec{u}_1^\varepsilon, \vec{u}_2^\varepsilon\}$ *of (12)-(13). Moreover*

$$\|\vec{u}^{\varepsilon}\|_{L^\infty(0,t;L^2(G))} + \|\vec{u}^{\varepsilon}\|_{L^2(0,t;J^1(G))} \leq C\|\vec{u}_0\|_{L^2(G)}\{1 + \|\vec{v}\|_{L^2(0,t;J_0(G))}\}$$

and

$$\|M_1\vec{u}_1^{\varepsilon}\|_{L^2(0,T;L^2(G))} + \|M_2\{\rho \mid \vec{u}_1^{\varepsilon} - \vec{u}_2^{\varepsilon} \mid\}^{1/2}\vec{u}_1^{\varepsilon}\|_{L^2(0,T;L^2(G))} + \|\mathcal{M}\vec{u}_2^{\varepsilon}\|_{L^2(0,T;L^2(G))} \leq C\varepsilon$$

Furthermore

$$\|(\rho\vec{u}^{\varepsilon})'\|_{L^2(0,T;(J_0^1(G)\cap H^2(G))*)} \leq C\{1 + \|\vec{v}\|_{L^2(0,T;J_0(G))}\}$$

The different constants are independent of ε.

Proof. Let $\vec{v}$ be in $L^2(0,T;J_0(G))$,then there exists $\vec{v}_n$ in $C(0,T;C_0^\infty(G))$ with $\nabla\cdot\vec{v}_n = 0$ such that

$$\vec{v}_n \to \vec{v} \text{ in } L^2(0,T;J_0(G)).$$

With $\vec{v}_n$, we have a weak solution $\vec{u}_n^{\varepsilon}$ of (12)-(13), given by Theorem 4.2. Since the estimates of Theorem 4.2 depend only on the $L^2(0,T;J_0(G))$-norm of $\vec{v}_n$, an argument as in the proof of Theorem 4.2 gives the stated result. □

Proof of Theorem 4.1.

(1) Let $\vec{u}^{\varepsilon}$ as in Theorem 4.2, then we have, by taking subsequences

$$\{\vec{u}^{\varepsilon}, M_1\vec{u}_1^{\varepsilon}, M_2(\rho \mid \vec{u}_1^{\varepsilon} - \vec{u}_2^{\varepsilon} \mid)^{1/2}\vec{u}_1^{\varepsilon}, \mathcal{M}\vec{u}_2^{\varepsilon}\} \to \{\vec{u}, 0, 0, 0\}$$

in

$$(L^2(0,T;J^1(G)))_{weak} \cap (L^\infty(0,T;L^2(G)))_{weak^*} \times\{L^2(0,T;L^2(G^+)))\}^2 \times L^2(0,T;L^2(G^-))).$$

(2) It is clear that $\{M_1\vec{u}_1^{\varepsilon}, \mathcal{M}\vec{u}_2^{\varepsilon}\} \to \{M_1\vec{u}_1, \mathcal{M}\vec{u}_2\}$ in the weak topology of

$$L^2(0,T;H^1(G^+))) \times L^2(0,T;H^1(G^-))).$$

Therefore

$$M_1\vec{u}_1 = 0 \ i.e.\ \vec{u}_1 = 0 \text{ in } \{G^+ \times (0,T)\}/\overline{\bigcup_t (G_+(\vec{v}) \times \{t\})}$$

and

$$\mathcal{M}\vec{u}_2 = 0 \ i.e.\ \vec{u}_2 = 0 \text{ in } G \times (0,T)/\overline{\cup_t G^-(\vec{v}) \times \{t\})}$$

(3) We now show that $M_2(\rho \mid \vec{u}_1 - \vec{u}_2) \mid)^{1/2}\vec{u}_1 = 0$ *i.e.*

$$\vec{u}_1 = \vec{u}_2 \text{ on } \bigcup_t (\Gamma_t \times \{t\}).$$

An application of Aubin's theorem gives

$$\rho\vec{u}^\varepsilon \to \rho\vec{u} \text{ in } L^2(0,T;H^{-1}(G))) \cap (L^2(0,T;L^4(G)))_{weak}$$

Since

$$\vec{u}_1^\varepsilon \to \vec{u}_1 \text{ in } (L^2(0,T;H^1(G^+)))_{weak}$$

and since

$$\rho \mid \vec{u}_1^\varepsilon - \vec{u}_2^\varepsilon \mid \to \rho \mid \vec{u}_1 - \vec{u}_2 \mid \text{in} L^2(0,T;H^{-1}(G)) \cap (L^2(0,T;L^2(G^+)))_{weak}$$

it follows from the compensated compactness theorem of Murat that

$$\rho \mid \vec{u}_1^\varepsilon - \vec{u}_2^\varepsilon \mid \vec{u}_1^\varepsilon \to \rho \mid \vec{u}_1 - \vec{u}_2 \mid \vec{u}_1$$

in the distribution sense in $G \times (0,T)$ and therefore

$$M_2(\rho \mid \vec{u}_1 - \vec{u}_2 \mid \vec{u}_1)(x,t) = 0$$

Thus,

$$\vec{u}_1(x,t) = \vec{u}_2(x,t) \text{ on } \bigcup_t (\Gamma_t(\vec{v}) \times \{t\})$$

(4) Suppose that $\vec{u}, \hat{\vec{u}}$ are two solutions of (10)-(11). Then a calculation as in Lemma 4.1 gives

$$\frac{d}{dt}\|\sqrt{\rho}(\vec{u} - \hat{\vec{u}})\|^2_{L^2(G)} + \|\nabla(\vec{u} - \hat{\vec{u}})\|^2_{L^2(G)} = 0$$

and hence $\vec{u} = \hat{\vec{u}}$. All the other assertions of the theorem are trivial to prove. □

5. The nonlinear case

The main result of the paper is the following theorem.

Theorem 5.1. *Let* $\{\vec{u}_0, \rho_0\}$ *be in* $J_0(G) \times L^\infty(G)$ *with* $\rho_0^+ = \rho_0^-$ *on* Γ_0 *and let* $\rho_0 = \rho_0^\pm$ *in* $G_0^\pm$ *with*

$$\rho_0^+ = \rho_0^- \, on \Gamma_0, \rho_0 \in L^\infty(G); 0 < a \le \rho_0(x), \quad \forall x \in G.$$

Then there exists $\{\vec{u}, \rho\} \in L^2(0,T;J_0^1(G)) \times L^\infty(0,T;L^\infty(G))$, *weak solution of (1)-(5) in the sense of Definition 2.2.*

Let

$$\mathcal{B} = \{\vec{v} : \|\vec{v}(\cdot,t)\|_{J_0(G))} \leq C(a)\exp(ct) \quad \forall t \in [0,T]$$
$$\|\vec{v}\|_{L(0,T;J_0^1(G))} \leq C(a)\exp(cT)\}$$

and let $\mathcal{A}$ be the mapping of $\mathcal{B}$, considered as a bounded convex subset of $L^2(0,T;J^{-1}(G))$, into $L^2(0,T;J^{-1}(G))$ defined by

$$\mathcal{A}(\vec{v}) = \vec{u} \tag{19}$$

where $\vec{u}$ is the unique solution of (10)-(11) given by Theorem 4.1 and $\rho(v)$ is the unique solution of the initial value problem

$$\rho' + \vec{v} \cdot \nabla\rho = 0, \rho(\cdot,0) = \rho_0 \geq a \text{ in } G.$$

We now show that $\mathcal{A}$ satisfies the hypotheses of Schauder's theorem and thus, has a fixed point.

Lemma 5.1. *Suppose all the hypotheses of Theorem 4.1 are satisfied and let $\mathcal{A}$ be as in (19). Then $\mathcal{A}$ maps $\mathcal{B}$ into $\mathcal{B}$.*

Proof. With $\vec{v} \in \mathcal{B}$ it is trivial to check that $\mathcal{A}(\vec{u}) \in \mathcal{B}$. □

Lemma 5.2. *Suppose all the hypotheses of Theorem 4.1 are satisfied, then $\mathcal{A}$ is a completely continuous mapping of $\mathcal{B}$ into $L^2(0,T;J^{-1}(G))$.*

Proof.

(1) Let $\vec{v}_n \in \mathcal{B}$ and let ρ_n be the solution of (6) and let $\vec{u}_n$ be the unique solution of (10)-(11) given by Theorem 4.1. With the estimates of Theorem 3.1, we have by taking subsequences $\{\rho_n, \rho_n'\} \to \{\rho, \rho'\}$ in

$$\{(L^\infty(0,T;L^\infty(G)))_{weak^*} \cap L^2(0,T;H^{-1}(G))\} \times (L^2(0,T;H^{-1}(G)))_{weak}$$

Since

$$\vec{v}_n \to \vec{v} \text{ in } (L^2(0,T;J_0^1(G)))_{weak}$$

it follows from the compensated compactness theorem of Murat that

$$\rho_n\vec{v}_n \to \rho\vec{v} \text{ in } \mathcal{D}'(G \times (0,T))$$

It is now trivial to check that ρ is the unique solution of

$$\rho' + \vec{v} \cdot \nabla\rho = 0, \rho(\cdot,0) = \rho_0.$$

(2) From the estimates of Theorem 4.1, we have

$$\|\vec{u}_n\|_{L^\infty(0,T;J_0(G))} + \|\vec{u}_n\|_{L^2(0,T;J^1(G))} \le C$$

Since $\vec{v}_n \to \vec{v}$ weakly in $L^2(0,T;J_0(G))$, it follows from Lemma 2.1 that

$$G^+(\vec{v}) = \bigcap_{n\ge n_0} G^+(\vec{v}_n); G^-(\vec{v}) = \bigcap_{n\ge n_0} G^-(\vec{v}_n)$$

and hence

$$\|\vec{u}_{1,n}\|_{L^2(0,T;H^1(G^+(\vec{v})))} \le \|\vec{u}_{1,n}\|_{L^2(0,T;H^1(G^+(\vec{v}_n))}$$

Therefore there exists a subsequence such that

$$\vec{u}_{1,n} \to \vec{u}_+ \text{ weakly in } L^2(0,T;H^1(G^+(\vec{v}))); \nabla\cdot\vec{u}_+ = 0 \text{ in } G^+(\vec{v})$$

Similarly for $\vec{u}_{2,n}$ with $\vec{u}_- = 0$ on ∂G and

$$\vec{u}_n \to \vec{u} \text{ in } (L^\infty(0,T;L^2(G)))_{weak^*}.$$

Again with the estimates of Theorem 4.1 we have

$$\|(\rho_n\vec{u}_n)'\|_{L^2(0,T;(J_0^1(G)\cap H^2(G))^*)} \le C$$

An application of Aubin's theorem gives

$$\rho_n\vec{u}_n \to \vec{\chi} \text{ in } L^2(0,T;J^{-1}(G)) \cap (L^2(0,T;L^2(G)))_{weak}$$

and as in the first part $\vec{\chi} = \rho\vec{u}$ in the distribution sense in $G\times(0,T)$ by the compensated compactness theorem.

(3) We now show that

$$\vec{u}_n \to \vec{u} \text{ in } L^2(0,T;J^{-1}(G)).$$

We have

$$\|\rho_n(\vec{u}_n - \vec{u})\|_{L^2(0,T;J^{-1}(G))} \le \|\rho_n\vec{u}_n - \rho\vec{u}\|_{L^2(0,T;J^{-1}(G))} + \|(\rho_n-\rho)\vec{u}\|_{L^2(0,T;J^{-1}(G))} \tag{20}$$

Let $\varphi \in L^2(0,T;J_0^2(G))$, then

$$\left|\int_0^T((\rho_n-\rho)\vec{u},\varphi)dt\right| \le \left|\int_0^T(\rho_n-\rho,\vec{u}\varphi)dt\right|$$
$$\le \|\rho_n-\rho\|_{L^\infty(0,T;H^{-1}(G))}\{\|\vec{u}\|_{L^2(0,T;L^4(G))}\|\varphi\|_{L^2(0,T;H_0^2(G))}$$
$$+\|\vec{u}\|_{L^2(0,T;H^1(G))}\|\varphi\|_{L^2(0,T;L^\infty(G))}\}$$
$$\le C\|\rho_n-\rho\|_{L^\infty(0,T;H^{-1}(G))}\|\vec{u}\|_{L^2(0,T;H^1(G))}\|\varphi\|_{L^2(0,T;H_0^2(G))}$$

Hence

$$\|(\rho_n-\rho)\vec{u}\|_{L^2(0,T;J^{-2}(G))} \le C\|\vec{u}\|_{L^2(0,T;H^1(G))}\|\rho_n-\rho\|_{L^\infty(0,T;H^{-1}(G))}$$

Since $(\rho_n-\rho)\vec{u}$ is uniformly bounded in the $L^\infty(0,T;L^2(G))$-norm, we have

$$\begin{aligned}\|(\rho_n(\cdot,t)-\rho(\cdot,t))\vec{u}(\cdot,t)\|_{J^{-1}(G)} &\le \varepsilon\|\rho_n(\cdot,t)-\rho(\cdot,t))\vec{u}(\cdot,t)\|_{J_0(G)}\\ &\quad+K_\varepsilon\|(\rho_n(\cdot,t)-\rho(\cdot,t))\vec{u}(\cdot,t)\|_{J^{-2}(G)}\end{aligned}$$

for almost all $t\in[0,T]$. Thus,

$$\begin{aligned}&\|(\rho_n-\rho)\vec{u}\|_{L^2(0,T;J^{-1}(G))}\\ &\quad\le \varepsilon\|(\rho_n-\rho)\vec{u}\|_{L^2(0,T;J_0(G))}+K_\varepsilon\|(\rho_n-\rho)\vec{u}\|_{L^2(0,T;J^{-2}(G))}\\ &\quad\le C\varepsilon+K_\varepsilon\|\vec{u}\|_{L^2(0,T;J^1(G))}\|\rho_n-\rho\|_{L^\infty(0,T;H^{-1}(G))}\end{aligned}$$

we get

$$(\rho_n-\rho)\vec{u}\to 0 \text{ in } L^2(0,T;J^{-1}(G)). \tag{21}$$

It follows from (20) and (21) that

$$\rho_n(\vec{u}_n-\vec{u})\to 0 \text{ in } L^2(0,T;J^{-1}(G)). \tag{22}$$

Let φ be in $L^2(0,T;J_0^1(G))$, then

$$\begin{aligned}\left|\int_0^T(\vec{u}_n-\vec{u},\varphi)dt\right| &\le \left|\int_0^T((\vec{u}_n-\vec{u})\rho_n,\rho_n^{-1}\varphi)dt\right|\\ &\le a^{-1}\int_0^T\int_G|(\vec{u}_n-\vec{u})\rho_n|\,|\varphi|\,dxdt\\ &\le a^{-1}\|\rho_n(\vec{u}_n-\vec{u})\|_{L^2(0,T;J^{-1}(G))}\|\varphi\|_{L^2(0,T;J_0^1(G))}\end{aligned}$$

Hence

$$\|\vec{u}_n-\vec{u}\|_{L^2(0,T;J^{-1}(G))}\le a^{-1}\|\rho_n(\vec{u}_n-\vec{u})\|_{L^2(0,T;J^{-1}(G))}$$

It follows from (22) that

$$\vec{u}_n\to\vec{u} \text{ in } L^2(0,T;J^{-1}(G)).$$

(4) We have

$$|((\rho_n\vec{u}_n)(\cdot,t)-(\rho\vec{u})(\cdot,t),\phi)|\le\|\rho_n\vec{u}_n-\rho\vec{u}\|_{C(0,T;J^{-1}(G))}\|\phi\|_{J_0^1(G)}$$

Hence

$$\begin{aligned}\lim((\rho_n\vec{u}_n(\cdot,t),\phi) &= ((\rho\vec{u})(\cdot,t),\phi)\\ &= ((\rho_+\vec{u}_+)(\cdot,t),\phi)_{G^+(\vec{v})}+((\rho_-\vec{u}_-)(\cdot,t),\phi)_{G^-(\vec{v})}\end{aligned}$$

for all $t \in [0, T]$ and for all $\phi \in J_0^1(G) \cap H^2(G)$. We have

$$\begin{aligned}
&\int_0^T \{(\nabla \vec{u}_{1,n}, \nabla\phi)_{G^+(\vec{v}_n)} - (\nabla \vec{u}_+, \nabla\phi)_{G^+(\vec{v})}\}dt \\
&= \int_0^T (\nabla \vec{u}_{1,n}, \nabla\phi)_{G^+(\vec{v}_n)/G^+(\vec{v})}dt + \int_0^T (\nabla\{\vec{u}_{1,n} - \vec{u}_+\}, \nabla\phi)_{G^+(\vec{v})}dt
\end{aligned}$$

with ϕ as before. Thus,

$$\begin{aligned}
&\left|\int_0^T (\nabla \vec{u}_{1,n}, \nabla\phi)_{G^+(\vec{v}_n)/G^+(\vec{v})}dt\right| \\
&\leq C\|\vec{u}_{1,n}\|_{L^2(0,T;H^1(G^+)}\text{meas}(G^+(\vec{v}_n)/G^+(\vec{v})) \\
&\leq C_1\text{meas}(G^+(\vec{v}_n)/G^+(\vec{v})).
\end{aligned}$$

for all $\phi \in C(0, T; C^1(G))$. Hence

$$\lim \int_0^T (\nabla \vec{u}_{1,n}, \nabla\phi)_{G^+(\vec{v}_n)}dt = \int_0^T (\nabla \vec{u}_+, \nabla\phi)_{G^+(\vec{v})}dt$$

Similarly for the term involving $\vec{u}_{2,n}$. We have

$$\begin{aligned}
&\lim \int_0^T ((\rho_n\vec{v}_n \cdot \nabla)\phi, \vec{u}_n)dt \\
&= \lim \int_0^T (\{(\rho_n\vec{v}_n - \rho\vec{v}) \cdot \nabla\}\phi, \vec{u}_n)dt + \lim \int_0^T ((\rho\vec{v} \cdot \nabla)\phi, \vec{u}_n)dt \\
&= \int_0^T ((\rho\vec{v} \cdot \nabla)\phi, \vec{u})dt \\
&= \int_0^T \{((\rho_+\vec{v}_+ \cdot \nabla)\phi, \vec{u}_+)_{G^+(\vec{v})} + ((\rho_-\vec{v}_- \cdot \nabla)\phi, \vec{u}_-)_{G^-(\vec{v})}\}dt
\end{aligned}$$

as

$$\begin{aligned}
&\left|\int_0^T ((\rho_n\vec{v}_n - \rho\vec{v}) \cdot \nabla\phi, \vec{u}_n)dt\right| \\
&\leq C\int_0^T \int_G |(\rho_n\vec{v}_n - \rho\vec{v})\vec{u}_n|\, dxdt \\
&\leq C\|\rho_n\vec{v}_n - \rho\vec{v}\|_{L^2(0,T;J^{-1}(G))}\|\vec{u}_n\|_{L^2(0,T;J_0^1(G))} \\
&\leq C_1\|\rho_n\vec{v}_n - \rho\vec{v}\|_{L^2(0,T;J^{-1}(G))}
\end{aligned}$$

With f as in Assumption 2.1 and with

$$\vec{u}_n \to \vec{u} \text{ in } L^2(0, T; J^{-1}(G)) \cap (L^2(0, T; J_0(G)))_{weak}$$

it follows from Lemma 2.2 that

$$\lim \int_0^T (\vec{f}(\cdot, \vec{u}_n), \phi)dt = \int_0^T (\vec{f}(\cdot, \vec{u}), \phi)dt$$
$$= \int_0^T \left\{ (\vec{f}_+(\cdot, \vec{u}_+), \phi)_{G^+(\vec{v})} + (\vec{f}_-(\cdot, \vec{u}_-), \phi)_{G^-(\vec{v})} \right\} dt$$

Thus, $\mathcal{A}$ is a completely continuous mapping of $\mathcal{B}$, considered as a subset of $L^2(0,T;J^{-1}(G))$ into $L^2(0,T;J^{-1}(G))$ and the lemma is proved. □

Proof of Theorem 5.1. Since $\mathcal{A}$ maps the closed, bounded convex set $\mathcal{B}$ of $L^2(0,T;H^{-1}(G))$ into a compact part of $\mathcal{B}$, it satisfies all the hypotheses of Schauder's theorem,(e.g.cf. M. Krasnoselkii,[5] p.124) and thus there exists $\vec{u} \in \mathcal{B}$ such that

$$\mathcal{A}(\vec{u}) = \vec{u}.$$ □

References

1. H. Beirao da Viega and A. Valli, *J. Math. Anal. Appl.* **71**, 338 (1980).
2. Y. Y. Belov and N.N. Yanenko, *Math. Notes*, **10**, 480 (1978).
3. B.Gustafsson and A.Sundstrom, *SIAM J.Appl.Math.* **35**, 242 (1978).
4. V. A. Kajikov, *Dokl.Akad.Nauk USSR*, **216**, 1008 (1974).
5. M. A. Krasnoselkii, *International Series of Monographs in Pure and Applied Mathematics*, (Pergamon Press Book, The Macmillan Company, New York 1964).
6. J. L. Lions, *On some problems connected with the Navier Stokes equations*, in *Nonlinear Evolution Equations*, Ed., M.Crandall, (Academic Press, New York, 1978) pp. 59–84.
7. Bui An Ton, *Abstract and Applied Analysis*, **6**, 619 (2005).
8. Bui An Ton, *Nonlinear Analysis, Theory, Methods and Applications* (in press)
9. Bui An Ton, *On a free boundary problem for a stochastic hybrid nonlinear hyperbolic parabolic system* (Submitted for publication).

Harmonic, Wavelet and p-Adic Analysis
Eds. N. M. Chuong *et al.* (pp. 175–209)

§9. SAMPLING IN PALEY-WIENER AND HARDY SPACES

VU KIM TUAN† and AMIN BOUMENIR‡

Department of Mathematics
University of West Georgia
Carrollton, GA 30118
†*E-mail: vu@westga.edu*
‡*E-mail: boumenir@westga.edu*

In the first part of this paper we study the problem of recovering bandlimited signals by irregular sampling, that is from their values at irregularly distributed points. We obtain a sufficient condition for a sampling sequence to fit any bandlimited signal with an arbitrary bandwidth. The sampling functions are made of eigensolutions of a singular Sturm-Liouville operator whose spectrum is the set of sample points. The main tool is the Gelfand-Levitan theory which is used to produce such an operator. The second part deals with sampling in Hardy spaces. There we prove a new sampling formula and show that Shannon type sampling formulae cannot hold for functions in the Hardy space $\mathcal{H}^2_+$. As an example, we work out a new series representation of the Riemann zeta function in the half-plane $\Re(s) > \frac{1}{2}$. Upper bounds for the truncation and amplitude errors for the sampling formulae are also provided.

Keywords: Shannon Sampling Theorem, Irregular Sampling, Inverse Spectral Problem, Entire Function, Interpolation, Hardy Space, Paley-Wiener Space
1991 *Mathematics Subject Classification.* 4A20, 41A05.

1. Sampling in Paley-Wiener Spaces

In view of the rapidly growing area of applications of sampling theory in digital signal processing, electrical engineering, medical imaging and data processing we would like to present a brief summary of results, old and new, and also describe the various connections between the areas of irregular sampling, function theory, differential equations, and inverse spectral theory.

Let $F \in L_2(\mathbb{R})$ and f be its Fourier transform

$$f(\lambda) = \frac{1}{\sqrt{2\pi}} \int_{-\infty}^{\infty} F(t) e^{i\lambda t}\, dt.$$

In signal processing F is called a finite-energy signal while f is its frequency content. If f vanishes outside an interval $(-w, w)$ then F is said to be a bandlimited signal and its bandwidth T is the smallest possible w. Most of the sounds, such as human voice and music, are bandlimited signals. A bandlimited signal $F \in L_2(\mathbb{R})$ with bandwidth T can also be described by the Paley-Wiener theorem as the restriction to the real line of an entire function of order at most one and of type at most T, see Ref. 2. An entire function $F(t)$ is said of order at most ρ and type at most T, if and only if, for every positive, but no negative ϵ,

$$|F(t)| = O\left(e^{(T+\epsilon)|t|^\rho}\right), \quad t \in \mathbb{C}.$$

Denote the set of all signals with bandwidth at most T by PW_T^1 and the Paley-Wiener space, the set of all bandlimited signals, by

$$PW^1 = \bigcup_{T>0} PW_T^1. \tag{1}$$

Part 1 is concerned with the very important problem of recovering bandlimited signals (Paley-Wiener functions) by irregular sampling and, as usual in any recovery problem, we need to address two issues: uniqueness and the recovery algorithm.

For uniqueness observe that bandlimited signals can be determined uniquely by their zeros in the complex plane.[40] Therefore, they can be uniquely recovered from their sampled values only if the sampling rate is not less than their density of zeros. If $n(R)$ denotes the number of zeros of F inside the disc $|t| \le R$, then for a nontrivial bandlimited signal F, the density of zeros $\frac{n(R)}{2R}$ is proportional to its bandwidth T:[40]

$$\frac{n(R)}{2R} \sim \frac{T}{\pi}. \tag{2}$$

So if we denote by $N(R)$ the number of sample points on the interval $[-R, R]$, then we can recover a signal F with bandwidth T exactly only if

$$\overline{\lim}_{R\to\infty} \frac{N(R)}{2R} \ge \frac{T}{\pi}.$$

The sampling rate $\frac{T}{\pi}$ per second is called the Nyquist sampling rate,[54] which is the minimum rate at which a signal with bandwidth T needs to be sampled for its exact reconstruction.

The problem of recovering a bandlimited signal from its values at regularly spaced points is classical. The celebrated Shannon sampling formula, a fundamental result in digital signal processing,[54]

$$F(t) = \sum_{n=-\infty}^{\infty} F\left(\frac{n\pi}{T}\right) \frac{\sin(Tt - n\pi)}{Tt - n\pi}, \qquad F \in PW_T^1, \tag{3}$$

recovers any bandlimited signal with bandwidth at most T from its equidistantly spaced $\frac{\pi}{T}$ apart samples. Formula (3) is also known as the Whittaker-Shannon-Kotelnikov (WSK) theorem.[54] Clearly, the Shannon sampling formula yields

$$\frac{N(R)}{2R} \sim \frac{T}{\pi},$$

which means the sampling rate in the Shannon sampling formula is optimal.

The sample points in the Shannon formula are uniformly spaced. Non-uniform sampling, that is sampling with samples at non-uniformly distributed points, occurs frequently in practice.[10,33,54] For examples, uniform sampling but with some missing sample points or with time jitters can be considered as simple cases of non-uniform sampling. Also when a signal varies rapidly it is more natural to sample at higher rate than when it is varying slowly.

The first result in the direction of non-uniform sampling was given by Paley and Wiener. Given a sequence of sample points $\{t_n\}_{n\in\mathbb{Z}}$, such that

$$\sup_{n\in\mathbb{Z}} \left| t_n - n\frac{\pi}{T} \right| < \frac{\pi}{4T}, \tag{4}$$

we can define the entire function $G(t) = (t-t_0)\prod_{n=1}^{\infty}\left(1-\frac{t}{t_n}\right)\left(1-\frac{t}{t_{-n}}\right)$. Then for any $F \in PW_T^1$ we have

$$F(t) = \sum_{n\in\mathbb{Z}} F(t_n) S_n(t), \tag{5}$$

where the sampling functions are

$$S_n(t) = \frac{G(t)}{G'(t_n)\,(t-t_n)}.$$

The constant $\frac{\pi}{4T}$ in (4) is known as the Kadec's constant.[24] Condition (4) cannot be relaxed due to the fact that signals with bandwidth T are Fourier transforms of functions from $L^2(-T,T)$, and only under (4), is the set $\{e^{it_n x}\}_{n\in\mathbb{Z}}$ complete in $L^2(-T,\ T)$ as shown by Levinson.[29]

A powerful tool in non-uniform sampling and a generalization of the WSK theorem is Kramer's theorem which we outline now.[25,54] Let $y(x,t)$ be a continuous function in t that belongs to $L^2(0,T)$, $0 < T < \infty$, for any fixed t and such that $\{y(x,t_n)\}_{n\in\mathbb{N}}$ is a complete orthogonal family of $L^2(0,T)$. Then any function F, which is the transform with the kernel $y(x,t)$ of a function $f \in L^2(0,T)$

$$F(t) = \int_0^T f(x)y(x,t)dx, \quad f \in L^2(0,T), \tag{6}$$

can be interpolated from its values at $\{t_n\}_{n\in\mathbb{N}}$ by

$$F(t) = \sum_{n=1}^{\infty} F(t_n) \frac{\int_0^T y(x,t)\overline{y(x,t_n)}\,dx}{\int_0^T |y(x,t_n)|^2\,dx}. \tag{7}$$

A simple application of Kramer's idea is given by a regular Sturm-Liouville (S-L) operator

$$\begin{cases} L(y) := -y''(x,t) + q(x)y(x,t) = ty(x,t), \qquad 0 \le x \le T < \infty, \\ y'(0,t) - hy(0,t) = 0, \\ y'(T,t) - Hy(T,t) = 0, \end{cases} \tag{8}$$

which gives rise to a generalized Fourier transform of the form (6). If t_n are the eigenvalues of (8), then $\{y(x,t_n)\}_{n\in\mathbb{N}}$ is a complete orthogonal family of $L^2(0,T)$. Thus, F defined by (6) with $y(x,t)$ being a solution of (8) can be recovered by formula (7).

However there are two main drawbacks to this approach: First, we do not sample bandlimited signals but only functions of the form (6), with no obvious relation to bandlimited signals. Secondly, the eigenvalues are imposed and cannot be changed once a regular Sturm-Liouville operator (8) has been chosen.

The Gelfand-Levitan inverse spectral theory[21] brings a quick remedy to both issues. In Ref. 6, Boumenir has shown that if the sampling sequence has the asymptotics

$$\sqrt{t_n} \sim \frac{\pi}{T} n \quad \text{as} \quad n \to \infty, \tag{9}$$

then with the help of the Gelfand-Levitan theory the sample points can be used to tailor a regular Sturm-Liouville problem (8), whose spectrum is precisely the given sample points. Moreover, the space of functions of the form (6) is related with the space of bandlimited signals PW_T^1 in a very obvious way. This is easily proved with the help of transformation operators.

From the practical point of view, (9) means that the sample points can be randomly distributed except at infinity, where they should be close to "regular" as specified by (9). One of the objectives of this paper is to improve on the condition at infinity (9) so to allow more general distributions of sample points.

Another important feature of our result is blind sampling which we explain now. Observe that all classical sampling formulae, such as Shannon, Paley-Wiener, and Kramer,[54] need a priori information of the bandwidth in order to set the sampling rate. Suppose now we receive a bandlimited signal from an unknown source with no information about its bandwidth. What would be the sampling rate for such a signal and more importantly how can we recover it? Since we do not know the bandwidth, we call it a blind recovery. A necessary condition for the sampling rate in case the bandwidth of the signal is unknown can be easily derived. Since it is at least $\frac{T}{\pi}$ for a signal with bandwidth T, which in our case is arbitrary, the sampling rate must satisfy

$$\overline{\lim}_{R\to\infty} \frac{N(R)}{2R} = \infty. \tag{10}$$

In other words, we must oversample to be on the safe side.

We shall exhibit a universal sampling sequence $\{t_n\}$ satisfying condition (10) that is appropriate for blind sampling. The fact that (9) is not applicable means that we cannot use a regular S-L problem as done in Ref. 6. Instead we should use the Gelfand-Levitan (G-L) inverse spectral theory to its full extent to construct a singular S-L operator. In other words, to generate a sampling formula valid for all bandwidths, the sampling sequence must fit the spectrum of a singular self-adjoint S-L problem. The issue reduces to the summability of a certain series of sampling points in order to satisfy the conditions of the Gelfand-Levitan-Gasymov theorem. In the last section of Part 1 we derive estimates for the truncation and amplitude errors of our sampling formula.

Since the famous paper by Shannon appeared, and driven by applications, the subject of sampling has evolved in so many directions that it is impossible to cite them all in few pages. It is worth mentioning, in the area of Kramer's theorem, that the string differential operator offers more freedom for its eigenvalues distribution, as shown by M.G. Krein.[14,28] Indeed, its eigenvalues do not obey any particular distribution law at infinity. Thus a sampling formula for strings would provide us with more interesting irregular sampling results, especially when sampling on a finite interval.[8] For application of sampling in parallel-beam tomography we refer to.[17] The interaction of interpolation, spline and adaptive irregular sampling can be found in Ref. 4. For computational driven methods there are works by Feichtinger and Grochenig,[18,23] where coefficient identification is used to express a bandlimited signal by a trigonometric polynomial. This leads to a linear system with a Toeplitz structure which can be solved by superfast algorithms of numerical linear algebra. For an overview of recent progress and applications in the area let us mention Refs. 3,4,7,9,10,36,37,53,54 and the references therein.

1.1. *Inverse spectral problem*

Consider a singular S-L problem

$$\begin{cases} L(y) := -y''(x,t) + q(x)y(x,t) = ty(x,t), & 0 \le x < \infty, \\ y'(0,t) - hy(0,t) = 0, \end{cases} \tag{11}$$

where q is locally integrable on $[0,\infty)$ and $h \in \mathbb{R}$. The end point $x = \infty$ is said to be limit point (LP) if there exists, for all complex number t, at least one solution y such that $y(\cdot,t) \notin L^2(1,\infty)$. If the end point $x = \infty$ is in the LP case then the operator L in (11) is self-adjoint and there is no need for an extra boundary condition. Thus we assume that the S-L operator in (11) is in the LP case at infinity and regular at $x = 0$, so it generates a generalized Fourier transform mapping $L^2(0,\infty)$ onto $L^2(\mathbb{R}, d\Gamma)$ which is defined by

$$F(t) = \int_0^\infty f(x)y(x,t)\,dx \tag{12}$$

$$\int_0^\infty |f(x)|^2\,dx = \int_{-\infty}^\infty |F(t)|^2\,d\Gamma(t), \tag{13}$$

where $\Gamma(t)$ is a monotone increasing, right-continuous function (unique up to an additive constant). The inverse transform takes the form

$$f(x) = \int_{-\infty}^\infty F(t)y(x,t)\,d\Gamma(t), \tag{14}$$

see Ref. 32. For example in case $q(x) = h = 0$ the generalized Fourier transform reduces to the classical Fourier cosine transform

$$F_c(f)(t) = \int_0^\infty f(x)\cos(x\sqrt{t})\,dx.$$

The function Γ is called a spectral function associated with the normalized eigensolutions $y(x,\lambda)$ and has the following asymptotics[21] at infinity

$$\Gamma(t) = \frac{2(1+h)^2}{\pi}\sqrt{t} + \Gamma(-\infty) - h\sqrt{1+h^2} + o(1), \quad t \to \infty. \tag{15}$$

Recall that in 1951, in their celebrated paper Ref. 21, Gelfand and Levitan gave separately the necessary and the sufficient conditions for the solvability of the inverse spectral problem. To close the gap, M.G. Krein in 1953, Ref. 26, announced two necessary and sufficient conditions for Γ to be a spectral function of a regular or singular S-L problem, which he then revised by adding the third condition in 1957, Ref. 27:

Theorem 1.1 (M. G. Krein). *In order for* Γ *to be a spectral function of*

$$\begin{cases} L(y) := -y''(x,t) + q(x)y(x,t) = ty(x,t), & 0 \le x < l \le \infty \\ y'(0,t) - hy(0,t) = 0 \end{cases}$$

for a given $l \le \infty$ *it is necessary and sufficient that*

(1) The function $\Pi(\eta) = \int_{-\infty}^{\infty} \frac{1-\cos(\eta\sqrt{t})}{t} d\Gamma(t)$ *where* $0 \le \eta \le 2l$ *is finite and has two absolute continuous derivatives on every interval* $[0, r]$ *where* $r < 2l$

(2) $\Pi'(0) = 1$

(3) $\liminf_{R\to\infty} sup\ n(R)/\sqrt{R} \ge l/\pi$ *where* $n(R)$ *represents the number of points in the spectrum that are also contained in the interval* $[0, R]$.

The issue of whether Krein's result needed two or three conditions was settled down by Marchenko[32] for the case $l = \infty$ and Yavryan[52] for the case $l < \infty$. They have shown that the third condition is superfluous.

Earlier, in 1964, Gasymov and Levitan Ref. 20 closed the gap of the Gelfand-Levitan result[21] by showing that the following two conditions are both necessary and sufficient for the solvability of the inverse spectral problem.

Theorem 1.2 (Gelfand-Levitan-Gasymov). *For a monotone increasing and right-continuous function* Γ *to be the spectral function of a self-adjoint singular S-L problem (11) with a real and locally integrable potential* $q(x)$ *over* $[0, \infty)$ *it is necessary and sufficient that:*
[A] (Existence) For any $f \in L^2(0, \infty)$ *with compact support*

$$\int_{-\infty}^{\infty} |F_c(f)(t)|^2 \, d\Gamma(t) = 0 \quad \Rightarrow \quad f = 0 \ \textit{almost everywhere.} \tag{16}$$

[B] (Smoothness) The function

$$\Phi_N(x) := \int_{-\infty}^{N} \cos\left(x\sqrt{t}\right) d\left(\Gamma(t) - \frac{2}{\pi}\sqrt{t_+}\right) \tag{17}$$

converges boundedly to a differentiable function Φ.

Here t_+ is the cut-off function which is equal to t if $t > 0$ and 0 otherwise. As we are concerned with the irregular sampling problem, we need to consider only an S-L problem (11) whose spectrum is purely discrete. In this case the Gelfand-Levitan-Gasymov theorem takes the form:

Theorem 1.3 (Gelfand-Levitan-Gasymov: discrete spectrum case). *For a given sequence* $\{t_n\}_{n\ge1}$ *to be the set of eigenvalues of (11),* $\alpha_n =$

$\int_0^\infty |y(x,t_n)|^2\,dx$ while $q(x)$ is a real and locally integrable potential it is necessary and sufficient that:
A1) For any $f \in L^2(0,\infty)$ with compact support

$$F_c(f)(t_n) = 0 \ \ \textit{for all} \ \ n \in \mathbb{N} \Rightarrow \ f = 0 \ \ \textit{almost everywhere.}$$

B1) The function

$$\Phi_N(x) := \sum_{t_n \le N} \alpha_n \cos\left(x\sqrt{t_n}\right) - \frac{1}{x}\sin\left(\sqrt{N}x\right) \tag{18}$$

converges boundedly to a differentiable function Φ.

Thus both Gelfand-Levitan-Gasymov's and Krein's theorems require two conditions. The major differences between two theorems is in the required smoothness and whether the measure is $\Gamma(t)$ or $\sigma(t) = \Gamma(t) - \frac{2}{\pi}\sqrt{t_+}$. We need also to mention that in Ref. 32,Theorem 2.3.1., p. 142, Marchenko has a similar theorem that falls in between Gelfand-Levitan-Gasymov and Krein theorems, where the smoothness condition is: $\Psi(x) = \left(\frac{1-\cos(tx)}{t^2},\ R\right)$ should be at least three times continuously differentiable. Note here that Ψ uses t instead of $\sqrt{t}$ while R is a distribution, and the reconstructed potential is only continuous. It is clear that the Gelfand-Levitan-Gasymov paper Ref. 20 gives the weakest possible smoothness on the potential, namely $q \in L^{1,loc}(0,\infty)$. Before going to sampling using the inverse spectral theory we revisit the Gelfand-Levitan-Gasymov theorem and show that in fact only one, namely the second condition is needed.

Theorem 1.4 (Gelfand-Levitan-Gasymov Revisited). *For a monotone increasing function Γ to be the spectral function of a problem (11) where q has m locally integrable derivatives it is necessary and sufficient that the sequence of functions Φ_N converges boundedly to a function Φ that has $m+1$ locally integrable derivatives.*

The essence of the inverse spectral problem is to recover the potential q and the initial condition h from the knowledge of Γ. The main idea is based on the existence of a transformation operator which expresses the solution $y(x,t)$ of (11) in terms of the $\cos\left(x\sqrt{t}\right)$ which is also a solution of (11) but in the particular (unperturbed) case $q = h = 0$. These transformation operators and their inverse are of Fredholm type and given by

$$y(x,t) = \cos\left(x\sqrt{t}\right) + \int_0^x K(x,\eta)\cos\left(\eta\sqrt{t}\right)dt, \tag{19}$$

$$\cos\left(x\sqrt{t}\right) = y(x,t) + \int_0^x H(x,\eta)y(\eta,t)\,d\eta,$$

and

$$K(x,x) = \frac{1}{2}\int_0^x q(\eta)d\eta + h.$$

The connection between Γ and q is established through the integral equation

$$F(x,\eta) + K(x,\eta) + \int_0^x K(x,s)F(s,\eta)ds = H(\eta,x) \tag{20}$$

where $F(x,\eta) = \int \cos\left(x\sqrt{t}\right)\cos\left(\eta\sqrt{t}\right)d\sigma(t)$ which in fact is the limit of the sequence $\frac{1}{2}\left[\Phi_N(x+\eta) + \Phi_N(x-\eta)\right]$ as $N \to \infty$. The main idea in Gelfand-Levitan-Gasymov proof is that condition $[A]$ implies uniqueness of the solution $K(x,.)$ for the integral equation (20) and by the Fredholm alternative uniqueness implies the existence. Condition $[B]$ is for smoothness only. Here we have two important remarks. In practice it is difficult to verify when condition $[A]$ holds. Next, once we have recovered q, we need to verify that no boundary condition is needed at $x = \infty$, i.e. the problem is in the LP case and so Γ is indeed a spectral function of a singular self adjoint problem.

Theorem 1.5. *Assume that* $\Phi_N(x) := \int_{-\infty}^N \cos\left(x\sqrt{t}\right)d\sigma(t)$ *converges boundedly to a differentiable function* Φ *as* $N \to \infty$, *then condition* $[A]$ *holds.*

Proof. Let f have compact support, then by the Paley-Wiener theorem[2] the Fourier cosine transform $F_c(f)$ of f in condition $[A]$ is an entire function of order $\frac{1}{2}$. We shall distinguish two cases. First assume that the continuous spectrum is not empty, say it contains an interval $[t_0 - \delta,\ t_0 + \delta]$ with $\delta > 0$. Then from condition $[A]$

$$0 = \int_{-\infty}^{\infty} |F_c(f)(t)|^2\, d\Gamma(t) \geq \int_{t_0-\delta}^{t_0+\delta} |F_c(f)(t)|^2\, d\Gamma(t)$$

and the fact that Γ is increasing about t_0 it implies that $F_c(f)(t) = 0$ for $t \in [t_0 - \delta, t_0 + \delta]$. Since $F_c(f)$ is entire we must have $F_c(f) = 0$ and the inverse Fourier cosine transform leads to $f = 0$. The same holds true if the spectrum of Γ includes a sequence with a finite accumulation point. Clearly if $\{t_n\}_{n=1}^{\infty}$ belong to the spectrum of Γ and $\lim_{n\to\infty} t_n = t_0$ then from

$$0 = \int_\sigma |F_c(f)(t)|^2\, d\Gamma(t) \geq \sum_{n\geq 1} |F_c(f)(t_n)|^2\, \alpha_n$$

with $\alpha_n > 0$, it follows that the entire function $F_c(f)$ vanishes on a sequence $\{t_n\}_{n=1}^{\infty}$ with a finite accumulation point. Hence $F_c(f)(t) = 0$ and again $f = 0$.

The second case is when we have a purely discrete spectrum with no finite accumulation point, i.e. Γ is a step function. Since Φ_N converges boundedly to a differentiable function, then necessarily $\Phi_N(0) \to \Phi(0)$ and

$$\begin{aligned}\Phi_N(0) &= \int_{-\infty}^{N} d\sigma(t)\\ &= \sigma(N) - \Gamma(-\infty)\\ &= \Gamma(N) - \frac{2}{\pi}\sqrt{N} - \Gamma(-\infty)\\ &= \Phi\,(0) + o\,(1)\end{aligned}$$

yields the asymptotic behavior of Γ at infinity

$$\Gamma(t) = \frac{2}{\pi}\sqrt{t} + \text{ const } + o\,(1) \qquad \text{as} \quad t \to \infty.$$

Since Γ is a step function with jumps α_n at t_n then Γ is constant on $(t_n,\ t_{n+1})$ and for any t_α and t_β such that $t_n < t_\alpha < t_\beta < t_{n+1}$ we have

$$\Gamma(t_\beta) - \Gamma(t_\alpha) = \frac{2}{\pi}\left(\sqrt{t_\beta} - \sqrt{t_\alpha}\right) + o\,(1)\,,$$

and since $\Gamma(t_\alpha) = \Gamma(t_\beta)$ we have for large $t \to \infty$

$$\sqrt{t_\beta} - \sqrt{t_\alpha} = o\,(1)\,.$$

Now let $t_\beta \to t_{n+1}$ while $t_\alpha \to t_n$ to obtain

$$\sqrt{t_{n+1}} - \sqrt{t_n} = o\,(1) \quad \text{as } n \to \infty. \tag{21}$$

One can see now that (21) forces the density of its eigenvalues to satisfy

$$\overline{\lim_{R\to\infty}}\, \frac{n\,(R)}{\sqrt{R}} = \infty. \tag{22}$$

Indeed (21) implies that for any given $M > 0$ there is a $K > 0$ such that for all $t_n > K$ we have

$$\sqrt{t_{n+1}} - \sqrt{t_n} < \frac{1}{M},$$

and thus for any $t_n,\ t_{n+1} \in [K,\ R]$ we have

$$t_{n+1} - t_n = \left(\sqrt{t_{n+1}} - \sqrt{t_n}\right)\left(\sqrt{t_{n+1}} + \sqrt{t_n}\right) < \frac{2}{M}\sqrt{R}.$$

Thus in the interval $[K,\ R]$ there are at least

$$\frac{R-K}{\frac{2\sqrt{R}}{M}} = \frac{(R-K)}{2\sqrt{R}}M$$

points t_n, i.e. $n(R) \geq \frac{(R-K)}{2\sqrt{R}}M$ which yields a lower bound on the density

$$\overline{\lim_{R\to\infty}} \frac{n(R)}{\sqrt{R}} \geq \frac{M}{2}.$$

Since M is arbitrary the limit (22) follows. We now show that condition $[A]$ is satisfied when (22) holds. If f has compact support on the interval $[0, a]$ then from the Paley-Wiener theorem[2]

$$F_c(f)(t^2) = \int_0^a f(x)\cos(xt)\,dx$$

is an entire function of exponential type at most a and according to Titchmarsh,[40] the distribution of its zeros should have the asymptotics

$$\overline{\lim_{R\to\infty}} \frac{n(R)}{R} \leq \frac{2a}{\pi}.$$

Hence the zero distribution of $F_c(f)(\lambda)$ is

$$\overline{\lim_{R\to\infty}} \frac{n(R)}{\sqrt{R}} \leq \frac{2a}{\pi}.$$

But since Γ is a step function with jumps at t_n, condition $[A]$ means that $F_c(f)(t_n) = 0$. Recall that the distribution of its zeros satisfies (22) which says that $F_c(f)$ has too many zeros and so can only be the trivial function. Thus $F_c(f)(t) = 0$ for all $t \in \mathbb{C}$ and consequently $f = 0$ in $L^2(0, \infty)$. □

1.2. *Sufficient conditions for spectral functions*

We now obtain sufficient, but verifiable, conditions on Γ for the bounded convergence of Φ_N to a differentiable function Φ. It is also clear that the convergence of the integral $\Phi_N(x) = \int_{-\infty}^{N} \cos(x\sqrt{t})\,d\sigma(t)$ depends solely on the behavior of the function σ for large values of t. First the case where $t \to -\infty$ is easily settled down by observing that $\cos(x\sqrt{t}) = \cosh(x\sqrt{-t})$ and for any $n \geq 0$ and $\varepsilon > 0$

$$t^n \cosh(x\sqrt{-t}) = O\left(\cosh\left((x+\varepsilon)\sqrt{-t}\right)\right)$$

holds for large $-t$, thus the existence of $\Phi_0(x)$ for all $x > 0$ is enough for the existence of its higher derivatives. In other words adding a continuous spectrum that is bounded from above while $\Phi_N(x)$ is defined for all $x \geq 0$ would not affect the convergence of the sequence Φ_N. Thus only the behavior of Γ when $t \to \infty$ matters and we shall examine only two cases.

If Γ is absolutely continuous at infinity, i.e. there is $K > 0$ such that

$$\sigma(t) = \Gamma(t) - \frac{2}{\pi}\sqrt{t_+} \quad \text{is differentiable almost everywhere for } t > K, \tag{23}$$

then for $N > K$ write

$$\Phi_N(t) = \Phi_K(t) + \int_K^N \cos\left(x\sqrt{t}\right) d\sigma(t). \tag{24}$$

Since the convergence depends on the integral we have for $N > K$

$$\begin{aligned}\frac{d}{dx}\int_K^N \cos\left(x\sqrt{t}\right) d\sigma(t) &= -\int_K^N \sqrt{t}\sin\left(x\sqrt{t}\right)\sigma'(t)\,dt \\ &= -2\int_{\sqrt{K}}^{\sqrt{N}} \sin(x\tau)\,\tau^2\sigma'\left(\tau^2\right)d\tau. \end{aligned}\tag{25}$$

For the existence of the Fourier sine transform (25) it is sufficient that

$$\tau^2\sigma'\left(\tau^2\right) \in L^2(\sqrt{K}, \infty),$$

or equivalently, $t^{3/4}\sigma'(t)$ is square integrable at infinity, which would then imply that $\Phi' - \Phi_K'$ belongs to $L^2(0,\infty)$ and is then certainly locally integrable. Since Φ_K' is locally integrable, then so is Φ'.

Another sufficient condition for the convergence of (25) as $N \to \infty$, and therefore, of Φ_N to a differentiable function Φ is : σ' is of bounded variation at infinity and

$$\sigma'(t) = o\left(\frac{1}{t}\right) \quad \text{as } t \to \infty. \tag{26}$$

In this case $\tau^2\sigma'\left(\tau^2\right)$ is of bounded variation and approaches 0 as $\tau \to \infty$. Thus it can be written as the difference of two monotone decreasing functions,

$$\tau^2\sigma'\left(\tau^2\right) = \psi_1(\tau) - \psi_2(\tau)$$

where $\psi_i(\tau) \downarrow 0$ as $\tau \to \infty$ and

$$\int_{\sqrt{K}}^{\sqrt{N}} \sin(x\tau)\,\tau^2\sigma'\left(\tau^2\right)d\tau = \int_{\sqrt{K}}^{\sqrt{N}} \sin(x\tau)\,\psi_1(\tau)d\tau - \int_{\sqrt{K}}^{\sqrt{N}} \sin(x\tau)\,\psi_2(\tau)\,d\tau. \tag{27}$$

The fact that we have a monotone decreasing function allows us to use the second mean value theorem, see Ref. 41, Theorem 6, to obtain

$$\begin{aligned}\left|\int_{\sqrt{K}}^{\sqrt{N}} \sin(x\tau)\,\psi_1(\tau)d\tau\right| &= \left|\psi_1\left(\sqrt{K}\right)\int_{\sqrt{K}}^{\sqrt{M}} \sin(x\tau)\,d\tau\right| \\ &\le \psi_1\left(\sqrt{K}\right)\frac{2}{x}, \quad K < M < N.\end{aligned}$$

Thus $\int_{\sqrt{K}}^{\sqrt{N}} \sin(x\tau)\,\psi_1(\tau)d\tau$ converges uniformly in any compact interval of x that does not contain 0 and similarly for the second integral in (27).

Thus as $N \to \infty$, $\int_{\sqrt{K}}^{\sqrt{N}} \sin(x\tau)\, \tau^2 \sigma'(\tau^2)\, d\tau$ converges uniformly in any compact interval not containing zero. Since the main assumptions are (23) and (26) we arrive at

Theorem 1.6. *Let Γ be a monotone increasing, right continuous function such that it is absolutely continuous at $t = \infty$ with bounded variation derivative and*

$$\Gamma'(t) = \frac{1}{\pi\sqrt{t}} + o\left(\frac{1}{t}\right) \qquad \text{as } t \to \infty,$$

then there exists a unique Sturm-Liouville operator (11) for which Γ is its spectral function.

We now move to the second case where we have a discrete spectrum at $t = \infty$. Without loss of generality by (24) we can assume that the spectrum is all discrete. Thus Φ_N reduces to a trigonometric sum and if $t_k < N \le t_{k+1}$ then

$$\Phi_N(x) = \sum_{n=1}^{k} \alpha_n \cos\left(x\sqrt{t_n}\right) - \frac{2}{\pi x}\sin\left(x\sqrt{N}\right). \tag{28}$$

In order to simplify the proof for convergence we first assume that

$$\alpha_n = \frac{2}{\pi}\left(\sqrt{t_{n+1}} - \sqrt{t_n}\right), \tag{29}$$

and decompose (28) as a telescope sum

$$\begin{aligned}
\frac{\pi}{2}\Phi_N(x) &= \sum_{n=1}^{k}\left(\sqrt{t_{n+1}} - \sqrt{t_n}\right)\cos\left(x\sqrt{t_n}\right) - \frac{1}{x}\sin\left(x\sqrt{N}\right) \\
&= \sum_{n=1}^{k}\left[\left(\sqrt{t_{n+1}} - \sqrt{t_n}\right)\cos\left(x\sqrt{t_n}\right) - \frac{1}{x}\sin\left(x\sqrt{t_{n+1}}\right) + \frac{1}{x}\sin\left(x\sqrt{t_n}\right)\right] \\
&\quad + \frac{1}{x}\sin\left(x\sqrt{t_{k+1}}\right) - \frac{1}{x}\sin\left(x\sqrt{N}\right) - \frac{1}{x}\sin\left(x\sqrt{t_1}\right).
\end{aligned} \tag{30}$$

Using the mean value theorem we successively obtain

$$\begin{aligned}
&\frac{\pi}{2}\Phi_{t_{k+1}}(x) + \frac{1}{x}\sin\left(x\sqrt{t_1}\right) \\
&= \sum_{n=1}^{k}\left(\sqrt{t_{n+1}} - \sqrt{t_n}\right)\left(\cos\left(x\sqrt{t_n}\right) - \cos\left(x\sqrt{\tau_n}\right)\right) \\
&= -x\sum_{n=1}^{k}\left(\sqrt{t_{n+1}} - \sqrt{t_n}\right)\left(\sqrt{t_n} - \sqrt{\tau_n}\right)\sin\left(x\sqrt{u_n}\right),
\end{aligned} \tag{31}$$

where

$$t_n \leq u_n \leq \tau_n \leq t_{n+1}. \tag{32}$$

Similarly we have

$$\begin{aligned}
&\frac{d}{dx}\left[\frac{\pi}{2}\Phi_{t_{k+1}}(x) + \frac{1}{x}\sin\left(x\sqrt{t_1}\right)\right] \\
&= -\sum_{n=1}^{k}\left(\sqrt{t_{n+1}} - \sqrt{t_n}\right)\left(\sqrt{t_n} - \sqrt{\tau_n}\right)\left(\sin\left(x\sqrt{u_n}\right)\right) \\
&\quad -x\sum_{n=1}^{k}\sqrt{u_n}\left(\sqrt{t_{n+1}} - \sqrt{t_n}\right)\left(\sqrt{t_n} - \sqrt{\tau_n}\right)\cos\left(x\sqrt{u_n}\right).
\end{aligned} \tag{33}$$

The sums in (31) and (33) converge uniformly if

$$\sum_{n=1}^{\infty}\sqrt{u_n}\left(\sqrt{t_{n+1}} - \sqrt{t_n}\right)\left(\sqrt{t_n} - \sqrt{\tau_n}\right) < \infty.$$

Because

$$\sqrt{u_n}\left(\sqrt{t_{n+1}} - \sqrt{t_n}\right)\left(\sqrt{t_n} - \sqrt{\tau_n}\right) < \frac{(t_{n+1} - t_n)^2}{2\sqrt{t_n}}$$

we arrive at

Theorem 1.7. *Assume that Γ is a step function at ∞ with jumps at $\{t_n\}$ such that*

$$\sum_{n=1}^{\infty}\frac{(t_{n+1} - t_n)^2}{\sqrt{t_n}} < \infty, \tag{34}$$

and for large n, $\alpha_n = \frac{2}{\pi}\left(\sqrt{t_{n+1}} - \sqrt{t_n}\right)$, then there exists a unique potential q and an initial condition h such that Γ is the spectral function of (11).

We can enlarge the class of normalizing constants α_n by adding an extra term β_n

$$\alpha_n = \frac{2}{\pi}\left(\sqrt{t_{n+1}} - \sqrt{t_n}\right) + \beta_n$$

where we clearly should choose $\beta_n \to 0$ so the previous analysis holds. It is readily seen from (28) that the new function can be expressed through the previously studied function $\Phi_N(x)$ defined in (30) as follows

$$\widetilde{\Phi}_N(x) = \sum_{n=1}^{k} \alpha_n \cos\left(x\sqrt{t_n}\right) - \frac{2}{\pi x}\sin\left(x\sqrt{N}\right)$$

$$= \sum_{n=1}^{k} \frac{2}{\pi}\left(\sqrt{t_{n+1}} - \sqrt{t_n}\right)\cos\left(x\sqrt{t_n}\right) - \frac{2}{\pi x}\sin\left(x\sqrt{N}\right) + \sum_{n=1}^{k} \beta_n \cos\left(x\sqrt{t_n}\right)$$

$$= \Phi_N(x) + \sum_{n=1}^{k} \beta_n \cos\left(x\sqrt{t_n}\right).$$

By differentiating we arrive at

$$\widetilde{\Phi}'_N(x) = \Phi'_N(x) - \sum_{n=1}^{k} \beta_n \sqrt{t_n}\sin\left(x\sqrt{t_n}\right),$$

and choosing $N = t_{k+1}$ reduces the above sum to

$$\widetilde{\Phi}'_{t_{k+1}}(x) = \Phi'_{t_{k+1}}(x) - \sum_{n=1}^{k} \beta_n \sqrt{t_n}\sin\left(x\sqrt{t_n}\right).$$

Thus for the new sum to converge boundedly to a differentiable function, we only need the series containing β_n to do so, i.e.

$$\sum_{n=1}^{k} \beta_n \sqrt{t_n}\sin\left(x\sqrt{t_n}\right) = \sum_{n=1}^{k} \frac{\beta_n t_n}{\left(t_{n+1} - t_n\right)^2} \frac{\left(t_{n+1} - t_n\right)^2}{\sqrt{t_n}} \sin\left(x\sqrt{t_n}\right).$$

Since the partial sums $\sum_{n=1}^{k} \frac{(t_{n+1}-t_n)^2}{\sqrt{t_n}} \sin\left(x\sqrt{t_n}\right)$ are already uniformly bounded, by Abel's theorem[55] convergence will follow if $\frac{\beta_n t_n}{(t_{n+1}-t_n)^2} \to 0$ as $n \to \infty$, i.e.

$$\beta_n = o\left(\frac{\left(t_{n+1} - t_n\right)^2}{t_n}\right).$$

Thus we have proved

Theorem 1.8. *Let Γ be a monotone increasing function, right-continuous that is a step function as $t \to \infty$, then if its jumps α_n satisfy*

$$\alpha_n = \frac{2}{\pi}\left(\sqrt{t_{n+1}} - \sqrt{t_n}\right) + o\left(\frac{\left(t_{n+1} - t_n\right)^2}{t_n}\right), \quad \text{as } n \to \infty, \tag{35}$$

where t_n satisfies (34), then Γ is the spectral function of a singular Sturm-Liouville operator (11) in the LP case at $x = \infty$.

The previous theorem gives a simple description of possible singular isospectral operators. Although the jumps can vary, they must be related to the eigenvalues. Formula (35) gives such a behavior at infinity.

Remark 1.1. While the condition (21) gives a necessary condition for the eigenvalues of a singular Sturm-Liouville problem, the condition $\sum_{n=1}^{\infty} \frac{(t_{n+1}-t_n)^2}{\sqrt{t_n}} < \infty$ is sufficient for reconstruction of a singular Sturm Liouville operator in the LP case. In the regular case the asymptotics is $t_n \approx cn^2$ as $n \to \infty$. For example the Laguerre differential operator is LP at $x = \infty$ and its eigenvalues $t_n = 4n + 3$. The necessary condition (21) obviously holds, but the sufficient condition (34) does not satisfy.

Now that $\Gamma(t)$ qualifies to be a spectral function, the S-L operator can be recovered by the G-L inverse spectral theory.[21] To this end we define

$$L(x,\eta) = \int_0^\infty \cos(x\sqrt{t}) \cos(\eta\sqrt{t})\, d\left(\Gamma(t) - \frac{2}{\pi}\sqrt{t}\right), \tag{36}$$

and solve the Fredholm integral equation

$$L(x,\eta) + K(x,\eta) + \int_0^x K(x,s)L(s,\eta)ds = 0, \qquad 0 \le \eta \le x,$$

to find $K(x,\eta)$, which is also differentiable. The potential q is then given by $q(x) = \frac{1}{2}\frac{d}{dx}K(x,x)$, the boundary condition h by $h = K(0,0)$, and the solution $y(x,t)$ by

$$y(x,t) = \cos\left(x\sqrt{t}\right) + \int_0^x K(x,\eta)\cos\left(\eta\sqrt{t}\right) d\eta. \tag{37}$$

1.3. *Sampling formula for $PW^{1/2}$*

Before the study of bandlimited signals can begin, we first consider signals F that can be expressed in the form (6) with $y(x,t)$ being a normalized solution of a S-L problem (11). Observe that a solution $y(x,t)$ of an S-L problem, as a function of complex variable t cannot grow faster than $e^{|x|\sqrt{|t|}}$ in the complex plane.[6] Thus F defined by (6) is an entire function of order at most 1/2 and normal type, which leads us to first consider the sampling problem for entire functions of order $\frac{1}{2}$.

Let $PW_T^{1/2}$ be the set of functions $F : [0,\infty) \to C$ such that $F(t^2)$ can be extended as an even function in PW_T^1, and

$$PW^{1/2} = \bigcup_{T>0} PW_T^{1/2}.$$

$F \in PW_T^{1/2}$ can be described as a function on $(0, \infty)$ that can be analytically extended onto the whole complex plane as an entire function of order 1/2, type at most T and such that

$$\int_0^\infty |F(t)|^2 t^{-1/2} dt < \infty.$$

Also the Paley-Wiener theorem[2] allows us to express $F \in PW_T^{1/2}$ through a Fourier cosine transform

$$F(t) = \int_0^T f(x) \cos(x\sqrt{t})\, dx, \quad f(x) \in L^2(0, T). \tag{38}$$

Since $F(t^2) \in PW_T^1$ the zeros of $F(t^2)$ satisfy the distribution law (2). Therefore the zeros of $F \in PW_T^{1/2}$ have the density

$$\frac{n(R)}{\sqrt{R}} \sim \frac{2T}{\pi}. \tag{39}$$

Because $F(t^2) \in PW_T^1$ is completely determined by its zeros, up to a multiple constant,[40] the same is also true for $F(t) \in PW_T^{1/2}$. Thus if the sampling rate is less than the density of its zeros, we lose uniqueness but if the sampling rate is more than the density of its zeros, we have a full recovery of the function. In other words to recover a function from $PW_T^{1/2}$ one needs to sample at the rate

$$\overline{\lim}_{R\to\infty} \frac{N(R)}{\sqrt{R}} \geq \frac{2T}{\pi}.$$

This is why to recover a function from $PW^{1/2}$ with a finite but unknown type, the sampling rate must obey

$$\overline{\lim}_{R\to\infty} \frac{N(R)}{\sqrt{R}} = \infty. \tag{40}$$

Clearly if the eigenvalues $\{t_n\}$ of a regular S-L on a finite interval $[0, a]$ are used as sample points, then instead of (40), we would have[32]

$$\frac{N(R)}{\sqrt{R}} \sim \frac{a}{\pi}.$$

Therefore one must consider a singular S-L problems which we now construct.

Let $y(x,t)$ be the normalized eigensolution of (11), $y(0,t) = 1$ and $y'(0,t) = h$.

Theorem 1.9. *Assume that q is locally integrable over $[0, \infty)$. Then $F \in PW_T^{1/2}$ if and only if $F(t) = \int_0^T f(x) y(x,t) dx$, where $f \in L^2(0, T)$.*

Proof. If $F \in PW_T^{1/2}$, then by (38) there exists $g \in L^2(0,T)$ such that $F(t) = \int_0^T g(x)\cos(x\sqrt{t})dx$. Now use the fact that[21] $\cos(x\sqrt{t})$ and $y(x,t)$ are transmuted by

$$\cos\left(x\sqrt{t}\right) = y(x,t) + \int_0^x H(x,\eta)y(\eta,t)d\eta \tag{41}$$

to write

$$\begin{aligned} F(t) &= \int_0^T g(x)\cos(x\sqrt{t})dx \\ &= \int_0^T g(x)\left[y(x,t) + \int_0^x H(x,\eta)y(\eta,t)d\eta\right]dx \\ &= \int_0^T g(x)y(x,t)dx + \int_0^T g(x)\int_0^x H(x,\eta)y(\eta,t)d\eta\, dx \\ &= \int_0^T g(x)y(x,t)dx + \int_0^T y(\eta,t)\int_\eta^T g(x)H(x,\eta)dx\, d\eta \\ &= \int_0^T \left[g(\eta) + \int_\eta^T g(x)H(x,\eta)dx\right]y(\eta,t)d\eta. \end{aligned}$$

Observe that $H(.,.)$ is a continuous function and so $\int_\eta^T g(x)H(x,\eta)dx \in L^2(0,T)$. Thus if $F \in PW_T^{1/2}$ then there exists $f(\eta) = g(\eta) + \int_\eta^T g(x)H(x,\eta)dx$ from $L^2(0,T)$ such that $F(t) = \int_0^T f(\eta)y(\eta,t)d\eta$. The converse is similar and uses the transmutation formula (37). □

Problem (11) has now a set of eigensolutions $y(x,t_n) \in L^2(0,\infty)$. If $F \in PW^{1/2}$, then by Theorem 1.9

$$F(t) = \int_0^\infty f(x)y(x,t)dx, \tag{42}$$

where $f \in L^2(0,\infty)$ has compact support. Therefore F, as the generalized Fourier transform of $f \in L^2(0,\infty)$, belongs to $L^2(\mathbb{R}, d\Gamma)$:

$$\sum_{n=1}^{\infty} |F(t_n)|^2(\sqrt{t_{n+1}} - \sqrt{t_n}) < \infty,$$

and

$$f(x) = \int_{-\infty}^{\infty} F(t)y(x,t)\, d\Gamma(t) = \frac{2}{\pi}\sum_{n\in\mathbb{N}} F(t_n)y(x,t_n)(\sqrt{t_{n+1}} - \sqrt{t_n}).$$

Thus we have proved

Theorem 1.10. *Let $F \in PW^{1/2}$ be sampled at $\{t_n\}_{n\in\mathbb{N}}$, where the sequence $\{t_n\}_{n\in\mathbb{N}}$ satisfies condition (34). Then F can be recovered by formula (42), where*

$$f(x) = \frac{2}{\pi} \sum_{n\in\mathbb{N}} F(t_n) y(x, t_n) \left(\sqrt{t_{n+1}} - \sqrt{t_n}\right),$$

and $y(x,t)$ is defined by (37).

Note that if the type T of F is known, then we can combine both the generalized and inverse generalized Fourier transforms to get

$$\begin{aligned} F(t) &= \int_0^T f(x) y(x,t) dx \\ &= \frac{2}{\pi} \sum_{n=1}^{\infty} F(t_n) \left(\sqrt{t_{n+1}} - \sqrt{t_n}\right) \int_0^T y(x, t_n) y(x,t) dx, \end{aligned}$$

i.e.

$$F(t) = \sum_{n\in\mathbb{N}} F(t_n) S_n^T(t), \tag{43}$$

where

$$S_n^T(t) = \frac{2}{\pi} (\sqrt{t_{n+1}} - \sqrt{t_n}) \int_0^T y(x,t_n) y(x,t) dx. \tag{44}$$

Observe that we can use the same points t_n for all $\mathrm{PW}_T^{1/2}$ and thus we have proved the sampling theorem:

Theorem 1.11. *Given a sequence $\{t_n\}$ satisfying condition (34) and any type $T > 0$, then there exists a sequence of sampling functions $S_n^T(t)$ such that for any $F \in PW_T^{1/2}$ we have the sampling formula (43).*

1.4. *Sampling formula for bandlimited signals*

Having shown in the previous section how to recover functions from $PW^{1/2}$, now we proceed to sample functions from PW^1. Recall that if $F(t)$ is an even function from PW_T^1, then $F(\sqrt{t}) \in PW_T^{1/2}$, and therefore, can be reconstructed using the technique of the previous section. Now if F is any function from PW_T^1, then

$$F(t) = F_1(t) + tF_2(t),$$

where F_1 and F_2 are two even bandlimited signals with bandwidth T defined by

$$F_1(t) = \frac{F(t) + F(-t)}{2}, \; F_2(t) = \frac{F(t) - F(-t)}{2t}.$$

It is crucial to observe that we can find the values of both F_1 and F_2 at t_n only if F is given at both t_n and $-t_n$. So to recover any function from PW^1 one must sample on a symmetric sequence of points. Denote by $\mathbb{Z}^* = \mathbb{Z} - \{0\}$.

Let $\{\tau_n\}_{n\in\mathbb{Z}^*}$ be a doubly infinite, unbounded, and strictly monotone increasing sequence $\cdots < \tau_{-2} < \tau_{-1} < 0 < \tau_1 < \tau_2 < \cdots$, that is also symmetric $\tau_n = -\tau_{-n}$, and satisfies condition

$$\sum_{n=1}^{\infty} \tau_n(\tau_{n+1} - \tau_n)^2 < \infty. \tag{45}$$

For simplicity we omit the index 0 in the sequence, and an example of symmetric sequences satisfying (45) is given by

$$\tau_n = (n + o(1))^{\delta}, \quad n \geq 1, \;\; \tau_n = -\tau_{-n}, \;\; 0 < \delta < \frac{1}{3}.$$

If F is known at τ_n, $n \in \mathbb{Z}^*$, then both F_1 and F_2 are known at τ_n, $n \in \mathbb{N}$. We now show that we can reconstruct F_1 and F_2. To this end define

$$G_1(t) = F_1(\sqrt{t}) \;\text{ and }\; G_2(t) = F_2(\sqrt{t}), \tag{46}$$

then both of them belong to $PW_T^{1/2}$ and are known at $t_n = \tau_n^2$, $n \geq 1$. Obviously condition (45) translates into (34), which allows us to define the spectral function as follows

$$\Gamma(t) = \frac{2}{\pi} \sum_{t_n \leq t} (\sqrt{t_{n+1}} - \sqrt{t_n}).$$

Using (36) we can construct the eigensolution $y(x,t)$ by

$$y(x,t) = \cos(x\sqrt{t}) + \int_0^x K(x,\eta)\cos(\eta\sqrt{t})\, d\eta.$$

We now show how to recover $F \in PW^1$. Denote

$$\begin{aligned} g_i(x) &= \frac{2}{\pi} \sum_{n=1}^{\infty} G_i(t_n) y(x,t_n)(\sqrt{t_{n+1}} - \sqrt{t_n}) \\ &= \frac{2}{\pi} \sum_{n=1}^{\infty} F_i(\tau_n) y(x,\tau_n^2)(\tau_{n+1} - \tau_n), \quad i = 1,2. \end{aligned}$$

Since $G_i \in PW_T^{1/2}$, the support of g_i is a subset of $[0, T]$. Thus we can reconstruct F_i from their values $F_i(\tau_n)$, i.e.

$$F_i(t) = G_i(t^2) = \int_0^\infty g_i(x) y(x, t^2)\, dx, \quad i = 1, 2,$$

and we arrive at

$$\begin{aligned} F(t) &= F_1(t) + tF_2(t) \\ &= \frac{1}{\pi} \int_0^\infty \sum_{n=1}^{\infty} (\tau_{n+1} - \tau_n) y(x, \tau_n^2) y(x, \tau^2) \\ &\quad \times \left[\left(1 + \frac{t}{\tau_n}\right) F(\tau_n) + \left(1 - \frac{t}{\tau_n}\right) F(\tau_{-n}) \right] dx. \end{aligned}$$

In case the bandwidth T of the signal F is known, the above sampling formula reduces to

$$F(t) = \sum_{n=1}^{\infty} \left[\left(1 + \frac{t}{\tau_n}\right) F(\tau_n) + \left(1 + \frac{t}{\tau_{-n}}\right) F(\tau_{-n}) \right] \tilde{S}_n^T(t), \tag{47}$$

where

$$\tilde{S}_n^T(t) = \frac{1}{\pi} (\tau_{n+1} - \tau_n) \int_0^T y(x, \tau_n^2)\, y(x, t^2)\, dx.$$

Thus we have proved

Theorem 1.12. *Given a symmetric sequence $\{\tau_n\}_{n \in \mathbb{Z}^*}$ satisfying condition (45), then any bandlimited signal F can be recovered from $\{F(\tau_n)\}_{n \in \mathbb{Z}^*}$ by the formula*

$$\begin{aligned} F(t) = \; & \tfrac{1}{\pi} \int_0^\infty \sum_{n=1}^{\infty} (\tau_{n+1} - \tau_n) y(x, \tau_n^2) y(x, \tau^2) \\ & \times \left[\left(1 + \tfrac{t}{\tau_n}\right) F(\tau_n) + \left(1 - \tfrac{t}{\tau_n}\right) F(\tau_{-n}) \right] dx. \end{aligned}$$

Moreover if the bandwidth T of the signal is known, then the sampling formula (47) can be used.

1.5. *Error analysis*

We already have proved that if $F \in PW_T^{1/2}$ then $F \in L^2_{d\Gamma}(0, \infty)$. We now show that if additionally $t^{k-1/4} F(t) \in L^2(0, \infty)$ then $t^k F(t) \in L^2_{d\Gamma}(0, \infty)$. Recall that $F(t), t^k F(t) \in L^2_{d\Gamma}(0, \infty)$ if and only if

$$F(t) = \int_0^\infty f(x) y(x, t) dx,$$

where f is in the domain of the operator L^k. Moreover, the generalized Fourier transform of $L^k f$ is $t^k F(t)$.

First if $k = 1$ then $t^{3/4}F(t) \in L^2(0,\infty)$ means $\int_0^\infty |tF(t)|^2 \, d\sqrt{t} < \infty$. Since $\frac{2}{\pi}\sqrt{t}_+$ is the spectral function for the operator L with $q(x) = h = 0$, and $y(x,t) = \cos(x\sqrt{t})$ we have

$$F(t) = \int_0^\infty \cos\left(x\sqrt{t}\right) g(x)\, dx,$$

where g is in the domain of the operator $L = -D^2$, which means that g is twice differentiable, $g'' \in L^2(0,\infty)$, and $g'(0) = 0$. Moreover from $F \in PW_T^{1/2}$ we have $\operatorname{supp}(g) \subset [0,T]$ and so

$$-tF(t) = \int_0^T \cos\left(x\sqrt{t}\right) g''(x)\, dx.$$

Recall that from the Gelfand-Levitan theory[21] we have the transmutation operator defined by

$$\cos(x\sqrt{t}) = (Vy(.,t))(x),$$

where

$$(Vf)(x) = f(x) + \int_0^x H(x,\eta) f(\eta) d\eta,$$

and H is a continuous kernel. Thus V is a bounded operator acting in $L^2(0,T)$, its adjoint V^* is also bounded in $L^2(0,T)$ and we have

$$\begin{aligned} -tF(t) &= \int_0^T \cos(x\sqrt{t})\, g''(x)\, dx \\ &= \int_0^T (Vy(.,t))(x)\, g''(x)\, dx \\ &= \int_0^T y(x,t)\, (V^* g'')(x)\, dx. \end{aligned}$$

Because $g'' \in L^2(0,T)$ then $(V^*g'')(x) \in L^2(0,T)$. Thus $-tF(t)$ as the generalized Fourier transform (12) of a function from $L^2(0,T)$, must belong to $L^2_{d\Gamma}(0,\infty)$. Repeating this process by induction we can prove it for any k. The converse is also true. Thus we have arrived at

Theorem 1.13. *Let $F \in PW^{1/2}$. Then $t^{k-1/4}F(t) \in L^2(0,\infty)$ if and only if $t^k F(t) \in L^2_{d\Gamma}(0,\infty)$.*

We now use the above result to find a bound for the truncation error. Let

$$Y(t,\tau) = \int_0^T y(x,t)y(x,\tau)\,dx.$$

For a fixed τ, $Y(.,\tau)$ can be seen as the generalized Fourier transform (12) of the square integrable function $y(.,\tau)$, supported between 0 and T. Parseval's formula (13) then yields

$$\begin{aligned}\int_0^T |y(x,\tau)|^2\,dx &= \int_{-\infty}^{\infty} |Y(t,\tau)|^2\,d\Gamma(t)\\ &= \frac{2}{\pi}\sum_{n=1}^{\infty} |Y(t_n,\tau)|^2\left(\sqrt{t_{n+1}}-\sqrt{t_n}\right)\\ &= \sum_{n=1}^{\infty}\frac{1}{\alpha_n}\left|S_n^T(\tau)\right|^2.\end{aligned}$$

Hence,

$$\begin{aligned}\left|F(t)-\sum_{n\le N}F(t_n)S_n^T(t)\right|^2 &= \left|\sum_{n>N}|F(t_n)S_n^T(t)|\right|^2\\ &\le \sum_{n>N}\alpha_n|F(t_n)|^2\sum_{n>N}\frac{1}{\alpha_n}\left|S_n^T(t)\right|^2\\ &\le \sum_{n>N}\alpha_n|F(t_n)|^2\sum_{n\in\mathbb{N}}\frac{1}{\alpha_n}\left|S_n^T(t)\right|^2\\ &= \sum_{n>N}\alpha_n|F(t_n)|^2\int_0^T|y(x,t)|^2\,dx.\end{aligned}$$

Now, if $t^{k-1/4}F(t)\in L^2(0,\infty)$, then by Theorem 1.13, $t^kF(t)\in L^2_{d\Gamma}(0,\infty)$ and so

$$\begin{aligned}\sum_{n>N}\alpha_n|F(t_n)|^2 &\le \int_{t_N}^{\infty}|F(t)|^2\,d\Gamma(t)\\ &\le t_N^{-2k}\int_{t_N}^{\infty}t^{2k}F^2(t)\,d\Gamma(t)\\ &\le t_N^{-2k}\int_0^{\infty}t^{2k}F^2(t)\,d\Gamma(t).\end{aligned}$$

In other words

$$\sum_{n>N}\alpha_n|F(t_n)|^2\le E_k^2\,t_N^{-2k},$$

where $E_k^2 = \int_0^\infty t^{2k} F^2(t)\, d\Gamma(t)$. Thus we have proved

Theorem 1.14. *Let $F \in PW_T^{1/2}$ such that $t^{k-1/4}F(t) \in L^2(0,\infty)$. Then the truncation error for the sampling formula (43) has the order*

$$\left| F(t) - \sum_{n \le N} F(t_n) S_n^T(t) \right| \le E_k \sqrt{\int_0^T |y(x,t)|^2\, dx}\; t_N^{-k}.$$

Now assume that the signal F has been sampled with a random error ϵ, the so called amplitude error. So instead of exact values $F(t_n)$ only perturbed samples $F_\epsilon(t_n)$ have been recorded

$$|F(t_n) - F_\epsilon(t_n)| \le \epsilon.$$

The truncation error in this case first decreases as the number of sample points N increases, and then it deteriorates as N is taken very large. Thus an important issue is to find the optimal number of sampling points N. From Theorem 1.14 we have

$$\begin{aligned} &\left| F(t) - \sum_{n \le N} F_\epsilon(t_n) S_n^T(t) \right| \\ &\le \left| \sum_{n \le N} \left[F(t_n) - F_\epsilon(t_n) \right] S_n^T(t) \right| + \left| \sum_{n > N} F(t_n) S_n^T(t) \right| \\ &\le \epsilon \sum_{n \le N} \left| S_n^T(t) \right| + E_k \sqrt{\int_0^T |y(x,t)|^2\, dx}\; t_N^{-k}. \end{aligned}$$

The Hölder inequality yields

$$\begin{aligned} \sum_{n \le N} \left| S_n^T(t) \right| &\le \sqrt{\sum_{n \le N} \alpha_n} \sqrt{\sum_{n \le N} \frac{1}{\alpha_n} \left| S_n^T(t) \right|^2} \\ &\le t_{N+1}^{1/4} \sqrt{\frac{2}{\pi} \sum_{n \in \mathbb{N}} \frac{1}{\alpha_n} \left| S_n^T(t) \right|^2} \\ &\le t_{N+1}^{1/4} \sqrt{\frac{2}{\pi} \int_0^T |y(x,t)|^2\, dx}. \end{aligned}$$

Consequently,

$$\left| F(t) - \sum_{n \le N} F_\epsilon(t_n) S_n^T(t) \right|$$
$$\le \epsilon t_{N+1}^{1/4} \sqrt{\frac{2}{\pi} \int_0^T |y(x,t)|^2 \, dx} + E_k \sqrt{\int_0^T |y(x,t)|^2 \, dx} \; t_N^{-k}.$$

Therefore, if N is chosen such that $t_N = O\left(\epsilon^{-\frac{4}{4k+1}}\right)$, then

$$\epsilon O(t_N^{1/4}) + O(t_N^{-k}) = O\left(\epsilon^{\frac{4k}{4k+1}}\right).$$

Hence we arrive at

Theorem 1.15. *Let $F \in PW_T^{1/2}$ such that $t^{k-1/4}F(t) \in L^2(0,\infty)$. Let the sampled values $F_\epsilon(t_n)$ have random error ϵ. Then if N is chosen so that $t_N \sim \epsilon^{-\frac{4}{4k+1}}$ then*

$$F(t) - \sum_{n \le N} F_\epsilon(t_n) S_n^T(t) \sim \sqrt{\int_0^T |y(x,t)|^2 \, dx} \left\{ \sqrt{\frac{2}{\pi}} + E_k \right\} \epsilon^{\frac{4k}{4k+1}}.$$

For bandlimited signals we have

Theorem 1.16. *Let $F \in PW_T$ such that $t^{2k}F(t) \in L^2(0,\infty)$. Then the truncation error for the sampling formula (47) has the order*

$$F(t) - \sum_{n=1}^{N} \left[\left(1 + \frac{t}{\tau_n}\right) F(\tau_n) + \left(1 + \frac{t}{\tau_{-n}}\right) F(\tau_{-n}) \right] \tilde{S}_n^T(t) = O\left(\tau_N^{-2k}\right).$$

If the sampled values $F_\epsilon(\tau_n)$ have random error ϵ, and if N is chosen so that $\tau_N \sim \epsilon^{-\frac{2}{4k+1}}$ then

$$F(t) - \sum_{n=1}^{N} \left[\left(1 + \frac{t}{\tau_n}\right) F_\epsilon(\tau_n) + \left(1 + \frac{t}{\tau_{-n}}\right) F_\epsilon(\tau_{-n}) \right] \tilde{S}_n^T(t) = O\left(\epsilon^{\frac{4k}{4k+1}}\right).$$

The above results in sampling are new.

2. Sampling in Hardy Space

The same questions can be asked for the sampling in Hardy space $\mathcal{H}_+^2$, the Hilbert space of analytic functions in the upper half-plane which is defined by

$$\mathcal{H}_+^2 = \Bigg\{ F(x+iy) \text{ analytic for } y>0, \\ \|F\|_2 := \sup_{y>0} \sqrt{\int_{-\infty}^{\infty} |F(x+iy)|^2\, dx} < \infty \Bigg\}.$$

The Hardy space plays an important role in analysis, control theory, and differential equations. Here we shall address, in the context of regular sampling, a crucial question for the practitioner: is a sequence of values $\{F(s_0+ibn)\}_{n\geq 0}$ enough to interpolate $F(\lambda) \in \mathcal{H}_+^2$ for $\Im(\lambda) > 0$. Also how close can the recovery formula be from the WSK theorem given by (3), i.e. can we find a sequence of sampling functions S_n such that for $F \in \mathcal{H}_+^2$

$$F(\lambda) = \sum_{n\geq 0} F(z_0+ibn) S_n(\lambda)? \tag{48}$$

Such a formula would extend the WSK theorem from the Paley-Wiener spaces to the Hardy space. We shall see very soon that, unfortunately such a formula, as (48), is impossible. Nevertheless a recovery formula, based on the expansion of the kernel of the Fourier transform, is possible. This would be the main result of this part. At the end of the paper we provide estimates for the truncation error, and a new series representation for the Riemann zeta function $\xi(z)$. Recall that the Fourier transform, restricted to the positive half-line,[12]

$$F(\lambda) = \mathcal{F}(f)(\lambda) = \int_0^{\infty} e^{i\lambda t} f(t)dt, \quad \lambda = x+iy, \quad y>0,$$

is a bijection between $L^2(0,\infty)$ and $\mathcal{H}_+^2$

$$f \in L^2(0,\infty) \text{ if and only if } \mathcal{F}(f) \in \mathcal{H}_+^2, \tag{49}$$

and moreover $\|F\|_2 = \sqrt{2\pi}\, \|f\|_{L^2(0,\infty)}$.

We first show that $\{s_0+ibn\}_{n\geq 0}$, where $b>0$ and $\Im(s_0) > 0$, is a regular sampling sequence, by proving uniqueness of the recovery.

Theorem 2.1. *Let $F_1,\ F_2 \in \mathcal{H}_+^2$, $\Im(s_0) > 0$, $b>0$ and $F_1(s_0+ibn) = F_2(s_0+ibn)$ for $n \geq 0$ then $F_1(s) = F_2(s)$.*

Proof. Since $F_1 - F_2 \in \mathcal{H}_+^2$, there exists $f \in L^2(0,\infty)$ such that $F_1 - F_2 = \mathcal{F}(f)$. Observe that $0 = F_1(s_0+ibn) - F_2(s_0+ibn)$ can be considered as the nth moment of the function $\dfrac{t^{(-is_0/b-1)}}{b} f\left(\frac{-\ln t}{b}\right)$ over $(0,\ 1)$, then a standard result in the theory of moments yields $f = 0$, Ref. 13, Theorem 5.3,p. 22. Thus $F_1(\lambda) = F_2(\lambda)$ for any $\Im(\lambda) > 0$, and so the values $\{F(s_0 + ibn)\}_{n \geq 0}$ are enough to determine a unique F in $\mathcal{H}^2_+$. □

We will show now that WKS-type sampling formula (48) is impossible.

Theorem 2.2. *There are no sampling functions $S_n(\lambda)$ such that (48) holds for functions in H^2_+.*

Proof. Assume that there are sampling functions S_n such that (48) holds for $F \in H^2_+$. Then for any $f \in L^2(0,\infty)$ with compact support

$$\int_0^\infty e^{i\lambda t} f(t)dt = F(\lambda) = \sum_{n\geq 1} S_n(\lambda) \int_0^\infty e^{-nt} f(t)dt$$

$$= \int_0^\infty \left(\sum_{n\geq 1} S_n(\lambda) e^{-nt} \right) f(t)dt.$$

Since the subset of functions with compact support is dense in $L^2(0,\infty)$ we must have for $\Im(\lambda) > 0$

$$e^{i\lambda t} = \sum_{n\geq 1} S_n(\lambda) e^{-nt} \quad \text{in} \quad L^2(0,\infty).$$

Set $e^{-t} = x$ to obtain

$$x^{-i\lambda} = \sum_{n\geq 1} S_n(\lambda) x^n \ \text{ in } \ L^2_{\frac{1}{x}}(0,1), \tag{50}$$

or equivalently

$$x^{-i\lambda - \frac{1}{2}} = \sum_{n\geq 1} S_n(\lambda) x^{n-\frac{1}{2}} \ \text{ in } \ L^2(0,1).$$

The convergence of the series in $L^2(0,1)$ implies

$$\left\| S_n(\lambda) x^{n-\frac{1}{2}} \right\|_{L^2(0,1)} \to 0 \ \text{ as } \ n \to \infty.$$

In other words, we have

$$|S_n(\lambda)| \sqrt{\int_0^1 x^{2n-1} dx} = |S_n(\lambda)| \frac{1}{\sqrt{2n}} = o(1) \Rightarrow |S_n(\lambda)| \leq M\sqrt{n}.$$

Hence, the series $\sum_{n\geq 1} S_n(\lambda) x^n$ is analytic in the disk $|x| < 1$, Ref. 1. From (50), both the series $\sum_{n\geq 1} S_n(\lambda) x^n$ and $x^{-i\lambda}$ coincide on $(0,1)$ and by the principle of analytic continuation, $x^{-i\lambda}$ should also be analytic in the unit disk, which is impossible and therefore (48) cannot hold. □

2.1. *Sampling formula*

We now present the main result of this part. Without loss of generality, we consider the cases $s_0 = \frac{1}{2}i$ or i, and $b = 1$, so the sampling points are either $\left\{\left(n+\frac{1}{2}\right)i\right\}_{n\geq 0}$ or $\{(n+1)\,i\}_{n\geq 0}$. The key idea is to expand the kernel $e^{i\lambda t}$ in the Fourier transform in terms of e^{-nt}. Because the system $\{e^{-nt}\}_{n\geq 0}$ is not orthogonal in $L^2(0,\infty)$, we need to use the Gram-Schmidt orthogonalization process, which amounts to expanding $x^{i\lambda}$ in terms of Legendre polynomials. We first recall the formula for the shifted factorial

$$(a)_k = a\,(a+1)\ldots(a+k-1) = \frac{\Gamma\,(a+k)}{\Gamma\,(a)}.$$

We have

Theorem 2.3. *Let $F \in \mathcal{H}^2_+$. Then*

$$F(\lambda) = \sum_{k=0}^{\infty} \frac{(2k+1)\left(\frac{1}{2}+i\lambda\right)_k}{\left(\frac{1}{2}-i\lambda\right)_{k+1}} \sum_{n=0}^{k} \frac{(-k)_n\,(k+1)_n}{(n!)^2} F\left(in+\frac{i}{2}\right), \tag{51}$$

where the series converges uniformly on any compact subset of the upper half plane, and

$$\sum_{k=0}^{\infty} \left| \sum_{n=0}^{k} \frac{(-k)_n\,(k+1)_n}{(n!)^2} F\left(in+\frac{i}{2}\right)\right|^2 < \infty.$$

Conversely, if $\{f_n\}$ is a sequence of complex numbers such that

$$\sum_{k=0}^{\infty} \left| \sum_{n=0}^{k} \frac{(-k)_n\,(k+1)_n}{(n!)^2} f_n\right|^2 < \infty, \tag{52}$$

then the series

$$\sum_{k=0}^{\infty} \frac{(2k+1)\left(\frac{1}{2}+i\lambda\right)_k}{\left(\frac{1}{2}-i\lambda\right)_{k+1}} \sum_{n=0}^{k} \frac{(-k)_n\,(k+1)_n}{(n!)^2} f_n,$$

converges uniformly on any compact subset of $\Im(\lambda) > 0$ to a function $F \in \mathcal{H}^2_+$, and moreover

$$F\left(in+\frac{i}{2}\right) = f_n$$

for any $n \in \mathbb{N}$.

Proof. First observe that, by setting $t = -\ln(x)$, we have $f(t) \in L^2(0,\infty)$ if, and only if, $g(x) := f(-\ln x)x^{-1/2} \in L^2(0,1)$. Thus $F \in \mathcal{H}^2_+$ if, and only if,

$$F(\lambda) = \int_0^1 g(x)x^{-i\lambda-1/2}dx, \quad \Im(\lambda) > 0, \qquad g \in L^2(0,1). \tag{53}$$

Let $P_k(x)$ be Legendre polynomials.[39] Then $P_k^*(x) = \sqrt{2k+1}\,P_k(1-2x)$, yields an orthonormal system of polynomials in $L^2(0,1)$ Ref. 39, and from the fact $P_k(1-2x) = F(-k, k+1; 1, x)$ we have

$$P_k^*(x) = \sqrt{2k+1}\sum_{n=0}^{k}\frac{(-k)_n\,(k+1)_n}{(n!)^2}x^n = \sum_{n=0}^{k} a_{kn}x^n, \tag{54}$$

where

$$a_{kn} = \sqrt{2k+1}\,\frac{(-k)_n\,(k+1)_n}{(n!)^2}.$$

In order to use (53), we expand $x^{-i\lambda-\frac{1}{2}}$ in Fourier-Legendre series of Legendre polynomials $P_k^*(x)$

$$x^{-i\lambda-\frac{1}{2}} = \sum_{k=0}^{\infty} c_k(\lambda)P_k^*(x) = \sum_{k=0}^{\infty} c_k(\lambda)\sum_{n=0}^{k} a_{kn}x^n, \quad 0 < x < 1, \tag{55}$$

where convergence holds in $L^2(0,1)$, and uniformly on any compact subset of $\mathbb{R}^2_+$, the upper half plane. Here

$$c_k(\lambda) = \int_0^1 x^{-i\lambda-1/2}P_k^*(x)dx = \frac{\sqrt{2k+1}\;(1/2+i\lambda)_k}{(-i\lambda+1/2)_{k+1}}.$$

Having expressed $x^{-i\lambda-1/2}$ in terms of x^n in (55), we now use (53) to go back to the Fourier transform

$$\begin{aligned}
F(\lambda) &= \sum_{k=0}^{\infty} c_k(\lambda)\int_0^1 g(x)P_k^*(x)\,dx \\
&= \sum_{k=0}^{\infty} c_k(\lambda)\int_0^1 g(x)\sum_{n=0}^{k} a_{kn}x^n\,dx \\
&= \sum_{k=0}^{\infty} c_k(\lambda)\sum_{n=0}^{k} a_{kn}\int_0^1 g(x)x^n\,dx
\end{aligned}$$

$$= \sum_{k=0}^{\infty} c_k(\lambda) \sum_{n=0}^{k} a_{kn} \int_0^1 g(x) x^{-i(in+i/2)-1/2}\, dx$$
$$= \sum_{k=0}^{\infty} c_k(\lambda) \sum_{n=0}^{k} a_{kn} F(in + i/2).$$

The convergence is uniform on any compact subset of $\Im(\lambda) > 0$. Since $g \in L^2(0,1)$, then $F \in H_+^2$ and $\left\{\sum_{n=0}^{k} a_{kn} F(in+i/2)\right\}_{k\geq 0}$ being the Fourier-Legendre coefficients of $g(x) \in L^2(0,1)$ in the Legendre expansion, must be in ℓ^2

$$\sum_{k=0}^{\infty} \left| \sum_{n=0}^{k} a_{kn} F(in + i/2) \right|^2 = \int_0^1 |g(x)|^2\, dx < \infty.$$

Conversely, let $\{f_n\}_{n=0}^{\infty}$ be a sequence of complex numbers satisfying

$$\sum_{k=0}^{\infty} \left| \sum_{n=0}^{k} a_{kn} f_n \right|^2 < \infty.$$

Denote

$$g_k = \sum_{n=0}^{k} a_{kn} f_n. \tag{56}$$

Then obviously $\sum_{k=0}^{\infty} |g_k|^2 < \infty$ and the function

$$g(x) := \sum_{k=0}^{\infty} g_k\, P_k^*(x)$$

belongs to $L^2(0,1)$. Use (53) to define $F \in \mathcal{H}_+^2$, where its Fourier-Legendre coefficients g_k satisfy

$$g_k = \int_0^1 g(x) P_k^*(x) dx = \sum_{n=0}^{k} a_{kn} \int_0^1 g(x) x^n\, dx \tag{57}$$
$$= \sum_{n=0}^{k} a_{kn} F(in + i/2).$$

So (56) and (57) lead to a triangular system

$$\sum_{n=0}^{k} a_{kn} F(in + i/2) = \sum_{n=0}^{k} a_{kn} f_n, \quad k \in \mathbf{N}.$$

From the fact that $a_{kk} \neq 0$ for any k, the system leads to

$$F(in + i/2) = f_n, \quad n \in \mathbf{N}. \tag{58}$$

Since $F \in \mathcal{H}_+^2$ and its values at $in + i/2$ agrees with the f_n, by (58), we have

$$\begin{aligned} F(\lambda) &= \sum_{k=0}^{\infty} \frac{(2k+1)\left(\frac{1}{2}+i\lambda\right)_k}{(-i\lambda+\frac{1}{2})_{k+1}} \sum_{n=0}^{k} \frac{(-k)_n\,(k+1)_n}{(n!)^2} F\,(in+i/2) \\ &= \sum_{k=0}^{\infty} \frac{(2k+1)\left(\frac{1}{2}+i\lambda\right)_k}{\left(-i\lambda+\frac{1}{2}\right)_{k+1}} \sum_{n=0}^{k} \frac{(-k)_n\,(k+1)_n}{(n!)^2} f_n. \end{aligned}$$

□

By a simple translation we can sample at the integer $i(n+1)$ instead of $in + i/2$. To this end, we only need to translate functions by $i/2$ upward, i.e. if $F \in \mathcal{H}_+^2$ then $F\,(\lambda + i/2) \in \mathcal{H}_+^2$, $\Im\,(\lambda) > 0$ and (51) yields

$$F(\lambda + i/2) = \sum_{k=0}^{\infty} \frac{(2k+1)\,(i\lambda)_k}{(-i\lambda+1)_{k+1}} \sum_{n=0}^{k} \frac{(-k)_n\,(k+1)_n}{(n!)^2} F\,(in+i) \qquad \text{for } \Im\,(s) > 0,$$

and then change back $\lambda + i/2$ into λ to yield

Theorem 2.4. *For any $F \in \mathcal{H}_+^2$ we have for $\Im(\lambda) > \frac{1}{2}$*

$$F\,(\lambda) = \sum_{k=0}^{\infty} \frac{(2k+1)\,(i\lambda+1/2)_k}{(-i\lambda+1/2)_{k+1}} \sum_{n=0}^{k} \frac{(-k)_n\,(k+1)_n}{(n!)^2} F\,(in+i)\,. \tag{59}$$

and the series converges uniformly in any compact domain contained in $\Im(s) > \frac{1}{2}$ of the complex plane.

2.2. *Truncation error*

Recall that a function $f \in Lip_s$ if

$$|f(x) - f(y)|^s < M\,|x-y|^s\,,$$

where $M,\ s > 0$. It is easily seen that $x^{-i\lambda-1/2} \in Lip_{\Im(\lambda)-1/2}$ if $\Im\,(\lambda) > \frac{1}{2}$, see Ref. 11. Therefore we have an estimate for the remainder, Ref. 39

$$\left| x^{-i\lambda-\frac{1}{2}} - \sum_{k=0}^{N} \frac{(2k+1)\left(\frac{1}{2}+i\lambda\right)_k}{\left(-i\lambda+\frac{1}{2}\right)_{k+1}} \sum_{n=0}^{k} \frac{(-k)_n\,(k+1)_n}{n!n!} x^n \right| \le \frac{c}{\left|N^{\Im(\lambda)-1/2}\right|}, \tag{60}$$

where c is a certain constant, and the estimate is uniform for $x \in (0,1)$. Set $x = e^{-t}$, and multiply by $f(t)e^{-t\xi-t/2}$, and integrate over $(0,\infty)$, where $\xi \geq 0$, to obtain for $\Im(\lambda) > \frac{1}{2}$

$$\left|F(\lambda+i\xi) - \sum_{k=0}^{N} \frac{(2k+1)\left(\frac{1}{2}+i\lambda\right)_k}{\left(-i\lambda+\frac{1}{2}\right)_{k+1}} \sum_{n=0}^{k} \frac{(-k)_n (k+1)_n}{n!n!} F\left(in+\frac{i}{2}+i\xi\right)\right|$$
$$\leq \frac{c}{\left|N^{\Im(\lambda)-1/2}\right|} \int_0^\infty |f(t)|\, e^{-t\xi-t/2} dt.$$

If we replace $\lambda + i\xi$ by λ, we then get for $\Im(\lambda) > \xi + \frac{1}{2}$

$$\left|F(\lambda) - \sum_{k=0}^{N} \frac{(2k+1)\left(\frac{1}{2}+\xi+i\lambda\right)_k}{\left(-i\lambda-\xi+\frac{1}{2}\right)_{k+1}} \sum_{n=0}^{k} \frac{(-k)_n (k+1)_n}{n!n!} F\left(in+\frac{i}{2}+i\xi\right)\right|$$
$$\leq \frac{c}{\left|N^{\Im(\lambda)-\xi-1/2}\right|} \int_0^\infty e^{-\xi t-t/2} |f(t)|\, dt,$$

where we choose ξ so that $|f(t)|\, e^{-t\xi-t/2}$ is integrable.

Theorem 2.5. *If $F \in \mathcal{H}_+^2$ then the truncation error for $\Im(\lambda) > \xi + \frac{1}{2}$ and $\xi > 0$ is given by*

$$\left|F(\lambda) - \sum_{k=0}^{N} \frac{(2k+1)(\xi+i\lambda)_k}{(-i\lambda+1-\xi)_{k+1}} \sum_{n=0}^{k} \frac{(-k)_n (k+1)_n}{n!n!} F(in+i\xi)\right|$$
$$\leq \frac{c}{\sqrt{2\pi}\sqrt{2\xi+1}\left|N^{\Im(\lambda)-\xi-1/2}\right|} \|F\|_2 .$$

In the last estimate we used the fact that $\|F\|_2 = \sqrt{2\pi}\,\|f\|_{L^2(0,\infty)}$ and in all of the above formulae c is the constant in (60).

2.3. *Example*

Here we would like to show how to interpolate the Riemann zeta function $\xi(s)$. Recall the fact that $\frac{1}{s}\xi(s) = \int_0^\infty e^{-st}\chi(t)dt$, where

$$\chi(t) = n \quad \text{if} \quad \ln(n) < t < \ln(n+1)\,.$$

For convergence purposes we should recast the Laplace transform as a Fourier transform

$$\widetilde{\xi}(\lambda) = \frac{1}{-i\lambda+\frac{1}{2}}\xi\left(-i\lambda+\frac{1}{2}\right) = \int_0^\infty e^{i\lambda t}e^{-t/2}\chi(t)dt,$$

and thus $\widetilde{\xi} \in \mathcal{H}_+^2$ since χ has a slow growth. From (51), $\widetilde{\xi}$ can be sampled for $\Im(\lambda) > \frac{1}{2}$ by

$$\widetilde{\xi}(\lambda) = \sum_{k=0}^{\infty} \frac{(2k+1)(i\lambda + 1/2)_k}{(-i\lambda + 1/2)_{k+1}} \sum_{n=0}^{k} \frac{(-k)_n (k+1)_n}{(n!)^2} \widetilde{\xi}(in + i/2)$$

and setting $-i\lambda + \frac{1}{2} = s$ yields

$$\frac{1}{s}\xi(s) = \sum_{k=0}^{\infty} \frac{(2k+1)(1-s)_k}{(s)_{k+1}} \sum_{n=0}^{k} \frac{(-k)_n (k+1)_n}{n!n!(n+1)} \xi(n+1),$$

$$\xi(s) = \sum_{k=0}^{\infty} \frac{(2k+1)(1-s)_k}{(s+1)_k} \sum_{n=0}^{k} \frac{(-k)_n (k+1)_n}{n!(n+1)!} \xi(n+1). \qquad (61)$$

Recall that Euler has already computed $\xi(2)$ to $\xi(26)$ for even n, while Stieltjes determined the values of $\xi(2), \ldots, \xi(70)$ to 30 digits of accuracy in 1887.[38] While the series representation for the Riemann zeta function converges in the domain $\Re(s) > 1$, the sampling formula (61) gives a series representation that is convergent in a larger domain, namely $\Re(s) > \frac{1}{2}$.

References

1. N. Achieser, *Theory of Appoximation* (Dover, 1992).
2. R. P. Boas, *Entire Functions* (Academic Press, New York, 1954).
3. J. J. Benedetto and P. G. Ferreira, *Applied and Numerical Harmonic Analysis* (Birkhäuser, Boston, 2001).
4. J. J. Benedetto and A. Zayed, *Applied and Numerical Harmonic Analysis* (Birkhäuser, Boston, 2004).
5. A. Beurling and P. Malliavin, *Acta Math.*, **118**, 79 (1967).
6. A. Boumenir, *J. Fourier Anal. Appl.*, **5**, 377 (1999).
7. A. Boumenir, *Math. Comp.*, **68**, 1057 (1999).
8. A. Boumenir and A. Zayed, *J. Fourier Anal. Appl.*, **8**, 211 (2002).
9. P. L. Butzer, *J. Math. Res. Exposition*, **3**, 185 (1983).
10. P. L. Butzer and R. L. Stens, *SIAM Rev.*, **34**, 40 (1992).
11. P. J. Davis, *Interpolation and Approximation* (Dover, 1975).
12. V. A. Ditkin and A. P. Prudnikov, *Integral Transforms and Operational Calculus* (Pergamon Press, 1965).
13. G. Doetsch, *Introduction to the Theory and Application of the Laplace Transformation* (Springer, 1970).
14. H. Dym and H. P. McKean, *Gaussian Processes, Function Theory and Inverse Spectral Problem* (Academic Press, 1976).
15. Y. Eldar, *Sampling without input constraint: Consistent reconstruction in arbitrary spaces*, in *Sampling, Wavelets, and Tomography* (Birkhäuser, 2004), pp. 33–59.

16. A. Eremenko and D. Novikov, *J. Math. Pures Appl.*, **83**, 313 (2004).
17. A. Faridani, *Sampling theory and parallel-beam tomography,* in *Sampling, Wavelets, and Tomography,*(Birkhäuser, 2004) pp. 225–253.
18. H. G. Feichtinger and K. Gröchenig, SIAM J. Math. Anal., **23**, 244 (1992).
19. G. Freiling and V. Yurko, *Inverse Sturm-Liouville Problems and Their Applications* (Nova Science, 2001).
20. M. G. Gasymov and B. M. Levitan, *Russian Math. Surveys*, **19**, 1 (1964).
21. I. M. Gelfand and B. M. Levitan, *Amer. Math. Transl.*, **1**, 239 (1951).
22. F. Gesztesy and B. Simon, *Annals of Math.*, **152**, 593 (2000).
23. K. Gröchenig, *Math. Comp.*, **59**, 181 (1992).
24. M.I. Kadec, *Soviet Math. Dokl.*, **5**, 559 (1964).
25. H. P. Kramer, *J. Math. Phys.*, **38**, 68 (1959).
26. M. G. Krein, *Dokl. Akad. Nauk SSR*, **88**, 405 (1953).
27. M. G. Krein, *Dokl. Akad. Nauk SSR*, **113**, 970 (1957).
28. M. G. Krein and I. S. Kac, *Amer. Math. Soc. Transl*, **103**, 19 (1970).
29. N. Levinson, *Amer. Math. Soc. Colloq. Publs.*, **26**(1940).
30. B. M. Levitan, *Inverse Sturm-Liouville Problems* (VNU Science Press, Utrech, 1987).
31. B. Logan, *Properties of High-Pass Signals,* in *Thesis, Department of Electrical Engineering* (Columbia University, New York, 1965).
32. V. A. Marchenko, *Operator Theory: Advances and Applications 22* (Birkhäuser, 1986).
33. F. Marvasti, ed., *Nonuniform Sampling: Theory and Application* (Kluwer Academic Plenum, New York, 2001).
34. F. Natterer, *SIAM. J. Appl. Math.*, **53**, 358 (1993).
35. F. Natterer, *Computational Radiology and Imaging* (Minneapolis, MN, 1997) pp. 17–32.
36. K. Seip, *Interpolation and Sampling in Spaces of Analytic Functions,* ULECT 33, (American Mathematical Society, 2004).
37. K. Seip, *SIAM J. Appl. Math.*, **47**, 1112 (1987).
38. H. M. Srivastava, *J. Math. Anal. Appl.*, **246**, 331 (2000).
39. P. K. Suetin, *Classical Orthogonal Polynomials* (Nauka, Moscow, 1979).
40. E. C. Titchmarsh, *Proc. London Math. Soc.*, **25**, 283 (1926).
41. E. C. Titchmarsh, *Theory of the Fourier Integral* (Oxford University Press, 1948).
42. P. Vaidyanathan, *Sampling theorems for non-bandlimited signals,* in *Sampling, Wavelets, and Tomography*(Birkhäuser, 2004) pp. 115–135.
43. Vu Kim Tuan, *J. Fourier Anal. Appl.*, **4**, 315 (1998).
44. Vu Kim Tuan, *Numer. Funct. Anal. and Optimiz.*, **20**, 387 (1999).
45. Vu Kim Tuan, *Frac. Cal. & Appl. Anal.* **2**, 135 (1999).
46. Vu Kim Tuan and A. I. Zayed, *Results in Math.*, **38**, 362 (2000).
47. Vu Kim Tuan, *J. Fourier Anal. Appl.*, **7**, 319 (2001).
48. Vu Kim Tuan and A. I. Zayed, *J. Math. Anal. Appl.*, **266**, 200 (2002).
49. Vu Kim Tuan, *Adv. Appl. Math.*, **29**, 563 (2002).
50. Vu Kim Tuan,*Proceedings of the International Conference on Abstract and Applied Analysis 2002, held in Hanoi, Vietnam, August, 2002,* eds., Nguyen

Minh Chuong, L. Nirenberg et al., (World Scientific, 2004) pp. 561–567.
51. D. Walnut, *J. Fourier Anal. Appl.*, **2**, 435 (1996).
52. V. A. Yavryan, *J. Contemp. Math. Anal.*, **27**, 75 (1992).
53. R. M. Young, *An Introduction to Nonharmonic Fourier Series* (Academic Press, 1980).
54. A. Zayed, *Advances in Shannon's Sampling Theory* (CRC Press, 1993).
55. A. Zygmund, *Trigonometric Series* (Cambridge University Press, 2003).

Harmonic, Wavelet and p-Adic Analysis
Eds. N. M. Chuong *et al.* (pp. 211–227)

§10. QUANTIZED ALGEBRAS OF FUNCTIONS ON AFFINE HECKE ALGEBRAS*

DO NGOC DIEP

Institute of Mathematics, VAST,
18 Hoang Quoc Viet Road, Cau Giay District, 10307, Hanoi, Vietnam
E-mail: dndiep@math.ac.vn

The so called quantized algebras of functions on affine Hecke algebras of type A and the corresponding q-Schur algebras are defined and their irreducible unitarizable representations are classified.

Introduction

The algebras of functions on groups define the structure of the groups themselves: the algebras of continuous functions on topological groups define the structure of the topological groups. This essentially is the so called Pontryagin duality for Abelian locally compact groups and the Tannaka-Krein duality theory for compact groups. The smooth functions on Lie groups define the structure of Lie groups. It is the essential fact that in this case we can produce the harmonique analysis on genral Lie groups. The quantized algebras of functions on quantum groups defined the structure of quantum groups etc. In the same sense we define quantized algebras of functions which define the structure of quantum affine Hecke algebras. Let us discuss a little bites in more detail. Let us denote by $\mathfrak{g}$ a Lie algebra over the field of complex numbers, $U(\mathfrak{g})$ its universal eveloping algebra, $\lambda \in P^*$ a positive highest weight, $V_v(\lambda)$ the associated representation of type I, i.e. with a positive defined Hermite form $(.,.)$ and $(xv_1, v_2) = (v_1.x^*v_2), \forall v_1, v_2 \in V_v(\lambda)$, of the quantized universal enveloping algebra $U_v(\mathfrak{g})$. Let $\{v^\nu_\mu\}$ be an orth-

*The work was supported in part by National Foundation for Research in Fundamental Sciences, Vietnam, Alexander von Humboldt Foundation, Germany, and was completed during the visit of the author at the Department of Mathematics, The University of Iowa, U.S.A.
The author thanks the organizers of the conference and especially Professor DSc. Nguyen Minh Chuong for invitation to partcipate and give talk at the conference.

ogonal basis of $V_v(\lambda)$. Consider the matrix elements of the representation defined by

$$C^{\lambda}_{\nu,s;\mu,r}(x) := (xv^r_\mu, v^s_\nu)$$

and the linear span $\mathcal{F}_v(G) := \langle C^{\lambda}_{\nu,s;\mu,r}\rangle$. It was shown in L. Korogodski and Y. Soibelman[7] that indeed it is equipped with a structure of an Hopf algebra, the so called the quantized algebra of functions on the quantum group corresponding to G. It was shown also that this algebra is generalized by the matrix coefficients of the standard representation of G in the case $G = \mathrm{SL}_2$, i.e. the algebra of functions on quantum group SL_2 is generalized by the matrix coefficients $t_{11}, t_{12}, t_{21}, t_{22}$ with the relations

$$\begin{aligned}
&t_{11}t_{12} = v^{-2}t_{12}t_{11},\ t_{11}t_{21} = v^{-2}t_{21}t_{11}\\
&t_{12}t_{22} = v^{-2}t_{22}t_{12},\ t_{21}t_{22} = v^{-2}t_{22}t_{21}\\
&t_{12}t_{21} = t_{21}t_{12}, \qquad t_{11}t_{22} - t_{22}t_{11} = (v^{-2} - v^2)t_{12}t_{21}\\
&\qquad\qquad\qquad\qquad t_{11}t_{12} - v^{-2}t_{12}t_{21} = 1
\end{aligned}$$

From this presentation of the algebra, L. Korogodski and Y. Soibelman[7] obtained the description of all the irreducible (infinite-dimensional) unitarizable representations of the quantized algebra of functions $\mathcal{F}_v(G)$: For the particular case of $\mathcal{F}_v(\mathrm{SL}_2(\mathbb{C}))$, its complete list of irreducible unitarizable representations consists of:

- One dimensional representations $\tau_t, t \in \mathbb{S}^1 \subset \mathbb{C}$, defined by $\tau_t(t_{11}) = t, \tau_t(t_{22}) = t^{-1}, \tau_t(t_{12}) = 0, \tau(t_{21}) = 0$.
- Infinite-dimensional unitarizable $\mathcal{F}_v(\mathrm{SL}_2(\mathbb{C}))$-modules $\pi_t, t \in \mathbb{S}^1$ in $\ell^2(\mathbb{N})$, with an orthogonal basis $\{e_k\}_{k=0}^{\infty}$, defined by

$$\begin{aligned}
&\pi_t(t_{11}) : \begin{cases} e_0 \mapsto 0, \\ e_k \mapsto \sqrt{1-v}e_{k-1},\ k > 0, \end{cases}\\
&\pi_t(t_{22}) : e_k \mapsto \sqrt{1-v}e_{k+1},\ k \geq 0\,,\\
&\pi_t(t_{12}) : e_k \mapsto tv^{2k}e_k,\ k \geq 0\,,\\
&\pi_t(t_{21}) : e_k \mapsto t^{-1}v^{2k+1}e_k,\ k \geq 0\,.
\end{aligned}$$

For the general case of $\mathcal{F}_v(G)$, consider the algebra homomorphism $\mathcal{F}_v(G) \to \mathcal{F}_v(\mathrm{SL}_2(\mathbb{C}))$, dual to the canonical inclusion $\mathrm{SL}_2(\mathbb{C}) \hookrightarrow G_{\mathbb{C}}$. Then every irreducible unitarizable representation of the quantized algebra of functions $\mathcal{F}_v(G)$ is equivalent to one of the representations from the list:

- The representations $\tau_t, t = \exp(2\pi\sqrt{-1}x) \in \mathbb{T} = \mathbb{S}^1$,

$$\tau_t(C^{\lambda}_{\nu,s;\mu,r}) = \delta_{r,s}\delta_{\mu,\nu}\exp(2\pi\sqrt{-1}\mu(x)).$$

- The representations $\pi_i = \pi_{s_{i_1}} \otimes \cdots \otimes \pi_{s_{i_k}}$, if $w = s_{i_1}...s_{i_k}$ is the reduced decomposition of the element w into a product of reflections, where the representations π_{s_i} is the composition of the homomorphisms

$$\pi_{s_i} = \pi_{-1} \circ p : F_v(G) \twoheadrightarrow \mathcal{F}_v(\mathrm{SL}_2(\mathbb{C})) \longrightarrow \mathrm{End}\,\ell^2(\mathbb{N})$$

The purpose of this paper is to obtain the same kind results for the quantized algebras of functions on affine Hecke algebras and quantum Schur-Weyl algebras. [Remark that it should be more reasonable to name them as the quantized algebras of functions on *quantum affine Weyl groups*, but the "non-affine counterpart" - *the quantum Weyl group* terminology was reserved by L. Korogodski and Y. Soibelman for some objects of different kind - the algebras generated not only by the quantized reflections but also the quantized universal algebra.]

We start from the following fundamental remarks:

- The affine Hecke algebras $\mathbb{H}(v, W^r_{aff})$ and the v-Schur algebras $\mathbb{S}_{n,r}(v)$ are in a complete Schur-Weyl duality. It is therefore easy to conclude that the corresponding quantized algebras of functions, what we are going to define are also in a complete Schur-Weyl duality.
- The negative universal enveloping algebras $U_v^-(\hat{\mathfrak{sl}}_n) \otimes_A R$, where $A = \mathbb{C}[v, v^{-1}]$, R is the center of $U_v^-(\hat{\mathfrak{sl}}_n)$, is isomorphic to the Hall algebras $U_n^-(\mathbb{H}(v, W^r_{\mathrm{aff}}))$ and there is a natural map Θ from the last onto the v-Schur algebra $\mathbb{S}_{n,r}(v)$. From this we then have some maps between the quantized algebras of functions

$$\mathcal{F}(\mathbb{S}_{n,r}(v)) \to \mathcal{F}(U_v^-(\hat{\mathfrak{sl}}_n)) \to \mathcal{F}(U_v(\hat{\mathfrak{sl}}_n)).$$

  The irreducible representations of $\mathcal{F}(\mathbb{S}_{n,r}(v))$ could be found in the set of restrictions of irreducible untarizable representations of the quantized algebras $\mathbb{C}[\mathrm{SL}_r]_q, 0 < q < 1$, of functions on the quantum group of type SL_r.
- For complex algebraic groups G the irreducible unitarizable $\mathbb{C}[G]_q$-modules are completely described[7] for $0 < q < 1$.

Our main result describes the complete set of irreducible unitarizable $\mathcal{F}(\mathbb{H}(v, W^r_{\mathrm{aff}}))$-modules and $\mathcal{F}_v(\mathbb{S}(n,d))$-modules, Theorems 2.1, 2.2, 3.1, 3.2.

Let us describe the paper in more detail: Section 1 is a short introduction to the related subjects and we define the quantized algebras of functions $\mathcal{F}_v(W^r_{\mathrm{aff}}) = \mathcal{F}(\mathbb{H}(v, W^r_{\mathrm{aff}}))$ and $\mathcal{F}_v(\mathbb{S}(n,d)) := \mathcal{F}(\mathbb{S}_{n,r}(v))$. In §2 we give a full description of all irreducible unitarizable representations of $\mathcal{F}_v(W^r_{\mathrm{aff}})$. In §3 we do the same for the v-Schur algebras $\mathcal{F}(\mathbb{S}_{n,r}(v))$.

NOTATION. Let us fix some conventions of notation. Denote F a ground local field of characteristic p, $\mathbb{C}$ the field of complex numbers, $\mathbb{Z}$ the ring of integers, SL_r the special linear groups of matrices of sizes $r \times r$ with determinant 1, $\mathbb{G}$ an algebraic group, $G = \mathbb{G}(F)$ the group of rational F-points, $\mathbb{T}$ some maximal torus in G, $X^*(\mathbb{T})$ the root lattice, $X_*(\mathbb{T})$ the co-root lattice, $\mathbb{C}[G]_q$ the quantized algebra of complex-valued functions on quantum group associated to G, $\mathbb{S}_{n,r}(q)$ the q-Schur algebra, $\mathbb{S}_{n,r}(v)$ the v-Schur algebra, $\mathcal{F}(\mathbb{H}(v, W^r_{\text{aff}}))$ the quantized algebra of functions on affine Hecke algebra, $\mathcal{F}(\mathbb{S}_{n,r}(v))$ the quantized algebra of functions on quantum v-Schur algebra.

1. Definition of the Quantized Algebras of Functions

We introduce in this section the main objects - the quantized algebras $\mathcal{F}_v(W^r_{\text{aff}})$ of functions on affine Hecke algebras. As remarked in the introduction, it should be better to name the quantized algebras of functions on quantum affine Weyl groups, but we prefer in this paper this terminology in order to avoid any confusion with the terminology from L. Korogodski and Y. Soibelman.[7]

1.1.

1.1.1. *p-adic presentation*

Let us first recall the definition of Iwahori-Hecke algebras. Let F be a p-adic field, i.e. a finite extension of $\mathbb{Q}_q$, which is by definition the completion with respect to the ultra-metric norm of the rational field of the ring $\mathbb{Z}_p := \varprojlim \mathbb{Z}/p^n\mathbb{Z}$. Denote $\mathcal{O}$ the ring of integers in F, $\mathcal{O}^\times$ the group of units in $\mathcal{O}$, $G = \mathrm{SL}_2(F)$, $B = \{\begin{bmatrix} x & y \\ 0 & x^{-1} \end{bmatrix}; x, y \in F, x \neq 0\}$ the Borel subgroup of G, $T = \{\begin{bmatrix} x & 0 \\ 0 & x^{-1} \end{bmatrix}; x \in \mathcal{O}^\times\}$ the maximal torus, and $N = \{\begin{bmatrix} 1 & y \\ 0 & 1 \end{bmatrix}; y \in \mathcal{O}\}$ the unipotent radical of G. It is easy to check that $B = TN$. Define the so called *Iwahori-Hecke subgroup*

$$I = \{\begin{bmatrix} x & y \\ \varpi z & w \end{bmatrix}; x, w \in \mathcal{O}^\times, y, z \in \mathcal{O}\},$$

where ϖ is the generic presentative in the presentation of the principal ideal $\mathcal{P} = \varpi\mathcal{O}$. Let us denote $\mu(x)$ the Haar measure on the locally compact group $G = \mathrm{SL}_2(F)$, $\mu(I) = vol(I)$ the volume of I with respect to this Haar measure, χ_I the characteristic function of the set I, $e_I := \frac{1}{\mu(I)}\chi_I$ the idempotent, $e_I^2 = e_I = e_I$, defining a projector. The *Iwahori-Hecke algebra* $\mathbf{IH}(G,I)$ is defined as

$$\mathbf{IH}(G,I) = e_I \mathbf{H}(G) e_I$$

$$= \{f : \mathrm{SL}_2(F) \to \mathbb{C}; f(hxk) \equiv f(x), \forall h,k \in I, f \in \mathbf{H}(G) := C_c^\infty(G)\},$$

where $\mathbb{H}(G) := C_c^\infty(G)$ is the involutive algebra of smooth (i.e. locally constant) functions on $\mathrm{SL}_2(F)$ with compact support, with the well-known convolution product

$$(f * g)(x) := \int_G f(y) g(y^{-1}x) d\mu(y)$$

and involution as usually. Recall that the affine Weyl group W^1_{aff} is defined as $\tilde{W}/\{\pm 1\}$, where $\tilde{W} = \langle D, Dw \rangle$, the group generated by two generators $w := \begin{bmatrix} 0 & 1 \\ -1 & 0 \end{bmatrix}$ and $D := \begin{bmatrix} \varpi & 0 \\ 0 & \varpi^{-1} \end{bmatrix}$. It is coincided with the dihedral group. Let us choose the following generators $w_1 = w = \begin{bmatrix} 0 & 1 \\ -1 & 0 \end{bmatrix}$ and $w_2 = \Pi := \begin{bmatrix} 0 & -\varpi^{-1} \\ \varpi & 0 \end{bmatrix}$. It is well-known the relations

$$w_1 w_2 w_1^{-1} = w_2^{-1}, \quad w_1^2 = -1, \quad w_2^2 = -1,$$

or the standard braid relations

$$w_1 w_2 w_1 = w_2 w_1 w_2, \quad w_1^2 = -1, \quad w_2^2 = -1.$$

The group W^1_{aff} is discrete and infinite, and every element of W^1_{aff} can be presented as a reduced word in w_1 and w_2 , namely $w = w_{i_1} \dots w_{i_k}$. The group G can be presented as the union of the double coset classes $G = I.W^1_{\mathrm{aff}}.I$. Let us denote f_w the characteristic function of the coset class $IwI, w \in W_{\mathrm{aff}}$. If $w = w_{i_1} \dots w_{i_k}$ is a reduced presentation of $w \in W^1_{\mathrm{aff}}$ then $f_{w_{i_1}} * \dots * f_{w_{i_k}}$ is independent of the reduced presentation of w and $f_w = f_{w_{i_1}} * \dots * f_{w_{i_k}}$. Let us denote $f_i = f_{w_i}, i = 1,2$. We have therefore a correspondence

$$w \in W^1_{\mathrm{aff}} \mapsto f_w \in \mathbf{IH}(G,I),$$

subject to the relations

$$\begin{cases} f_i f_j f_i = f_j f_i f_j \\ \quad f_i^2 = (q-1) f_i + q, \text{ with } q = (\mathcal{O} : \mathcal{P}). \end{cases}$$

Let us do a change of variable $v := \frac{1}{\sqrt{q}}$ then we have

$$\begin{cases} f_i f_j f_i = f_j f_i f_j \\ (f_i + 1)(f_i - v^{-2}) = 0. \end{cases}$$

This is the so called *Coxeter presentation* of the Iwahori-Hecke algebra in SL_2 case.

For rank r groups of type A, i.e. SL_r we have the same picture, see for example.[1] Let us consider also the Hecke algebra $\mathbf{H}(G) = C_c^\infty(G)$, of smooth (i.e. locally constant) functions on G with compact support, under convolution product and involution. Corresponding to the map of rings

$$\mathbb{F}_q \longleftarrow \mathcal{O} \longrightarrow F, \text{ with } q = p^\ell = (\mathcal{O} : \mathcal{P}), \text{ for some integer } \ell$$

we have the maps of the groups of rational points

$$\mathbb{G}(\mathbb{F}_q) \longleftarrow \mathbb{G}(\mathcal{O}) \longrightarrow \mathbb{G}(F).$$

The preimage in $\mathbb{G}(\mathcal{O})$ of the Borel subgroup $B(\mathbb{F}_q)$ is called the *Iwahori subgroup*. It was shown that $G = \mathbb{G}(F) = I.W^r_{\text{aff}}.I$. The Iwahori-Hecke algebra $\mathbf{IH}(G, I)$ is defined as the algebra of smooth I-bi-invariant functions with compact support on $\mathbb{G}(F)$ under convolution and involution as a sub-algebra of the Hecke algebra $\mathbf{H}(G) = C_c^\infty(G)$. Denote by f_{s_i} the characteristic function of the double coset class $I.s_i.I$ in $G = \cup_{w \in W^r_{\text{aff}}} I.w.I$, and normalize as in the rank one case we also obtain the relations

$$f_{s_i} f_{s_j} f_{s_i} = f_{s_j} f_{s_i} f_{s_j},$$

$$(f_{s_i} + 1)(f_{s_i} - v^{-2}) = 0, \text{ where } v = \frac{1}{\sqrt{q}}, q = p^l = (\mathcal{O} : \mathcal{P}),$$

$$f_\sigma f_\gamma = f_{\sigma\gamma} \text{ if } \ell(\sigma\gamma) = \ell(\sigma) + \ell(\gamma).$$

1.1.2. *Affine hecke algebras* $\mathbb{H}(v, W^r_{\text{aff}})$

As usually let us denote v the formal quantum parameter. *(Abstract) Iwahori-Hecke algebras or affine Hecke* are defined in two equivalent ways: in *Coxeter presentation* as group algebras of affine Weyl groups and in *Bernstein presentation* as some abstract algebras presented by generators with relations. In Coxeter presentation:

Definition 1.1. An (abstract) Iwahori-Hecke or affine Hecke algebra is an $\mathbb{C}[v, v^{-1}]$-algebra generated by $T_\sigma, \sigma \in W^r_{\text{aff}}$, subject to the relations:

$$T_{s_i} T_{s_j} T_{s_i} = T_{s_j} T_{s_i} T_{s_j},$$

$$(T_{s_i}+1)(T_{s_i}-v^{-2})=0,$$

$$T_\sigma T_\gamma = T_{\sigma\gamma} \text{ if } \ell(\sigma\gamma)=\ell(\sigma)+\ell(\gamma).$$

Let us go to the Bernstein presentation of affine Hecke algebras as some abstract algebras presented by generators with relations.

Definition 1.2. An affine Hecke algebra in Bernstein presentation is an $\mathbb{C}[v,v^{-1}]$-algebra with generators $T_i^{\pm}, i=1,\dots r-1$, and $X_j^{\pm}, j=1,\dots,r$, subject to the relations:

$$\begin{array}{ll} T_iT_i^- = 1 = T_i^-T_i, & (T_i+1)(T_i-v^{-2})=0, \\ T_iT_{i+1}T_i = T_{i+1}T_iY_{i+1}, & T_iT_j = T_jT_i, \text{ if } |i-j|>1, \\ X_iX_i^- = 1 = X_i^-X_i, & X_iX_j = X_jX_i, \\ T_iX_iT_i = v^{-2}X_{i+1}, & X_jT_i = T_iX_j, \text{ if } J \neq i, i+1. \end{array}$$

In this definition we denoted T_i in place of T_i^+, X_i in place of X_i^+, etc.... we keep this agreements in the future use.

The isomorphism between two definitions can be established as follows. Associate $T_{s_i} \mapsto T_i$ and $\tilde{T}_\sigma^{-1} \mapsto X_1^{\sigma_1}\dots X_r^{\sigma_r}$, where $\tilde{T}_\sigma := v^{\ell(\sigma)}T_\sigma$, if $\sigma = (\sigma_1,\dots,\sigma_r)$ is dominant.

1.2.

1.2.1. *Admissible representations of p-adic groups*

Let us recall that a representation of p-adic group is called *supercuspidal* iff all its matrix coefficients have compact support modulo the center of the group. It is well-known the following fact: Given any irreducible representation π of G, there exists a Levi subgroup L and a supercuspidal representation σ of L such that π is a sub-quotient of the induced representation $\imath_P^G(\sigma) := \operatorname{Ind}_P^G \operatorname{infl}\sigma$. Every representation of the form $\imath_P^G(\sigma)$ has finite length for any irreducible representation of P and the other pair (L',σ') has the same properties as (L,σ) if and only if there exists an element $x \in G$ such that $L' = xLx^{-1}$ and $\sigma' = \sigma^x$, where $\sigma^x(h) := \sigma(xhx^{-1})$. The pair (L,σ) is called a *cuspidal pair* and the conjugacy class of (L,σ) is called the *support* of π. Two pairs (L,σ) and (L',σ') are called *innertially equivalent* iff there exist $x \in G$ and $\chi \in \mathcal{X}_L^{unr}$ such that $L' = xLx^{-1}$ and $\sigma' = (\sigma\otimes\chi)^x$. Given an innertially equivalent class $s = \langle(L,\sigma)\rangle$ one defines the sub-category $\mathcal{R}^s(G)$ of the category $\mathcal{R}(G)$ of smooth representations, consisting of all representations, all the sub-quotients of which have support in s. One of the well-known result of Bernstein is the fact that $\mathcal{R}(G) = \times_s \mathcal{R}^s(G)$ as the direct product of categories. The category $\mathcal{R}_{cusp}(G) := \times_s \mathcal{R}^s(G)$. Another well-known result of I. Bernstein, A. Borel and P. Kutzko is the fact that there is an equivalence from the category of unramified representations $\mathcal{R}^{unr}(G)$, for $G = \mathrm{SL}_2$, to the category of finite dimensional representations of the Iwahori-Hecke algebra $\mathbb{H}(G, I)$. The general case was treated in numerous works, see for example, Henniart.[5]

1.2.2. *Dipper-James construction of irreducible finite dimensional representations of* $\mathbb{H}(v, W^r_{\mathrm{aff}})$

For affine Hecke algebras of type A_{r-1} there are constructions of all irreducible finite dimensional representations parametrized by Young tableaux, or partitions. Let us recall it in brief form. For each Young diagram λ a so called Specht $\mathbb{H}(v, W^r_{\mathrm{aff}})$-module S^λ was defined in Ref. 2 and for the value $v = q$ not a root of unity provide a complete list of irreducible finite dimensional representations of $\mathbb{H}(q, W^r_{\mathrm{aff}})$ modules.

If q is a primitive ℓth root of unity, Dipper and James[2] constructed also a complete set of $\mathbb{H}(q, W^r_{\mathrm{aff}})$ modules D^μ, parametrized though all Young diagram with at most $\ell - 1$ rows of equal length. Let us describe this construction in more detail. Let $\lambda = (\lambda_1, \dots, \lambda_r)$, $Y_\lambda = \mathfrak{S}_{\lambda_1} \times \cdots \times \mathfrak{S}_{\lambda_r} \subset \mathfrak{S}_n$. Define the symmetrization

$$\mathrm{Sym}_\lambda := \sum_{w \in Y_\lambda} T_w$$

and the anti-symmetrization

$$A_\lambda := \sum_{w \in Y_\lambda} (-q)^{n(n-1)/2 - \ell(w)} T_w .$$

Let S^λ be the submodule of the induced $\mathbb{H}(q, W^r_{\mathrm{aff}})$ module $W^\lambda \cong \mathbb{H}(q, W^r_{\mathrm{aff}}) \otimes_{\mathbb{H}(\lambda)} \mathbb{C}$, [where $\mathbb{H}(\lambda)$ is the sub-algebra generated by T_i such that $s_i \in Y_\lambda$], generated by $A'_\lambda W^\lambda$ for λ' is obtained from λ by interchanging rows with columns,

$$S^\lambda = \mathbb{H}(q, W^r_{\mathrm{aff}}) A_{\lambda'} \mathbb{H}(q, W^r_{\mathrm{aff}}) \,\mathrm{Sym}_\lambda .$$

It was proven that there exists an explicit basis of the $\mathbb{H}(v, W^r_{\mathrm{aff}})$ modules

$$\mathbb{H}(v, W^r_{\mathrm{aff}}) A_{\lambda'} \mathbb{H}(v, W^r_{\mathrm{aff}}) \,\mathrm{Sym}_\lambda \subseteq \mathbb{H}(v, W^r_{\mathrm{aff}}),$$

which is evaluable at $q \in \mathbb{C}^\times$ and such that the basis elements evaluated at q remain linearly independent over $\mathbb{C}$ for all $q \in \mathbb{C}^\times$. Let $(.,.)$ be the bilinear form on the $\mathbb{H}(v, W^r_{\mathrm{aff}})$ module W^λ. Then the modules $D^\lambda = S^\lambda/(S^\lambda \cap$

$(S^\lambda)^\perp$) are either 0 or simple. The Young diagram is called ℓ-regular iff it has at most $\ell - 1$ rows of equal length. The module D^μ is nonzero if and only if μ is ℓ-regular. We refer the reader to the original work of Dipper and James[2] for a detailed exposition.

1.2.3. *The langlands correspondence*

Recall that a representation of p-adic group G is called smooth if the stabilizer of any vector is an open-closed subgroup in G. Let us denote $\tilde{V}$ the contragradient representation of V, Let $\rho : G = \mathbb{G}(F) \to \operatorname{End} V$ be an admissible (i.e. smooth and $\tilde{\tilde{V}} = V$) representation of G. One of the most important properties of admissible representations of p-adic groups is the fact that the space V^I of I-invariant vectors in an admissible representation V, is finite dimensional. For every element f from the Iwahori-Hecke algebra $\mathbf{IH}(G, I) \cong C_c^\infty(I \setminus G/I)$ we associate an operator in finite dimensional vector space V^I,

$$\rho(f) := \int_{\mathbb{G}(F)} f(x)\rho(x)dx.$$

It is not hard to see that this correspondence gives us a representation of the Iwahori-Hecke algebra $\mathbf{IH}(G, I)$ in the finite dimensional space V^I. It was proven that the correspondence $V \mapsto V^I$ provides a functor from, and is indeed an equivalence between the category of admissible representations of G generated by I-fixed vectors and the category of finite dimensional representations of the Iwahori-Hecke algebra $\mathbf{IH}(G, I) \cong \mathbb{H}(v, W^r_{\text{aff}})|_{v=q}$. This result was essential proven by A. Borel, P. Kutzko end Bernstein in rank one case and by Harris-Taylor[6] and Henniart[5] in the general (rank r) case. We refer the readers to Refs. 5,6 for more detailed exposition of the *local Langlands Correspondence.*

1.3.

We can define now our main objects - the quantized algebras of functions on quantum affine Hecke algebras.

1.3.1. *Quantized algebras of functions*

Let us consider the product of matrix coefficients, associated with the product of elements of the affine Hecke algebra, of finite dimensional representations, see Ref. 9. With respect to this product we have some noncommutative algebras.

Definition 1.3. The quantized algebra $\mathcal{F}(\mathbb{H}(v, W^r_{\text{aff}}))$ or $\mathcal{F}_v(W^r_{\text{aff}})$ of functions on the quantum affine Hecke algebra $\mathbb{H}(v, W^r_{\text{aff}})$ is by definition the algebra generated by matrix coefficients of all finite-dimensional representations of the quantum affine Hecke algebra $\mathbb{H}(v, W^r_{\text{aff}})$.

1.3.2. *Inclusion*

Proposition 1.1. *The natural inclusion $W^1_{\text{aff}} \hookrightarrow W^r_{\text{aff}}$ induces a natural projection of quantized algebras of functions*

$$\mathcal{F}(\mathbb{H}(v, W^r_{\text{aff}})) \twoheadrightarrow \mathcal{F}(\mathbb{H}(v, W^1_{\text{aff}})).$$

Proof. It easy an easy consequence from the corresponding inclusion of the affine Weyl groups, $W^1_{\text{aff}} \hookrightarrow W^r_{\text{aff}}$. □

2. Irreducible Representations

The main subject of this section is to describe all (up to unitary equivalence) inequivalent unitarizable representations of the quantized algebras of functions on affine Hecke algebras. We describe first in the rank 1 case and then use the projection $\mathcal{F}(\mathbb{H}(v, W^r_{\text{aff}})) \twoheadrightarrow \mathcal{F}(\mathbb{H}(v, W^1_{\text{aff}}))$ to maintain the general case.

2.1. *Rank 1 case*

Lemma 2.1. *The quantized algebra $\mathcal{F}_v(W^f \ltimes X_*(T))$ is generated by the restrictions $t_{11}|_{W^f \ltimes X_*(T))}$ and $t_{12}|_{W^f \ltimes X_*(T))}$ with some defining relations.*

Proof. It was proven in L. Korogodski and Y. Soibelman[7] that in every finite-dimensional representation of $\mathcal{F}[\mathrm{SL}_2(\mathbb{C})]_q$, there exists an action of quantum Weyl elements $\bar{w}$. For the groups of type A_1, the root and coroot lattices are isomorphic $X^*(\mathbb{T}) \cong X_*(\mathbb{T})$. We can therefore see $W_{\text{aff}} = W^f \ltimes X^*(\mathbb{T}) \cong \bar{W}_{\text{aff}} = W^f \ltimes X_*(\mathbb{T})$ as some subgroups of $\mathrm{SL}_2(\mathbb{C})$. Therefore we have the restrictions of the representations from the list of irreducible representations of $\mathrm{SL}_2(\mathbb{C})$. Two generators of W^1_{aff} are $w = \begin{bmatrix} 0 & 1 \\ -1 & 0 \end{bmatrix}$ and $D = \begin{bmatrix} \varpi & 0 \\ 0 & \varpi \end{bmatrix}$. In the representation described in Ref. 7, they are defined by two matrix elements t_{11} and t_{12}, restricted to our affine Weyl group. □

Lemma 2.2. *Every irreducible unitarizable representation of $\mathcal{F}_v(W^f \ltimes X_*(T))$ can be obtained by restricting some irreducible unitarizable representations of $\mathcal{F}_v(\mathrm{SL}_2(\mathbb{C})$.*

Proof. First remark that if V is a representation of $\mathcal{F}_v(W^f \ltimes X_*(T))$ and $\operatorname{Ind} V = \mathcal{F}_v(\mathrm{SL}_2(\mathbb{C})) \otimes_{\mathcal{F}_v(W^f \ltimes X_*(T))} V$ the induced representation of $\mathcal{F}_v(\mathrm{SL}_2(\mathbb{C}))$, then there is the well-known Frobenius duality

$$\operatorname{Hom}(\operatorname{Ind} V, W) \cong \operatorname{Hom}(V, W|_{\mathcal{F}_v(W^f \ltimes X_*(T))}).$$

Let us consider a $\mathcal{F}_v(W^f \ltimes X_*(T))$ module V. Taking induction $\operatorname{Ind} V = \mathcal{F}_v(\mathrm{SL}_2(\mathbb{C})) \otimes_{\mathcal{F}_v(W^f \ltimes X_*(T))} V$, we have a $\mathcal{F}_v(\mathrm{SL}_2(\mathbb{C}))$ module. The irreducible ones can be therefore obtained from the list of irreducible unitarizable reprenatations $\pi_{w,t}$ of $\mathcal{F}_v(\mathrm{SL}_2(\mathbb{C}))$. □

Let us denote the restrictions of representations of $\mathcal{F}[\mathrm{SL}_2]_q$ on $\mathcal{F}(\mathbb{H}(v, W^r_{\mathrm{aff}}))$ by the same letters.

Theorem 2.1. *Every irreducible unitarizable representations of $\mathcal{F}_v(W^1_{\mathrm{aff}})$ is equivalent to one of the unitarily inequivalent representation from the list:*

(1) The representations $\tau_t, t \in \mathbb{S}^1$, defined by

$$\tau(t_{11}) = t,\ \tau(t_{22}) = t^{-1},$$
$$\tau(t_{21}) = 0,\quad \tau(t_{12}) = 0,$$

(2) The representations $\pi_{w,t}, w \in W^f, t \in \mathbb{S}^1$, defined by

$$\pi_t(t_{11}) : \begin{cases} e_0 \mapsto 0, \\ e_k \mapsto \sqrt{1-v} e_{k-1},\ k > 0, \end{cases}$$
$$\pi_t(t_{22}) : e_k \mapsto \sqrt{1-v} e_{k+1},\ k \geq 0,$$
$$\pi_t(t_{12}) : e_k \mapsto t v^{2k} e_k,\ k \geq 0,$$
$$\pi_t(t_{21}) : e_k \mapsto t^{-1} v^{2k+1} e_k,\ k \geq 0.$$

Proof. It is directly deduced from Lemmas 2.1, 2.2 and the following fact. Let us now recall that L. Korogodski and Y. Soibelman[7] obtained the description of all the irreducible (infinite-dimensional) unitarizable representations of the quantized algebra of functions $\mathcal{F}_v(G)$: For the particular case of $\mathcal{F}_v(\mathrm{SL}_2(\mathbb{C}))$, its complete list of irreducible unitarizable representations consists of:

- One dimensional representations $\tau_t, t \in \mathbb{S}^1 \subset \mathbb{C}$, defined by $\tau_t(t_{11}) = t, \tau_t(t_{22}) = t^{-1}, \tau_t(t_{12}) = 0, \tau(t_{21}) = 0$.

- Infinite-dimensional unitarizable $\mathcal{F}_v(\mathrm{SL}_2(\mathbb{C}))$-modules $\pi_t, t \in \mathbb{S}^1$ in $\ell^2(\mathbb{N})$, with an orthogonal basis $\{e_k\}_{k=0}^{\infty}$, defined by

$$\pi_t(t_{11}) : \begin{cases} e_0 \mapsto 0, \\ e_k \mapsto \sqrt{1-v}e_{k-1},\ k > 0, \end{cases}$$
$$\pi_t(t_{22}) : e_k \mapsto \sqrt{1-v}e_{k+1},\ k \geq 0,$$
$$\pi_t(t_{12}) : e_k \mapsto tv^{2k}e_k,\ k \geq 0,$$
$$\pi_t(t_{21}) : e_k \mapsto t^{-1}v^{2k+1}e_k,\ k \geq 0.$$
□

2.2. *Rank r case*

Let us consider the representations which are the composition of the homomorphisms

$$\pi_{s_i} = \pi_{-1} \circ p : F_v(G) \twoheadrightarrow \mathcal{F}_v(\mathrm{SL}_2(\mathbb{C})) \longrightarrow \mathrm{End}\,\ell^2(\mathbb{N}).$$

Theorem 2.2. *Every irreducible unitarizable representation of $\mathcal{F}_v(W^r_{\mathrm{aff}})$ is equivalent to one of the unitarily inequivalent representations:*

- *The representations $\pi_{w,t} = \pi_{s_{i_1}} \otimes \dots \pi_{s_{i_1}} \otimes \tau_t$, $w = s_{i_1} \dots s_{i_k} \in W^f$ is a reduced decomposition of w, $t \in \mathbb{S}^1$.*

Proof. For the general case of $\mathcal{F}_v(G)$, consider the algebra homomorphism $\mathcal{F}_v(G) \twoheadrightarrow \mathcal{F}_v(\mathrm{SL}_2(\mathbb{C}))$, dual to the canonical inclusion $\mathrm{SL}_2(\mathbb{C}) \hookrightarrow G_{\mathbb{C}}$. Then every irreducible unitarizable representation of the quantized algebra of functions $\mathcal{F}_v(G)$ is equivalent to one of the representations from the list:

- The representations $\tau_t, t = \exp(2\pi\sqrt{-1}x) \in \mathbb{T} = \mathbb{S}^1$,

$$\tau_t(C^{\lambda}_{\nu,s;\mu,r}) = \delta_{r,s}\delta_{\mu,\nu}\exp(2\pi\sqrt{-1}\mu(x)).$$

- The representations $\pi_i = \pi_{s_{i_1}} \otimes \cdots \otimes \pi_{s_{i_k}}$, if $w = s_{i_1}...s_{i_k}$ is the reduced decomposition of the element w into a product of reflections, where the representations π_{s_i} is the composition of the homomorphisms

$$\pi_{s_i} = \pi_{-1} \circ p : \mathcal{F}_v(G) \twoheadrightarrow \mathcal{F}_v(\mathrm{SL}_2(\mathbb{C})) \longrightarrow \mathrm{End}\,\ell^2(\mathbb{N}).$$
□

3. Schur-Weyl Duality

The Schur-Weyl duality is well-known for finite-dimensional representations of quantum affine Hecke algebras and quantum v-Schur algebras. For (possibly infinite dimensional) representations of the quantized algebras of functions on them we also have this kind of duality. We use it then to describe (possibly infinite-dimensional) representations of q-Schur algebras. The main idea is to use the maps

$$U_n(\hat{sl}_r) \twoheadrightarrow U_n^-(\hat{sl}_r) \rightarrowtail U_n^-(\hat{sl}_r) \otimes_A R \twoheadrightarrow \mathbb{S}_{n,r}(v)$$

3.1.

3.1.1. *v-Schur algebras $\mathbb{S}_{n,r}(v)$*

We recall first the definition of the affine v-Schur algebras. Let $s \in \mathbb{N}$ be an nonnegative integer, and $r \in \mathbb{N}^* = \mathbb{N} \setminus \{0\}$ a positive integer. Denote

$$\mathcal{A}_r^n = \{(i_1, \dots, i_r); 1 \leq i_1 \leq \dots \leq i_r \leq n\}$$

be the fundamental domain of the both actions of $W_{\text{aff}}^r = \hat{\mathfrak{S}}_r$ on $\mathbb{Z}^r$ on the left by

$$s_j.(i_1, \dots, i_r) := (i_1, \dots, i_{j+1}, i_j, \dots, ir), 1 \leq j < r,$$

$$\lambda.(i_1, i_r) := (i_1 + s\lambda_1, \dots i_r + s\lambda_r), \lambda \in \mathbb{Z}^r$$

and on the right by

$$(i_1, \dots, i_r).s_j := (i_1, \dots, i_{j+1}, i_j, \dots, ir), 1 \leq j < r,$$

$$(i_1, i_r).\lambda := (i_1 + s\lambda_1, \dots i_r + s\lambda_r), \lambda \in \mathbb{Z}^r.$$

For an element $\mathbf{i} \in \mathcal{A}_r^n$, denote the stabilizer as $\mathfrak{S}_{\mathbf{i}}$. Let us consider the projector $e_{\mathbf{i}} := \sum_{\delta \in \mathfrak{S}_{\mathbf{i}}} T_\delta$. Define the affine v-Schur algebra as

$$\mathbb{S}_{n,r}(v) := \bigoplus_{\mathbf{i},\mathbf{j} \in \mathcal{A}_r^n} \mathbb{H}_{\mathbf{i},\mathbf{j}} = \bigoplus_{\mathbf{i},\mathbf{j} \in \mathcal{A}_r^n} e_{\mathbf{i}} \mathbb{H}(v, W_{\text{aff}}^r) e_{\mathbf{j}}.$$

It was proven that $\mathbb{H}_{\mathbf{i},\mathbf{j}} = e_{\mathbf{i}} \mathbb{H}(v, W_{\text{aff}}^r) e_{\mathbf{j}}$ is exactly the $\mathbb{C}[v, v^{-1}]$-linear span of the element $T_\sigma = \sum_{\sigma \in \mathfrak{S}_{\mathbf{i}} \setminus \hat{\mathfrak{S}} / \mathfrak{S}_{\mathbf{j}}} T_\sigma$. It was proven that this affine v-Schur algebra $\mathbb{S}_{n,r}(v)$ is a quotient of the modified quantum group $\dot{U}_v^-(\hat{gl}_n)$.

3.1.2. *v-Schur duality*

One defines

$$\mathbb{T}_{n,r} := \bigoplus_{\mathbf{i} \in \mathcal{A}_r^n} e_{\mathbf{i}} \mathbb{H}(v, W_{\text{aff}}^r).$$

Define $T_\sigma := \sum_{\delta\in\sigma} T_\delta$, for each coset class $\sigma \in \mathfrak{S}_{\mathbf{i}} \setminus \hat{\mathfrak{S}}_r$, then $\{T_\sigma\}$ form a basis of $\mathbb{T}(n,r)$. The algebra $\mathbb{H}(v, W^r_{\text{aff}})$ acts on $\mathbb{T}(n,r)$ by multiplication on the right and the algebra $\mathbb{S}_{n,r}(v)$ acts on $\mathbb{T}(n,r)$ on the left by multiplication

$$e_{\mathbf{i}}he_{\mathbf{j}}.e_{\mathbf{k}}h' := \delta_{\mathbf{j},\mathbf{k}}e_{\mathbf{i}}he_{\mathbf{j}}h', \forall h, h' \in e_{\mathbf{i}}\mathbb{H}(v, W^r_{\text{aff}}).$$

The Schur-Weyl duality for finite dimensional representations is as follows.

$$\mathbb{S}_{n,r}(v) \cong \operatorname{End}_{\mathbb{H}(v,W^r_{\text{aff}})} \mathbb{T}(n,r),$$

$$\mathbb{H}(v, W^r_{\text{aff}}) \cong \operatorname{End}_{\mathbb{S}_{n,r}(v)} \mathbb{T}(n,r).$$

Remark that a geometric realization of this Schur-Weyl duality is an important subject in the Deligne-Langlands interplay and was highly developed, see e.g. Ref. 1.

Theorem 3.1. *The unitarizable $\mathcal{F}(\mathbb{H}(v, W^r_{\text{aff}}))$-modules and $\mathcal{F}(\mathbb{S}_{n,r}(v))$ modules are in a complete Schur-Weyl duality*

$$\mathcal{F}(\mathbb{S}_{n,r}(v)) \cong \operatorname{End}_{\mathcal{F}(\mathbb{H}(v,W^r_{\text{aff}}))} \mathbb{T}(n,r),$$

$$\mathcal{F}(\mathbb{H}(v, W^r_{\text{aff}})) \cong \operatorname{End}_{\mathcal{F}(\mathbb{S}_{n,r}(v))} \mathbb{T}(n,r),$$

Proof. It is enough to recall that the quantum algebras of functions are consisting of matrix coefficients of all finite dimensional representations of the affine Hecke algebras and affine v-Schur algebras respectively. □

3.2.

3.2.1. *Restriction maps*

Let us first recall[9] the definition of the so called modified universal enveloping algebras $\dot{U}(\mathfrak{g})$. Denote as before $X^*(T)$ the weight lattice, $X_*(T)$ the co-weight lattice. For each $\lambda', \lambda'' \in X^*(T)$ define

$$_{\lambda'}U_{\lambda''} := U(\mathfrak{g})/(\sum_{\mu\in X_*(T)} (K_\mu - v^{\langle\mu,\lambda\rangle})U(\mathfrak{g}) + U(\mathfrak{g}) \sum_{\mu\in X_*(T)} (K_\mu - v^{\langle\mu,\lambda\rangle}))$$

and the natural projection

$$U(\mathfrak{g}) \twoheadrightarrow {}_{\lambda'}U_{\lambda''}.$$

By definition the modified universal enveloping algebra $\dot{U}(\mathfrak{g})$ is the direct sum

$$\dot{U}(\mathfrak{g}) := \bigoplus_{\lambda'\in X^*(T),\lambda''\in X_*(T)} {}_{\lambda'}U_{\lambda''}.$$

The v-Schur algebras can be considered as some quotient of the modified quantized universal enveloping algebras $\dot{U}_v(\mathfrak{g})$ which is different from $U(\mathfrak{g})$ replacing $U^0(\mathfrak{g}) = \mathbb{C}$ by the direct sum of infinite number of copies of $\mathbb{C}$, one for each element of the weight lattice $X^*(T)$, see G. Lusztig (Ref. 9, Chap.$\tilde{2}$3, 29). It was shown that the category of highest weight finite dimensional representations with weight decomposition of $U(\mathfrak{g})$ is equivalent to the category of highest weight representations of $\dot{U}(\mathfrak{g})$, but the algebras $U(\mathfrak{g})$ admit also the representations without weight decomposition.

Recall from the work of Schiffmann. The main idea is related with the maps

$$U_n(\hat{sl}_r) \twoheadrightarrow U_n^-(\hat{sl}_r) \rightarrowtail U_n^-(\hat{sl}_r) \otimes_A R \twoheadrightarrow \mathbb{S}_{n,r}(v)$$

3.2.2. *Description of irreducible representations*

Theorem 3.2. *The restrictions of irreducible unitarizable* $\mathcal{F}_v(U_n(\hat{sl}_r))$ *modules to* $\mathcal{F}_v(\mathbb{S}_{n,r}(v))$ *give a complete list of irreducible unitarizable* $\mathcal{F}_v(\mathbb{S}_{n,r}(v))$ *modules.*

Proof. The proof combines Lemmas 2.1, 2.2 and the following fact. In the particular case of $\mathbb{S}_{2,d}(v)$ Doty and Giaquinto[3] have a more presice description: The v-Schur-Weyl algebra is just the image of the quantized universal eveloping algebra $U_v(\mathfrak{sl}_2)$ in the d-tensor product power of the standard 2-dimensional representation. It is isomorphic to the algebra generated by elements E, F, K and K^{-1} subject to the relations:

(a) $KK^{-1} = K^{-1}K = 1$,
(b) $KEK^{-1} = v^2E, \quad KFK^{-1} = v^{-2}F$,
(c) $EF - FE = \frac{K-K^{-1}}{v-v^{-1}}$,
(d) $(K - v^d)(K - v^{d-2}) \dots (K - v^{-d+2})(K - v^{-d}) = 0$.

We use again the map $\mathcal{F}_v(\mathbb{S}(n,d)) \to \mathcal{F}_v(\mathbb{S}(2,d))$ associated with the natural inclusion of the Weyl groups $W^1_{\text{aff}} \hookrightarrow W^r_{\text{aff}}$ □

Remark 3.1. Denote

$$(V_{\varepsilon,t}, \Pi_{\varepsilon,t}) = \begin{cases} (\mathbb{C}, \tau_t), & \text{if } \varepsilon = 0, \\ (\ell_2(\mathbb{N}), \pi_{w,t}), w \in W^f, & \text{if } \varepsilon = 1 \end{cases}$$

and define

$$\widetilde{\mathbb{T}}_{n,r} := \sum_{\varepsilon=0,1}^{\oplus} \int_{t\in\mathbb{S}^1}^{\oplus} V^*_{\varepsilon,t} \otimes V_{\varepsilon,t} dt.$$

We have therefore the **Schur-Weyl Duality for unitarizable representations**: Every irreducible unitarizable representation of the quantum affine Hecke algebra $\mathbb{H}(v, W_{\mathrm{aff}}^r)$ is a sub-representation of the representation in the space of $\mathbb{S}_{n,r}(v)$-invariants $\widetilde{\mathbb{T}}_{n,r}^{\mathbb{S}_{n,r}(v)}$ and conversely, every irreducible unitarizable representation of the quantum v-Schur algebra $\mathbb{S}_{n,r}(v)$ is a sub-representation of the representation in the space of $\mathbb{H}(v, W_{\mathrm{aff}}^r)$-invariants $\widetilde{\mathbb{T}}_{n,r}^{\mathbb{H}(v, W_{\mathrm{aff}}^r)}$

Acknowledgments

This work was completed during the stay of the author as a visiting mathematician at the Department of mathematics, The University of Iowa. The author would like to express the deep and sincere thanks to Professor Tuong Ton-That and his spouse, Dr. Thai-Binh Ton-That for their effective helps and kind attention they provided during the stay in Iowa, and also for a discussion about the PBW Theorem and Schur-Weyl duality. The deep thanks are also addressed to the organizers of the Seminar on Mathematical Physics, Seminar on Operator Theory in Iowa and the Iowa-Nebraska Functional Analysis Seminar (INFAS), in particular the professors Raul Curto, Palle Jorgensen, Paul Muhly and Tuong Ton-That for the stimulating scientific atmosphere. The deep thanks are addressed to professors Phil Kutzko and Fred Goodman for the useful discussions during their seminar lectures on Iwahori-Hecke algebras and their representations.

The author would like to thank the University of Iowa for the hospitality and the scientific support, the Alexander von Humboldt Foundation, Germany, for an effective support.

References

1. N. Chriss and V. Ginzburg, *Representation Theory and Complex Geometry*, (Birkhäuser, Boston, 1997).
2. R. Dipper and G. James, *Proc. London Math. Soc.* **52**, 20 (1986).
3. S. Doty and A. Giaquinto, *Presenting quantum Schur algebras as quotients of the quantized universal enveloping algebra of* $\mathfrak{gl}_2$, (math.QA/0011164).
4. F. Goodman and H. Wenzl, *J. of Algebra*, **215**, 694 (1999).
5. G. Henniart, *Invent. Math.*, **139**, 339 (2000).
6. M. Harris and R. Taylor, *On the geometry and cohomology of some simple Shimura varieties*, (preprint, Harvard Univ., 1999).
7. L. Korogodski and Y. Soibelman, *Algebras of Functions on Quantum Groups: Part I*, in *AMS Math. Survey and Monographs*, Vol. **56**, 1998.
8. P. Kutzko, *Ann. of Math.* **112**, 381 (1980).

9. G. Lusztig, *Introduction to Quantum Groups*, (Birkhäuser, Boston-Basel-Berlin, 1993).
10. V. Nistor, *Higher orbital integrals, Shalika germs, and the Hochschild homology of the Hecke algebras*, (arXiv:math.RT/0008133, August 2000).
11. O. Schiffmann, *On the center of affine Hecke algebras of type A*, (arXiv:math.QA/0005182, May 2000).

Harmonic, Wavelet and p-Adic Analysis
Eds. N. M. Chuong *et al.* (pp. 229–265)

§11. ON THE *C*-ANALYTIC GEOMETRY OF Q-CONVEX SPACES

VO VAN TAN*

Suffolk University, Department of Mathematics,
Beacon Hill, Boston, Massachusetts. 02114, USA
E-mail: tvovan@suffolk.edu

This article surveys the investigations in the past 40 years, of certain global aspects of q-convex spaces, introduced by Andreotti and Grauert, as well as its recent developments. This expository account is self-contained and includes new results which did not appear elsewhere

Keywords: Plurisubharmonic functions, Levi convexity, complete intersections.
Primary 32 F10, 32 C15, 32 F 05 Secondary 32 C35.

1. Introduction

In 1962, Andreotti and Grauert, in a pioneering work Ref. 1, introduced the notion of q-convex spaces which generalized Stein and compact spaces and proved the following important

Theorem 1.1. *Let X be a C-analytic space of C-dim $X = n$. Then for any analytic coherent sheaf $\mathcal{F}$ on X and any integer q with $1 \leq q \leq n := C\,dim\,X$,*

- $dim\, H^i(X, \mathcal{F}) < \infty$ *for any $i \geq q$, provided X is q-convex, and*
- $H^i(X, \mathcal{F}) = 0$ *for any $i \geq q$ if X is q-complete*

Shortly before its appearance, results of this article were presented at a Bourbaki Seminar Ref. 60. However, the major *shortcoming* (or *challenge*) of this paper (as well as Ref. 60) stems from the fact that it did not offer a single non trivial example of q-convex spaces with $q > 1$.

On the other hand, 1-convex spaces were developed in full swing and were completely classified then in Refs. 34,58,59; yet they still generate,

*A sabbatical leave granted by the College of Arts and Sciences which allowed the author to complete this project is gratefully acknowledged.

seemingly endless inspirations, even now Refs. 103,104, due mainly to the discovery of highly electrifying, unexpected and concrete examples.

That glitch partly explained the hiatus, during the '60's, of any investigation into the C-analytic global structure of q-convex spaces, with few exceptions Refs. 3,4. In fact few years earlier, namely in 1959, Grauert Ref. 32 enunciated the following

Conjecture 1.2. Any C-analytic space of C-dim. $X = n$ is q-convex for some $q \leq n$, which was received skeptically, even then (Ref.60, Remarque, p. 193). Indeed we now know that it was actually not accurate.

The first non trivial example of q-complete manifolds with $q > 1$ occurred in the framework of Lie Group Theory in 1967 Ref. 76(see also Ref. 38, p. 295). In 1972, the final straw which broke the camel back should be attributed to Ref. 12, in which were exhibited a series of thought provoking examples of q-convex manifolds. This was a major breakthrough and catapulted booming ventures in this direction Refs. 28,29,55,78 to name a few.

Early on Refs. 94,96 it was realized that, unlike the case where $q = 1$, the disadvantage of investigating the global analytic structure of q-convex spaces, for $q > 1$, resided in the scarcely of global holomorphic functions and/or the lack of an effective operational system to control the compact analytic subvarieties of appropriate dimensions.

Naturally the strategy is to transplant such spaces into the framework of holomorph-convex (resp. holomorphically spreadable) spaces and to engage in some proxy crossfire. The outcomes turned out to be quite promising Refs. 77,89,94,96,102.

Therefore the main goal of this survey is to present the progress and achievements in this area by numerous experts in the past 4 decades, as well as its recent developments

So this paper is organized as follows:

In Sec. 2, we shall state the precise notion of q-convex spaces and analyze the main difficulties inherent to the central problems.

In Sec. 3 (resp. Sec. 4) we shall look at q-convex spaces within the context of holomorphically convex spaces (resp. K-complete spaces).

In Sec. 5, some aspect of the duality between algebraic and analytic geometry will be explored and various notions of q-convexity introduced by the German school following up the works of Behnke and Thullen and by the Japanese school along the footstep of Oka, will be discussed.

Finally in Sec. 6, we shall try to reallocate our resources and look upon the prospect of further research direction

From now on, unless the contrary is explicitly stated, all C-analytic spaces X with structural sheaf O_X are assumed to be reduced, of finite dimension and equipped with some countable topology. Also $Coh(X)$ will denote the category of analytic coherent sheaves and compact irreducible C-analytic subvarieties are assumed to be of positive dimensions

2. The Andreotti Grauert Legacy

2.1. *The initial challenge*

Definition 2.1 (Ref. 1). *Let X be a C-analytic space and let $\psi : X \to \mathbb{R}$. For any $x \in X$, let W_x be some neighborhood of x, isomorphic to some C-analytic subvariety V defined in an open set U of some $\mathbf{C}^N$. Let $\tau : W_x \cong V \subset U \subset \mathbf{C}^N$ be the isomorphism. Now ϕ is said to be strongly q-plurisubharmonic (resp. weakly q-plurisubharmonic) (or q-convex (resp. weakly q-convex) for short) if*

- *there exists a C^2 function $\psi : U \to \mathbb{R}$ such that, for any $\xi \in \mathbf{C}^N$ the Levi form*

$$L_z(\psi; \xi) = \sum_{1 \le i,j \le N} \frac{\partial^2 \psi(z)}{\partial z_i \partial z_j} \xi_i \xi_j$$

  has at most $q-1$ eigenvalues ≤ 0 (resp < 0) for any $z \in U$.
- *$\psi | V = \phi o \tau$.*

Remark 2.1. One can check that the above definition does not depend on the particular choice of the local isomorphism τ (see e.g. Ref. 59).

Definition 2.2 (Refs. 1,14,25). *Let X be a C-analytic space. X is said to be strongly q-pseudoconvex (or q-convex for short) if there exist*

(a) compact set $K \subset X$.
(b) an exhaustion function $\phi : X \to \mathbb{R}$, i.e. the sets $\{x | \phi(x) \le r\}$ are compact for any $r \in R$ such that ϕ is q-convex for any $x \in X \backslash K$.

In the special case where $K = \emptyset$, we say that X is *strongly q-complete* (or *q-complete* for short).

Definition 2.3. Let X be a C-analytic space. X is said to be *cohomologically q-convex* (resp. *cohomologically q-complete*) if C-dim $H^i(X, \mathcal{F}) < \infty$ (resp. $H^i(X, \mathcal{F}) = 0$) for any $\mathcal{F} \in Coh(X)$ and $i \ge q$.

Problem 2.1. *(**Characterization Problem**) Does the converse to Theorem 1.1 hold?*

The main impetus to this problem stems from the fact that it has a positive answer, provided $q = 1$(cf. Sec.3), and recently when $q = \dim X$ (cf. Sec. 6)

On the other hand, for $q > 1$, despite the bleak outlook, as was explained in the introduction, it was not a total lost; indeed, topologically, one is well informed in view of the following

Theorem 2.1 (Ref.86, see also Ref. 41). *Any q-complete manifold X has the same homotopy type as a CW complex of $\mathbb{R} - dim = n + q - 1$. In particular*

$$H_{n+i}(X, \mathbf{Z}) = 0 \quad \textit{for all} \quad i \geq q$$

and

$$H_{n+q-1}(X, \mathbf{Z}) \quad \textit{is free}$$

and its counterpart, namely

Proposition 2.1 (Ref. 56). *Let X be cohomologically q-complete space. Then*

$$H_{n+i}(X, \mathbf{C}) = 0 \quad \textit{for all} \ i \geq q.$$

It is fair to say that by the time of the appearance of Ref. 1 and Ref. 86, the following examples were well known, at least among the circle of experts:

Example 2.1. Let $X_1 := \mathbf{C}^{n-q+1} x \mathbf{P}_{q-1}$. Then one can check that X_1 is q-complete in view of the presence of $\psi(z) := \phi \circ \pi$ where $\phi(z) := \sum_{0 \leq j \leq n-q} |z_j|^2$ where $z := (z_0, \ldots, z_{n-q})$are coordinates in $\mathbf{C}^{n-q+1}$ and $\pi : \mathbf{C}^{n-q+1} x \mathbf{P}_{q-1} \rightarrow \mathbf{C}^{n-q+1}$ is the first projection. Notice that X_1 is holomorphically convex (see Definition 2.1 below).

Example 2.2. Let $X_2 := \mathbf{C}^n \setminus \mathbf{C}^{n-q}$, let $\phi(z) := \left(\sum_{1 \leq j \leq q} |z_j|^2 \right)^{-q}$ where $\mathbf{C}^{n-q} \cong \{z | z_1 = \cdots = z_q = 0\}$ and let $z := (z_1, \ldots, z_n) \in \mathbf{C}^n$.

Then $\phi(z) + \psi(z)$ will guarantee the q-completeness of X_2 where $\psi(z) := (\sum_{1 \leq i \leq n} |z_i|^2)$.

Notice that X_2 is not holomorphically convex.

Example 2.3. Let $X_3 := \mathbf{P}_n \backslash \mathbf{P}_{n-q}$ and let

$$\varphi(z) := \Big(\sum_{0 \leq i \leq q-1} |z_i|^2 \Big)^{-1} \Big(\sum_{0 \leq j \leq q-1} |z_i|^2 \Big)$$

where $z := (z_0 : \cdots : z_n)$ are the homogeneous coordinates of $\mathbf{P}_n$ and $\mathbf{P}_{n-q} \cong \{z | z_0 = \cdots = z_{q-1} = 0\}$. Then one can check that $\phi(z) := \log \varphi(z)$ is an exhaustion q-convex function on X_3.

Notice that $\Gamma(X_3, O) = \mathbf{C}$. In order to visualize the subtlety of Problem 2.1, we would like to mention 2 mini-puzzles which still baffle most observers.

Example 2.4. Let C be a connected non singular compact C-analytic curve in $\mathbf{P}_3$ and let $X_1 := \mathbf{P}_2 \setminus C$. Then it is known that Ref. 11 X_1 is cohomologically 2-complete

Question 2.1. Is X_1 always 2-complete?

Example 2.5. Let us consider the Veronese embedding Refs. 37,46.

$$\begin{array}{rcl} \tau : \mathbf{P}_2 & \longrightarrow & \mathbf{P}_5 \\ (x : y : z) & \longrightarrow & (x^2 : xy : y^2 : yz : z^2 : xz) \end{array}$$

where $(x : y : z)$ are homogeneous coordinates of $\mathbf{P}_2$.

Then $\nu := \tau(\mathbf{P}_2)$, the so called *Veronese* surface, is a compact surface of degree

$$d = c_1(\tau^* \mathcal{H})^2 = c_1(2\mathcal{H})^2 = 4 \quad \text{in} \quad \mathbf{P}_5$$

where $\mathcal{H}$ (resp.H) is a hyperplane line bundle on $\mathbf{P}_5$ (resp. $\mathbf{P}_3$). Now let $X_2 := \mathbf{P}_5 \backslash v$. From the following exact sequence.

$$\begin{array}{ccccc} H^2(\mathbf{P}_5, \mathbf{Z}) & \xrightarrow{d} & H^2(\nu, \mathbf{Z}) & \rightarrow & H^3(\mathbf{P}_5, \nu, \mathbf{Z}) \\ \mathbf{Z} & \longrightarrow & \mathbf{Z} & \rightarrow & \mathbf{Z}/d\mathbf{Z} \end{array}$$

and the Alexander-Lefschetz duality it follows readily that.

$$H_7(\chi_2, \mathbf{Z}) \cong H^3(\mathbf{P}_5, \nu \mathbf{Z}) \cong \mathbf{Z}/4\mathbf{Z}.$$

Hence X_2 is not 3-complete, in view of Theorem 2.1

Question 2.2. Is X_2 cohomologically 3-complete?

As was mentioned earlier, Problem 2.1 was motivated by the case where $q = 1$; thus we would like to come back and try to find out how our understanding of that special case came about and that will be the purpose of the next section.

2.2. *The prime time*

Let $J \subset O_X$ be an arbitrary analytic coherent ideal sheaf. Let us consider the following fundamental result due mainly to Serre (see e.g.Ref. 40)

Theorem 2.2. *Assume that $H^1(X, J) = 0$. Then X is 1-complete*

Proof. From the following exact sequence

$$0 \to J \to O_X - \kappa \to O_X/J \to 0$$

one obtains the surjectivity of κ^*

$$\Gamma(X, O_X) - \kappa^* \to \Gamma(X, O_X/J) - \delta \to H(X, J) \tag{1}$$

in view of the hypothesis.

Now for any sequence $\{x_k\} =: E \subset X$ without accumulation point, let J_1 be the ideal sheaf in O_X determined by E. Then (1) tells us that there exists some $f \in \Gamma(X, O_X)$ such that $f(E)$ is unbounded, i.e. X is holomorphically convex.

On the other hand let J_2 be the ideal sheaf in O_X determined by a pair of distinct points $\{x, y\} \subset X$. Then (1) implies the existence of an $f \in \Gamma(X, O_X)$ such that $f(x) \neq f(y)$ i.e. X is free of compact C-analytic subvarieties. Hence X is Stein. Then a main result in Ref. 58 tells us that X admits a real analytic exhaustion function ϕ which is 1-convex. □

This result certainly helped to push the envelope for 1-convex spaces which are completely characterized by the following important result

Theorem 2.3 (Ref. 59). *Assume that $\dim H^1(X, J) < \infty$. Then*

(i) X is a proper modification of a Stein space Y at finitely many points, i.e. there exist a Stein space Y, a finite set $T \subset Y$, and a proper and surjective morphism $\pi : X \to Y$, inducing a biholo-morphism $X \backslash S \cong Y \backslash T$ where $S := \pi^{-1}(T)$.

(ii) $\pi * O_X \cong O_Y$.

(iii) $H^i(X, \mathcal{F}) \cong H^i(S, \mathcal{F}|S)$ for any $i \geq 1$ and $\mathcal{F} Coh(X)$.

(Note that $\mathcal{F}|S$ stands for topological restriction. Thus $\mathcal{F}|S$ may not be coherent in S).

From now on S is referred to as the *exceptional set* of X

Proof.

Step 1. Let $\{x_\nu\} \subset X$ be an infinite discrete set and let us associate to each x_ν a constant, say $\nu^m \in \mathbf{C}$ (with $m = 1, \ldots$). This data gives rise to a global section $\omega_m \in \Gamma(X, O/J)$. From the exact sequence (1), let χ_m be the image of ω_m by δ in $H^1(X, J)$. By hypothesis, we obtain the following relation in $H^1(X, J)$

$$\sum_{1 \le m \le k} \alpha_m \chi_m = 0$$

where $0 \neq \alpha_m \in \mathbf{C}$.

In view of the exactness of (1), this implies the existence of some $f \in \Gamma(X, O_X)$, the image of which by κ^* is exactly the section

$$\sum_{1 \le m \le k} \alpha_m \omega_m \in \Gamma(X, O_X/J).$$

In other words, for every $\{x_v\}$, one can find a polynomial

$$P(t) := \sum_{1 \le m \le k} \alpha_m t^m \quad \text{with} \quad \alpha_m \neq 0,$$

such that

$$f(x_v) = P(v). \tag{2}$$

In particular $|f(x_v)| \to \infty$, i.e. X is holomorphically convex.

Step 2. Since X is holomorphically convex, we infer from Refs. 17,70, that there exist a Stein space Y and a surjective and proper morphism $\pi : X \to Y$ such that $\pi^{-1}\pi(x)$.

Now, let $S := \{x \in X | x$ is not an isolated in $\pi * O_X \cong O_Y$.

Claim $T := \pi(S)$ is finite.

Assume the contrary and let $\{x_v\}$ be a infinite discrete set in T. Let W_v, be an irreducible component of $\pi^{-1}(x_v)$. Let u_{2v-1}, u_{2v} be two distinct points of W_ν. Thus the sequence $\{u_v\}$ is a discrete set in X. In view of (2), for large v, $f(u_{v+1}) \neq f(u_v)$, sine $P(t+1) - P(t)$ has only finitely many zeroes. But $f|W_v$ is constant, which contradicts the fact that $f(u_{2v}) \neq f(u_{2v-1})$. Hence T is finite and S is compact. Consequently π induces an isomorphism

$$X/S \cong Y \setminus T \tag{3}$$

Step 3. Following Ref. 33 we have the following isomorphism.

$$H^k(X, \mathcal{F}) \cong H^0(Y, \mathcal{R}^k \pi * \mathcal{F}) \quad \text{for any} \quad k \ge 0. \tag{4}$$

On the other hand, we infer from (3) that

$$(\mathcal{R}^k \pi * \mathcal{F})_x = 0 \quad \text{for any } x \notin T \quad \text{and any } k \ge 1. \tag{5}$$

Consequently, one obtains from (4) and (5)

$$H^i(X, \mathcal{F}) \cong H^0(T, \mathcal{R}^i\pi * \mathcal{F}) \cong H^i(S, \mathcal{F}|S) \quad \text{for any } i \geq 1. \qquad \square$$

Corollary 2.1. *Assume that* $\dim H^1(X, J) < \infty$. *Then there exist a compact analytic subvariety* $S \subset X$ *and an exhaustion function* $\psi : X \to \mathbb{R}$ *which is real analytic and 1-convex for any* $x \in X \setminus S$.

Remark 2.2. Notice that for the above 3 results, the requirement that X is to be reduced, of finite dimension and to have a countable topology is redundant. This shows the privileged standing of cohomologically 1-convex spaces within this context.

Such results motivated the following classical

Definition 2.4 (Ref. 34). *Let* S *be a compact* C*-analytic space, let* V *be a holomorphic vector bundle on* S *and let us identify* S *with the zero section of* V. *Then* V *is said to be weakly negative if* S *admits a* 1*-convex neighborhood* $N \subset V$.

Similarly V *is said to be weakly positive if* V^* *is weakly negative.*

Remark 2.3. This definition is very convenient in Complex Analysis, since it coincides exactly with the notion of ampleness in Algebraic geometry. See e.g Refs. 29,43,78,99.

The hyperplane line bundle H on $\mathbf{P}_n$ is ample and hence weakly positive

Certainly the above key results direct us to the following

Problem 2.2 (Proper modification Problem). *Let* X *be a* q*-convex (resp. a cohomologically* q*-convex) space. Is* X *always the proper modification of some* q*-complete (resp. cohomologically* q*-complete) space* Y*?*

Concisely, can one find

(a) a q*-complete (resp. cohomologically* q*-complete) space* Y, *some compact* C*-analytic subvariety* $T \subset Y$ *and*
(b) a proper morphism $\pi : X \to Y$ *inducing a biholomorphism* $X \backslash S \cong Y \backslash T$ *where* $S := \pi^{-1}(T)$*?*

Now Problem 2.2 could be filtered out to 2 special versions, namely

Problem 2.3 (The Alteration Problem). *Let* X *be a* q*-convex space. Can one always find*

(a) some compact analytic subvariety $S \subset X$, *and*

(b) an exhaustion function ϕ which is q-convex for any $x \in X \backslash S$?

Problem 2.4 (The Supporting Problem). *Let X be a cohomologically q-convex space. Can one always find some compact analytic subvariety $S \subset X$ such that $H^i(X, \mathcal{F}) \cong H^i(S, \mathcal{F}|S)$ for any $i \geq q$ and any $\mathcal{F} \in Coh(X)$?*

In Ref. 102 we have

Theorem 2.4. *Let X be a 1-convex space with its exceptional set S. Assume that C-dim. $S < q$. Then X is q-complete.*

Consequently, one would like to raise the following

Problem 2.5 (The Obstruction Problem). *Assume that X is q-convex (resp. cohomologically q-convex) with no compact C-analytic subvarieties of C-dim $\geq k$ and let $p := \max\{k, q\}$. Is X p-complete (resp.cohomologically p-complete)?*

To round off this discussion let us mention the following

Example 2.6. Let γ be an irreducible hypersurface in $\mathbf{P}_{n+1}$ with only isolated singularites say $\{p_i\}$ and let $\pi : M \to \gamma$ be its non singular resolution. Let $Z := \cap_k \mathcal{H}_k$ where $\mathcal{H}_k$ are distinct irreducible hypersurface sections on γ with $1 \leq k \leq q$ such that $\{p_i\} \notin \mathcal{H}_k$ for any k and any i.

Certainly, $Y := \mathcal{Y} \backslash Z = \cup_k (\mathcal{Y} \backslash \mathcal{H}_k)$ as union of q Stein spaces is indeed q-complete, Ref. 66 (see also Ref. 90). Now let $X := M \backslash \pi^{-1}(Z)$. Then one can check that:

(1) X is a q-convex manifold
(2) X admits $S := \pi^{-1}(\cup_k p_k)$, as its q-maximal compact analytic subvariety (see Def. 2.6 below)
(3) $\pi|X : X \to Y$ is a proper modification, inducing a biholomorphism $X \backslash S \cong Y \backslash \cup_i p_i$.

2.3. *The shattered dreams*

We are now in a position to provide a series of counterexamples to some of the above mentioned problems. First of all let us mention the following

Theorem 2.5.

(a) If X is q-convex, then X admits only finitely many compact, irreducible components of C-dimension $\geq q$

(b) If X is q-complete, then X does not have any compact irreducible components of C-dimension $\geq q$.

Proof.

a) Since X is q-convex, the set $\mathbf{K} := \{x \in X | \phi(x) \leq \sup_K \phi\}$ is compact in X, for some exhaustion function $\phi : X \to \mathbb{R}$. Let A be a compact analytic subvariety in X with $m := C\text{-}\dim A \geq q$ and let us assume that A is *not* contained $\mathbf{K}$.

Let $x \in A$ with $\phi(x) = \sup_A \phi$. It is clear that $x \notin \mathbf{K}$. In view of the definition, there exists an open neighborhood W_x of x in X which can be realized as a closed analytic subvariety V of some open set $U \subset \mathbf{C}^N$, for some N. Let $\tau : W_x \cong V \subset U \subset \mathbf{C}^N$ be the isomorphism and let ψbe a q-convex function on U such that $\psi|V = \phi \circ \tau$. Let $\mathcal{W} := \tau(\mathcal{A})$ with $A := \mathcal{A} \cap U$ and let $z := \tau(x)$. In view of the q-convexity of ψ, one can find a subspace Σ through z in with C-dimension $\Sigma = N - q + 1$ such that $\psi|\Sigma$ is 1-convex and hence so does $\psi|\mathcal{W} \cap \Sigma$.

However

$$
\begin{aligned}
C.\dim \mathcal{W} \cap \Sigma &\geq (N - q + 1) + m - N \\
&\geq m - q + 1 \\
&\geq 1 \text{ since } m \geq q
\end{aligned}
$$

The maximality of ψ at z implies (see Ref. 40) the constancy of ψ on $\mathcal{W} \cap \Sigma$, and this will contradict the q-convexity of $\psi|\mathcal{W} \cap \Sigma$. Hence $A \subset \mathbf{K}$.

Now let Σ_m be the collection of m-dimensional compact irreducible components of X. Certainly Ref. 40 Σ_m is an analytic subvariety of X of pure C-dim $= m$. Let us assume that $m \geq q$. The previous argument tells us that the compact irreducible components of Σ_m lie in $\mathbf{K}$. Since $\mathbf{K}$is compact, those irreducible compact components are finite in number.

(b) This follows directly from part (a). □

Remark 2.4. This simple result already convinces us that Conjecture 1.2 was not accurate.

Now the following technical result is needed.

Lemma 2.1 (Ref. 94). *Let $\pi : X \to Y$ be a proper modification of C-analytic spaces, inducing a biholomorphism*

$$X \setminus S \cong Y \setminus T \tag{6}$$

for some compact C-analytic subvariety $T \subset Y$, where $S := \pi^{-1}(T)$. Assume that Y is cohomologically q-complete. Then

$$H_k(S, \mathbf{C}) \cong H_k(X; \mathbf{C}) \quad \textit{for all } k \geq n + q.$$

Proof. Let us consider the following exact sequence

$$H_k(Y, \mathbf{C}) \to H_k(Y, T; \mathbf{C}) \to H_{k-1}(T; \mathbf{C}). \tag{7}$$

Since Y and T are cohomologically q-complete, it follows readily from Pro. 2.1, that

$$H_k(Y, \mathbf{C}) = H_{k-1}(T; \mathbf{C}) \simeq 0 \text{ for all } k \geq n + q.$$

Consequently (7) tells us that

$$H_k(Y, T; \mathbf{C}) \cong 0 \quad \text{forall } k \geq n + q. \tag{8}$$

On the other hand, in view of Alexander-Lefschetz duality, for any $k \geq 0$

$$H_k(X, S, \mathbf{C}) \cong H^{2n-k}(X \setminus S; \mathbf{C}) \tag{9}$$

and

$$H^i(YT; \mathbf{C}) \cong H^{2n-k}(Y \setminus T; \mathbf{C}) \tag{10}$$

We infer from (6) that

$$H^i(X \setminus S; \mathbf{C}) \cong H^i(Y \setminus T; \mathbf{C}) \quad \text{for all } i \geq 0 \tag{11}$$

From the following exact sequence

$$H_{k+1}(X, S; \mathbf{C} \cong H_k(S, \mathbf{C} \to H_k(X; \mathbf{C}) \to H_k(X, S; \mathbf{C})$$

our desired conclusion will follow from (8), (9), (10) and (11). □

Example 2.7. Let T be a compact C-analytic threefold in $\mathbf{P}_8$ such that,

$$H^2(T, \mathbf{C}) \cong \mathbf{C}^2 \tag{12}$$

see e.g Ref. 63 and let $X_1 := \mathbf{P}_8 \backslash T$. Then X_1 is 5-convex Ref. 12 and free of compact C-analytic subvarieties of C-dim ≥ 5.

Claim. X_1 is not a proper modification of any cohomologically 5-complete space.

Indeed if it were, Lemma 2.1 tells us that there exists some compact analytic subvariety S such that

$$0 = H_{13}(S, \mathbf{C}) \cong H_{13}(X_1; \mathbf{C})$$

since by construction dim $S < 5$.

Now Poincare duality tells us that $H^3_c(X_1, \mathbf{C}) \cong H_{13}(X_1; \mathbf{C}) = 0$. On the other hand from the following exact sequence

$$\mathbf{C} \cong H^2(\mathbf{P}_8, \mathbf{C}) - \rho \to H^2(\mathcal{T}, \mathbf{C}) \to H^3_c(X_1, \mathbf{C})$$

we infer readily that ρ is surjective, contradicting (12) and our claim is proved.

Example 2.7' Let X be a compact C-analytic surface in $\mathbf{P}_4$ such that

$$H^1(X, \mathbf{C}) \neq 0$$

see e.g. Refs. 42,51 and let $X_2 := \mathbf{P}_4 \backslash X$. Then X_2 is 2-convex Ref. 12 and free of compact C-analytic subvarieties of C-dim ≥ 2. One can check Ref. 96 that X_2 is not cohomologically 2-complete. Dimensionwise, this counterexample is optimal, in view of results in Ref. 20.

Although immersed in such a hostile environment, one still can retrieve a small silver lining which has a differential geometric flavor:

Proposition 2.2 (Ref. 94). *Let M be a compact C-analytic manifold and let $\mathcal{E} \to M$ be a Griffiths positive, rank q holomorphic vector bundle. Let $Y := \{\Omega = 0\}$ for some non singular section $\Omega \in \Gamma(M, \mathcal{E})$. Then $X := M \backslash Y$ is q-complete.*

Proof. Let $\pi : Z \to M$ be the blow up of M along Y, inducing a biholomorphism

$$\chi := Z \setminus \Xi \cong M \setminus Y \tag{13}$$

where $\Xi := \mathbf{P}(N)$ and N is the normal bundle of Y in M. Let L be the line bundle on M determined by Ξ. Following Ref. 36, relative to some open coverings $\{U_i\}$ of Z, there is an induced hermitian metric $\{h_i, U_i\}$ on L such that

$$\Theta_{\mathcal{L}} | U_i \cong -\partial\partial \log h_i \text{ has at most } q-1 \text{ non positive eigenvalues, for any } i \tag{14}$$

where Θ_L is a $(1,1)$ curvature form of $\mathcal{L}$. By definition, $h_i : U_i \to \mathbb{R}$ are smooth functions such that

$$h_i |\sigma_1|^2 = h_j |\sigma_j|^2 \quad \text{on} \quad U_i \cap U_j \tag{15}$$

where $\{\sigma_i\}$ are local defining equations of Ξ om U_i. Then, we infer from (15) that the function $\phi | U_i := |\sigma_i|^2 h_i$ is well defined. Certainly $\Psi := -\log \phi$ is an exhaustion function on χ. Furthermore, in view of (14),

$$-\partial\partial \log \Psi | U_i = -\partial\partial \log h_i - \partial\partial \log |\sigma_i|^2$$

has at most $q-1$ non positive eigenvalues on $U_i \backslash (U_i \cap \Xi)$. It follows from (15) that χ and hence X is q-complete. □

Definition 2.5 (Ref. 37). *Let E be a holomorphic vector bundle on a C-analytic manifold M. The curvature operator $\Theta \in \Lambda^2(Hom(\mathcal{E}, \mathcal{E}))$ is said to be positive at $x \in M$, if*

(a) for any $0 \neq \lambda \in \mathcal{E}_x$ the multivector $(\lambda, \Theta\lambda) \in \Lambda^2 T_x^*(M)$ is positive of type $(1,1)$, where $T^*(M)$ is the complex cotangent bundle, or equivalently if
(b) for any vector $v \in T_x'(M)$, the hermitian matrix $-i\langle\Theta(x); v, v\rangle \in Hom(\mathcal{E}_x, \mathcal{E}_x)$ is positive definite where $T'(M)$ is the holomorphic tangent bundle

$\mathcal{E}$ is said to be Griffiths positive if Θ is everywhere positive.

Example 2.8. Let M be a compact C-analytic manifold and let $\mathcal{E} := \oplus_{1 \leq i \leq q} \mathcal{L}_i$ where each $\mathcal{L}_i$ is an ample line bundle. Then one can check that E is a rank q Griffiths positive vector bundle.

2.4. *The watchtower*

Theorem 2.5 admits the following generalization

Theorem 2.6 (Ref. 100). *(a) If X is cohomologically q-convex, then it admits only finitely many compact irreducible components of C-dim $\geq q$.*
(b) If X is cohomologically q-complete, then it is free of compact irreducible components of C-dim $\geq q$.

This result inspires the following:

Definition 2.6 (Ref. 94). *A compact analytic subspace S in X is said to be a q-maximal if*

(i) $\dim S_i \geq q$ for any irreducible components S_i of S and
(ii) If $T \subset X$ is an irreducible compact C-analytic subvariety, with $\dim T \geq q$, then necessarily $T \subset S$.

In the special case where $q = 1$, this is exactly the notion of maximality of Grauert Ref. 34. This naturally raises the following

Question 2.3. Does any q-convex space (resp. cohomologically q-convex space) X admits a q-maximal compact analytic subvariety S, if X contains at least one compact C-analytic subvariety of C-dim $\geq q$?

Remark 2.5. Certainly the answer to Question 2.3 is affirmative if $q = 1$ (cf. Theorem 2.3 or $q = n$ (cf. Theorem 2.6.(a)).

Now in the extreme case when $q = n$, in view of results in Refs. 68,82, combined with Theorem 2.6, we have the following classification:

Theorem 2.7. *Refs. 68,82*

(i) X is cohomologically n-convex iff X carries an n-maximal compact analytic subvariety S
(ii) X is cohomologically n-complete iff X is free of compact irreducible components of C-dim $= n$.

From Theorem 2.7 (ii), we deduce the following

Corollary 2.2. *Assume that X irreducible. Then X is compact iff there exists $\mathcal{H} \in Coh(X)$ such that*

$$H^n(X, \mathcal{H}) \neq 0.$$

We are now ready to prove the following

Theorem 2.8. *Assume that X is cohomologically $(n-1)$ convex. Then X admits an $(n-1)$-maximal compact subvariety S.*

Proof. Let X_i be the compact irreducible components of X with C-$\dim X_i = n$. Let $S_1 := \cup_i X_i$. We infer from Theorem 2.7(i), that S_1 is compact. Let χ_ν be the irreducible components of X with $X_i \neq \chi_v$ for any v and i. Let $X' := \cup_v X_v$. It follows from Theorem 2.7 (ii), that X' is cohomologically n-complete.

On the other hand, since

$$H^i(X', \mathcal{F}) \cong H^i(X, \mathcal{F}')$$

for any i and any $\mathcal{F}` \in Coh(X')$ where $\mathcal{F} \in Coh(X)$ is the trivial extension of $\mathcal{F}'$ to X, it follows readily that X' is also cohomologically $(n-1)$ convex. Consequently, we infer from the proof of Theorem 1 in Ref. 13 that X' admits only finitely many irreducible compact analytic subvarieties S_k of C-dim $= n-1$. Let $S_2 := \cup_k S_k$. Then one can check easily that $S := S_1 \cup S_2$ is the $(n-1)$ maximal compact analytic subvariety of X. □

2.5. *The bottom line*

By leaps and bounds, within the past decade, n-convex spaces were completely classified. Indeed, those results not only drastically improve Theorem 2.7, but also settle affirmatively Problem 2.1 when $q = n$; namely

Theorem 2.9. *Refs. 22,61 Let X be a C-analytic space with $\dim X = n$. Then*

(i) *X is q complete for some $q \leq n+1$.*
(ii) *X is n-convex if X carries an n-maximal compact C-analytic subvariety S.*
(iii) *X is n-complete if X is free of compact irreducible components of C-$\dim n$.*

Corollary 2.3 (Ref. 69). *$H^i(X, \mathcal{F}) = 0$ for all $\mathcal{F} \in Coh(X)$ and all $i > n$.*

Remark 2.6. When X is non singular, Theorem 2.9 was established earlier Ref. 35, the proof of which played a crucial role in this general case Ref. 22.

Complementing Theorem 2.9, we have

Theorem 2.10. *Any n-convex space X is a proper modification of some n-complete space Y.*

Proof. Let X_i be the compact irreducible components of X with C-$\dim X_i = n$ and let $S := \cup_i X_i$ be the n-maximal compact analytic subvariety of X. Let $Y := \cup_v \chi_v$ where χ_v are the irreducible components of X with $X_i \neq \chi_v$ for any v and i and let $T := Y \cap S$. Then one can check that

(1) Y is n-complete, and
(2) $\pi : X \to Y$ is the required proper modification morphism such that $X \backslash S \cong Y \backslash T$. □

Notice that Theorem 2.9 is derived from the following

Theorem 2.11 (Ref. 22). *Let S be an analytic subspace of X. Assume that S is q-complete. Then S admits a fundamental system of q-complete neighborhoods N in X.*

Corollary 2.4 (Ref. 9). *Let S be a compact C-analytic subvariety in X. Then S admits a fundamental system of q-complete neighborhoods N in X, provided C-$\dim S < q$.*

Corollary 2.5 (Ref. 9). *Let S be a compact C-analytic subvariety in X. Then S admits a fundamental system of q-complete neighborhoods N in X, provided C-dim $S < q$.*

Certainly Theorem 2.11 is of interest in its own right. The special case $q = 1$, was established in Ref. 84 the proof of which is quite difficult. Simplifications and generalizations of it can be found also in Refs. 19, 66.

From Ref. 102 one has the following

Theorem 2.12. *A C-analytic space is cohomologically q-complete iff each of its irreducible component is ones.*

Hence we have

Question 2.4. Does Theorem 2.12 hold if one replaces cohomologically q-completeness by q-completeness?

This question again is motivated by the fact that it has an affirmative answer, if $q = 1$ Ref. 2 or $q = n$ (cf. Theorem 2.9(iii)).
Warning. On the basis of Subsecs. 2.2 and 2.5, unless the contrary is explicitly stated, from now on, we are dealing exclusively with q-convex spaces X with $1 < q < n := C\text{-dim}\, X$.

3. The Holomorph-convex Spaces

3.1. *Preliminaries*

Definition 3.1. Let X be a C-analytic space. Then X is said to be *holomorphically convex*, if for any sequence $\{x_k\}$ without accumulation point, there exists a holomorphic function f such that

$$\limsup_{k\to\infty} |f(x_k)| = \infty.$$

As was seen above, holomorphically convex spaces played a crucial role in the classification of 1-convex spaces. This is due to the so called Remmert-Cartan reduction theory.

Theorem 3.1 (Refs. 17,27,70,71). *Let X be a given holomorphically convex space. For any pair of distinct points $x, y \in X$, let us define $x \sim y$ iff $f(x) = f(y)$ for any $f \in \Gamma(X, O_X)$. Then*

(i) $\sim$ is an equivalence relation,
(ii) $X/\sim =: Y$ is a Stein space,

(iii) *The natural projection* $\pi : X \to Y$ *is a proper surjective morphism with only connected fibres,*
(iv) *The natural map* $\pi^* : \Gamma(Y, O_Y) \to \Gamma(X, O_X)$ *is an isomorphism of function algebras, and*
(v) X *admits a countable topology*

Proposition 3.1 (Ref. 67). $H^i(X, \mathcal{F})$ *acquires a Hausdorff topology for any* $i \geq 0$ *and any* $\mathcal{F} \in Coh(X)$.

Remark 3.1. The morphism $\pi : X \to Y$ is referred to as the *Remmert-Stein reduction* of X. An attempt to classify holomorphically convex spaces was initiated in Ref. 95. Obviously, two special classes of holomorphically convex spaces stand out: Stein spaces and compact spaces. Since the former are completely characterized by Theorem 2.2, the next section will be devoted to the latter.

3.2. *The compactness*

One has the following fundamental result due to Cartan and Serre, see e.g. Ref. 40

Theorem 3.2. *Assume that* X *is compact. Then*

$$dim\, H^i(X, \mathcal{F}) < \infty \quad \text{for any } i \geq 0 \text{ and any } \mathcal{F} \in Coh(X).$$

It admits a milestone generalization

Theorem 3.3 (Ref. 33). *Let:* $X \to Y$ *be a proper holomorphic morphism of* C*-analytic spaces. Then for any* $\mathcal{F} \in Coh(X)$ *the higher direct images sheaves* $\mathcal{R}^k\pi * \mathcal{F} \in Coh(Y)$ *for any* $k \geq 0$.

Now let us introduce the following Refs. 42,43

Definition 3.2. For and any $F \in Coh(X)$, let

$$cd(X) := \quad \text{the smallest integer } k \geq 0 \text{ such that } H^i(X, F) = 0 \text{ for } i > k$$

and let

$$fd(X) := \quad \text{the smallest integer } k \geq 0 \text{ such that } C\text{-dim}H^i(X, F) < \infty \text{ for } i \geq$$

We are now in a position to prove the following classification result

Theorem 3.4. *Let* X *be an irreducible* C*-analytic space. Then the following conditions are equivalent:*

(a) X is compact,
(b) $cd(X) = n$,
(c) $fd(X) = 0$.

Proof. In view of Corollary 2.2 and Theorem 3.2, it remains to prove only the implication
(c) $\to$ (a). Assume that X is non compact. Let $T \subset X$ be an infinite discrete set of points. Let us define $\mathcal{F}$ by

$$\mathcal{F}_x = \begin{cases} 0 & \text{if } x \notin T \\ \mathbf{C} & \text{if } x \in T \end{cases}$$

Then one can check easily that $\mathcal{F} \in Coh(X)$ and $\dim\, H^0(X, \mathcal{F}) \cong \dim H^0(T, \mathcal{F}) = \infty$. Contradiction. □

Notice that for the equivalence (a) $\leftrightarrow$ (c), the hypothesis of irreducibility of X is not needed.

Throughout the remaining of this section all C-analytic spaces are assumed to be holomorphically convex.

3.3. *The cohomologically q-convexity*

Theorem 3.5.

(i) X is cohomologically q-complete iff X is free of compact C-analytic subvarieties of C-dim $\geq q$,
(ii) X is cohomologically q-convex iff X admits a q-maximal compact analytic subvariety S.

Proof.
(a) Let $\pi : X \to Y$ be the Remmert Stein reduction. Since $H^i(X, \mathcal{F}) \cong H^0(Y, \mathcal{R}^i\pi * \mathcal{F})$, Ref. 33 for any $i \geq 0$ and $\mathcal{F} \in Coh(X)$ and since (see e.g. Ref. 8)

$$\mathcal{R}^k\pi * \mathcal{F}_x \cong H^k(\pi^{-1}(x), \mathcal{F}).$$

Our desired conclusion will follow, in view of Corollary 2.3.
(b) *Step 1.* Let $\pi : X \to X'$ be a surjective proper morphism which contracts S to finitely many points. Certainly X' is holomorphically convex and free of compact C-analytic subvarieties of C-dim $\geq q$. Hence the first part of the proof tells us that X' is cohomologically q-complete. We infer from Ref. 100, Theorem 3, that

$$H^i(X, \mathcal{F}) \cong H^i(S, \mathcal{F}|S) \quad \text{for any } \ i \geq q \ \text{ and any } \mathcal{F} \in Coh(X).$$

In particular X is cohomologically q-convex.

Step 2. Assume that X is cohomologically q-convex. Let $S :=$ union of all compact C-analytic subvarieties of C-dim $\geq q$ and let $T := \pi(S)$. It is enough to show that T finite. Assume the contrary and let $T' \subset T$ be an infinite discrete set of points. Let $S' := \pi^{-1}(T')$. By hypothesis one can find an integer $m \geq q$ and infinitely irreducible components $S_v \subset S'$ with dim $S_v = m$ for any v. In view of Corollary 2.2 there exist $\mathcal{F}_v \in Coh(S_v)$ such that $H^m(S_v, \mathcal{F}_v) \neq 0$ i.e. the stalks $(\mathcal{R}^m\pi_*\mathcal{F}_v) \neq 0$ at x_v for any v. Let $\mathcal{G} := \cup\mathcal{G}_v$ where $\mathcal{G}_v$ is the trivial extension of $\mathcal{F}_v$ to X. Then one can check easily that $\mathcal{G} \in Coh(X)$ and

$$\dim H^m(X, \mathcal{G}) = \dim H^0(T, \mathcal{R}^m\pi_*\mathcal{G}) \geq \dim H^0(T', \mathcal{R}^m\pi_*\mathcal{G}) = \infty$$

contradicting the hypothesis of cohomologically q-convexity of X. □

Corollary 3.1. *Any cohomologically q-convex space X admits a proper modification morphism $\pi : X \to Y$ where Y is some cohomologically q-complete space.*

Corollary 3.2. *Let S be a q-maximal compact analytic subvariety of some cohomologically q-convex space X. Assume that dim S < p. Then X is cohomologically p-complete.*

From the results of this section and those in Sec. 2, we are now ready to completely classify Ref. 98 the holomorphically convex spaces, see Tab. 1.

3.4. *The q-convexity*

First of all let us mention the following

Lemma 3.1. *Refs. 85,89 Let $\pi : X \to Y$ be a holomorphic map of C-analytic spaces. Then there exists a decreasing chain of C-analytic subvarieties $\mathcal{A}_v$ in X with $v \leq n$.*

$$X = \mathcal{A}_v \supset \mathcal{A}_{v-1} \supset \cdots \supset \mathcal{A}_0 \supset \mathcal{A}_{-1} = \emptyset$$

such that for each v, we have

(1) $\dim \mathcal{A}_{v-1} < \dim \mathcal{A}_v$,
(2) sing $\mathcal{A}_v \subset \mathcal{A}_{v-1}$ and
(3) $\pi|\mathcal{A}_v \setminus \mathcal{A}_{v-1} : \mathcal{A}_v \setminus \mathcal{A}_{v-1} \to Y$ has constant rank.

We are now in a position to prove the following

Theorem 3.6. *X is q-complete iff X is free of compact analytic subvarieties of C-—dim $\geq q$.*

	X is ...	*cd(X)*		*fd(X)*	*X admits ...*	
	free of compact C-analytic subvarieties of C-dim $\geqslant 1$ & is holomorphically convex	0		n	a n-maximal compact C-analytic subvariety	
Furthermore X is assumed to be holomorphically convex	free of compact C-analytic subvarieties of C-dim $\geqslant 2$	1		n-1	a (n-1)- maximal compact C-analytic subvariety	*Furthermore X is assumed to be holomorphically convex*
	. . . .				. . . .	
	free of compact C-analytic subvarieties of C-dim $\geqslant$ q-1	q		q	a q- maximal compact C-analytic subvariety	
	. . . .				. . . .	
	free of compact C-analytic subvarieties of C-dim $\geqslant$ q-1	n-2		2	a 2- maximal compact C-analytic subvariety	
	free of compact C-analytic subvarieties of C-dim $\geqslant$ n	n-1		1	a 1- maximal compact C-analytic subvariety & is holomorphically convex	
		n	**X is compact** *then ...* *If X is ireductible*			

Table 1

Proof. The proof is by induction on $n := \dim X$. If $n = 1$, it follows from Theorem 2.9 that X is q-complete for some $q > 1$. Hence by induction we can assume that this Theorem does hold for spaces X with $\dim X < n$. By Lemma 3.1, there exists a C-analytic subvariety $X' \subset X$, such that

$$\dim X' < \dim X, \quad \text{and sing } X \subset X'$$

and

$$\pi | X \setminus X' : X \setminus X' \to Y \text{ has constant rank} \tag{16}$$

Since $Y' := \pi(X')$ is a Stein space therefore by induction assumption X' is q-complete. In view of Ref. ?, Satz 6.2, there exist a neighborhood U of X' in X and a q-convex function $\phi : U \to \mathbb{R}$. Let V be an open neighborhood of X' in X, such that $V \subset U$, and let $\rho : U \to \mathbb{R}$ be a smooth function such that

$$0 \leq \rho(x) \leq 1 \quad \rho(x) = 1 \text{ on } V \text{ and supp } \rho \subset U. \tag{17}$$

Since Y is Stein, there exists a 1-convex exhaustion function φ on Y. Now let

$$\Psi := \rho\phi + \chi(\varphi \circ \pi) \tag{18}$$

where $\chi : \mathbb{R} \to \mathbb{R}^+$ is a rapidly increasing smooth and convex function, with $\chi(t) \geq t$ for any $t \geq 0$, $\chi' > 0$ and $\chi'' > 0$. Notice that $\psi(x)$ is q-convex if

(i) $x \in V$ since $\chi(\phi \circ \pi)$ is weakly 1-convex on X,
(ii) $x \in X \backslash U$ in view of (17), (18) and (16).
(iii) $x \in U \backslash V$ since, in view of the choice of χ, χ' will be large enough to compensate the possible negative eigenvalues of the Levi form $L(\rho\phi)$.

Hence our proof is complete since ψ is an exhaustion function, by construction. □

Remark 3.2. Apparently, the original idea to tackle Theorem 3.6 by using Lemma 3.1 was due to Andreotti (see Ref. 85). Such an approach was initiated in Ref. 62 with only partial success, due to the lack of a crucial piece of hardware (Ref. 66, Satz 6.2). The completion of this program, first appeared in Ref. 89.

Corollary 3.3. *X is q-convex iff X admits a q-maximal compact analytic subvariety S.*

Proof. Assume that X admits a q-maximal compact analytic subvariety S. Let $\pi : X \to Y$ be a blowing down morphism which contracts S to finitely many points $T \subset Y$ Certainly Y is holomorphically convex and is free of compact analytic subvarieties of C-dim $\geq q$. We infer from Theorem 3.6 that Y is indeed q-complete and our desired conclusion will follow □

Corollary 3.4. *Assume that X is a q-convex space with its maximal compact analytic subvariety S. Then X is p-complete iff C-—dim$S < p$.*

4. The K-complete Spaces

4.1. *Preliminaries*

Definition 4.1. Let X be a C-analytic space. Then X is said to be *K-complete* (or *holomorphically spreadable*) if for any $x \in X$, there exist some neighborhood U of x in X and finitely many functions $\{f_1, \ldots, f_k\} \in \Gamma(U, O_U)$ such that $\{y \in U | f_i(y) = f_i(x)$ for all $1 \leq i \leq k\} = \{x\}$.

Notice that any open subset of a Stein space X is K-complete.

Theorem 4.1 (Refs. 31,52). *Any K-complete space admits a countable topology.*

Now one has the following important

Theorem 4.2. *(see e.g. Ref. 40, V.D.4) A pure n-dimensional C-analytic space X is K-complete iff X could be realized as a ramified domain $\pi : X \to \mathbf{C}^n$, i.e. the fibres of π are discrete.*

Adopting the same "touch- and-go" approach as above, we'll study q-convex spaces in this context; namely, throughout the rest of this chapter all C-analytic spaces X are assumed to be K-complete.

4.2. *The twin primes*

Theorem 4.3 (Ref. 102). *Assume that X is cohomologically q-convex Then X is cohomologically q-complete.*

Proof. In view of Theorem 2.12, one can assume, without loss of generalities that X is irreducible. Theorem 4.2 implies the existence of a holomorphic map $\pi : X \to \mathbf{C}^n$ with discrete fibres.

Certainly, the result is trivial if dim $X = 1$, since X is Stein. So, one can assume that the theorem does hold for C-analytic spaces with C-dim $X <$

n. Furthermore, it is known that X is cohomologically n-complete; hence by induction one can assume that X is already cohomologically $(q+1)$-complete.

Let us consider the following exact sequences

$$0 \to \mathcal{H} \to \mathcal{F} - h \to f\mathcal{F} \to 0 \tag{19}$$

$$0 \to f\mathcal{F} - \iota \to \mathcal{F} \to \mathcal{F}/f\mathcal{F} \to 0 \tag{20}$$

where $h : F \to F$ is a multiplication by $f := g \circ \pi, g \in \Gamma(\mathbf{C}^n, O)$ with $g \neq 0$ and $H := \ker h$. From (19), one obtains the surjectivity of

$$h^* : H^i(X, \mathcal{F}) \to H^i(X, f\mathcal{F})$$

for any fixed $i \geq q$, by induction hypothesis.

On the other hand, from (20), it follows readily that:

$$\iota^* : H^i(X, f\mathcal{F}) \to H^i(X, \mathcal{F}) \quad \text{is surjective,}$$

since $\mathrm{supp}(\mathcal{F}/f\mathcal{F}) = S := \{f = 0\}$ and $\dim S = n - 1$. So the morphism

$$\Phi_g := \iota^* \circ h^* : H^i(X, \mathcal{F}) \to H^i(X, \mathcal{F})$$

is surjective.

If X is cohomologically q-convex, i.e. dim $\mathrm{H}^i(X, \mathcal{F}) < \infty$, Φ_g is indeed bijective, for any fixed $g \neq 0$. Now let us assume that $H^i(X, \mathcal{F}) \neq 0$ and let ω be a non zero element in $H^i(X, \mathcal{F})$; for any such g as above, Φ_g will induce a monomorphism

$$\begin{aligned} \Gamma(\mathbf{C}^n, O) &\to H^i(X, \mathcal{F}) \\ g \quad &\longmapsto \quad \Phi_g(\omega) \end{aligned}$$

Thus $\dim \Gamma(\mathbf{C}^n, O) \leq \dim H^i(X, \mathcal{F})$ which contradicts the infinite dimensionality of $\Gamma(\mathbf{C}^n, O)$. Therefore X is cohomologically q-complete. □

In parallel with this result, one has

Theorem 4.4 (Ref. 102). *Assume that X is q-convex. Then X is q-complete.*

Proof. In view of the q-convexity of X, there exist a compact set $K \subset X$ and an exhaustion function $\phi : X \to \mathbb{R}$ which is q-convex at any $x \in X \backslash K$. For any real numbers $a > b > c > \sup_K \phi$, one has

$$K \subset U := \{x | \phi(x) < c\} \subset V := \{x | \phi(x) < b\} \subset W := \{x | \phi(x) < a\}.$$

Since X is K-complete, a result in Ref. 2 tells us that there exists a 1-convex function $f : W \to \mathbb{R}$. Now let ρ be a smooth function on X with $0 \leq \rho(x) \leq 1$ such that

$$\rho(x) = \begin{cases} 1 \text{ on } V \\ 0 \text{ on } X \setminus W \end{cases}$$

and let $g(x) := \rho(x)f(x)$. From the boundedness of g and the compactness of U we infer the existence of a real constant $A \gg 0$, such that $Ag + \phi$ is 1-convex on U. Again by virtue of the boundedness of f and the compactness of W, one can find a constant $B \gg 0$ such that

$$Ag + B\phi \text{ is } q\text{-convex on } X \setminus V = (X \setminus W) \cup (W \setminus V).$$

Now let $r : X \to \mathbb{R}$ be a smooth function with

$$r' > 0, \quad r'' > 0, \quad r(t) = t \text{ if } t < c \text{ and } r'(t) > B \text{ if } t > b.$$

Then one can check that $\Phi := Ag + r_0\phi : X \to \mathbb{R}$ is a q-convex function on X. Furthermore Φ is an exhaustion function since ϕ is and r is a convex function. □

Certainly this result tells us that within the framework of holomorphically spreadable spaces, Problem 2.1 is reduced to the following

Problem 4.1. *Let X be a K-complete space.*
Assume that X is cohomologically q-complete, is X always q-complete?

4.3. *An indentation*

Definition 4.2. An open subset $D \subset \mathbf{C}^n$ is said to have *boundary of class* C^α with $1 \leq \alpha \leq \infty$, at $x \in \partial D$, if there exist an open neighborhood U of x and a C^α function $\rho : U \to \mathbb{R}$ with the following property:

$$D \cap U = \{y \in U | \rho(y) < 0 \text{ and } d\rho(x)\} \neq 0. \tag{21}$$

∂D is said to be of class C^α if it is of class C^α at every point $x \in \partial D$. A function $r \in C^\alpha(U)$ satisfying (21) is called a *local defining function* for ∂D at x.

If U is a neighborhood of ∂D, a function $\rho \in C^\alpha(U)$ satisfying (21) is called a *global defining* function for ∂D.

Definition 4.3. Let $D \subset \mathbf{C}^n$ be a domain. D is said to be locally *weakly q-convex* (or *Levi q-convex*), if for every $z \in \partial D$ and $t \in H_z(\partial D)$, the Levi form

$$\mathcal{L}_z(\phi; t) \text{ has at most } q-1 \text{ negative eigenvalues,}$$

where ϕ is a local defining function for ∂D at z and $H_z(\partial D)$, the *holomorphic tangent space* to ∂D at z where

$$H_z(\partial D) := \Big\{ w = (w_1, \dots, w_n) \in \mathbf{C}^n | \sum_{1 \leq i \leq n} \partial\phi \setminus \partial x_i(z) w_i = 0 \Big\}.$$

Now we are in a position to present a small contribution toward Problem 4.1 :

Theorem 4.5. *Let M be a K-complete manifold and let $D \subset M$ be a bounded domain such that ∂D is of class C^2. Then D is locally weakly q-convex iff $H^i(D, O_D) = 0$ for all $i \geq q$.*

Proof.
(i) Assume that $H^i(D, O_D) = 0$ for all $i \geq q$. Then a result in Ref. 92, Theorem 1 (see also Ref. 23) tells us that D is locally weakly q-convex.
(ii) Assume that D is locally weakly q-convex. Then an argument in Ref. 1, Pro. 15 tells us that one can find a neighborhood U of ∂D and a global defining q-convex function $r : U \to \mathbb{R}$. Now let $\phi := -1/r$ and let V' be a relative compact neighborhood of ∂D such that $V' \subset U$. Let

$$\begin{aligned} \rho(x) &= 1 \text{ on } V' \\ &= 0 \text{ on } D \setminus (D \cap U) \end{aligned}$$

and let $\varphi := \rho\phi$.

Since M is K-complete, a result in Ref. 2 tells us that there exists a 1-convex function ψ on $D \cup V$. Since D is bounded, there exists a real constant $A \gg 0$, such that $A\psi + \phi$ is an exhaustion q-convex function on D, and our proof is complete. □

Remark 4.1. Notice that in the second part of the proof, the boundedness condition is crucial, due to a counterexample in Ref. 26. On the other hand, that hypothesis is superfluous for the first part of the proof. Also the notion of "test classes" introduced in Ref. 23 seems to be a good tool for further investigations.

5. The Cultural Diversity

5.1. *The GAGA duality*

As Serre Ref. 80 convinced us that the parallel development in algebraic and analytic geometry is, as always, a two way street, namely it benefits both discipline. This is no exception in this context. Indeed in Algebraic geometry still stands the following crucial

Question 5.1. Let $C \subset \mathbf{P}_3$ be a non singular connected C-analytic compact curve in $\mathbf{P}_3$. Is C a set theoretic complete intersection in $\mathbf{P}_3$?

This question is related to Question 2.1 in view of the following:

Theorem 5.1. *Let $Y \subset \mathbf{P}_N$ be a compact C-analytic subvariety of pure codimension q. Assume that Y is a set theoretic complete intersection. Then $X := \mathbf{P}_N \backslash Y$ is q-complete*

Proof. In view of the hypothesis, $Y = \bigcap_{\leq j \leq q} Y_j$ where each Y_j is an analytic hypersurface of degree d_j in $\mathbf{P}_N$. Let $d := l.c.m\{d_1, \ldots, d_q\}$ and let us consider the d-uple embedding Ref. 46

$$\begin{array}{rccc} \tau : & \mathbf{P}_N & \rightarrow & \mathbf{P}_v \\ & (z_0, \ldots, z_N) & \mapsto & (Z_0, \ldots, Z_v) \end{array}$$

where $v + 1 := \binom{N+d}{N}$ and the Z_k are monomials $(0 \leq k \leq v)$ of degree d in the $(N+1)$ variables $(z_0, \ldots, z_N)$. Let $W := \tau(\mathbf{P}_N)$. Then one can find Ref. 100 constants, say $\{\alpha_j^k\}$ with $0 \leq k \leq v$ and $1 \leq j \leq q$ such that Y_j are mapped onto hyperplanes

$$\mathcal{H}_j := \Big\{ \sum_{0 \leq k \leq v} \alpha_j^k Z_k = 0 \Big\} \text{ in } \mathbf{P}_v$$

i.e. $\tau^{-1}(\mathcal{H}_j \cap W) = \mathcal{Y}_j$. Since Y is of pure codimension q, one can check that the constants $\beta_1, \ldots, \beta_q$ are linearly independent, where $\beta_j := (\alpha_j^0, \ldots, \alpha_j^v) \in \mathbf{C}^{v+1} \setminus \{0\}$. Consequently $\bigcap_{\leq j \leq q} \mathcal{H}_j$ will determine a $\mathbf{P}_{v-q}$ in $\mathbf{P}_v$. Hence τ embeds $X := \mathbf{P}_N \backslash Y$ biholomorphically, as closed submanifold in $\mathbf{P}_v \backslash \mathbf{P}_{v-q}$ and thus X is q-complete in view of Example 2.3. □

As far as the set theoretic complete intersection is concerned, let us mention the following

Definition 5.1. Let X be a compact C-analytic subvariety in $\mathbf{P}_N$. Then X is said to be of *minimal degree* (see e.g.Ref. 37) if

(a) X is non degenerate, i.e. X does not contain in any $\mathbf{P}_{N'}$ with $N' < N$,
(b) deg X = codim $X + 1$.
On the other hand, it is known Ref. 21 that the only compact surfaces of minimal degree in $\mathbf{P}_N$ are

(1) Veronese surfaces
(2) *Rational normal scrolls*, Ref. 46 i.e. rational ruled surfaces which can be embedded in $\mathbf{P}_N$ by the complete linear system $|O_N(1)|$.

As we have seen in Example 2.5, Veronese surfaces are not set theoretic complete intersections in $\mathbf{P}_5$. However, it is known that rational normal scrolls are (see e.g Ref. 91).

Now the intertwining relationship between the above 2 problems is entangled in a web full of intrigues, due to the following

Conjecture 5.1 (Refs. 44,45). *Let $Y \subset \mathbf{P}_N$ be a connected compact C-analytic submanifold of C-dimension n. Then Y is a strict complete intersection provided $n > 2/3N$.*

Translated into our context, on the basis of Theorem 5.1, we have a weaker

Conjecture 5.2. *Let Y and $\mathbf{P}_N$ be as in Conjecture 5.1. Then $X := \mathbf{P}_N \backslash Y$ q-complete if $N > 3q$ where $q =$ codim Y.*

In fact this bound is quite sharp, due to the following

Example 5.1. Let $Y := \mathbf{G}(2,5)$ be the Grassmannian of 2-planes in $\mathbf{C}^5$. By using the Plucker embedding Ref. 37, Y can be realized as a closed analytic submanifold of codimension 3 in $\mathbf{P}_9$. Yet $X := \mathbf{P}_9 \backslash Y$ is not even 4-complete. Indeed in Ref. 16 it was explicitly shown that C-dim $H^4(X, \Omega^9) > 0$ where Ω^9 is the canonical sheaf of $\mathbf{P}_9$.

In the midst of this hotly contested debate, along came the following

Theorem 5.2 (Ref. 65). *Let Y be a connected compact C-analytic submanifold of codimension q in $\mathbf{P}_N$ and let $X := \mathbf{P}_N \backslash Y$. Then X is $(2q-1)$ complete.*

For small codimension ($\neq 2$), on the basis of explicit construction in the remarkable article Ref. 16, this result so far is quite sharp; and this turns out to be unwelcome news for the above conjectures. Indeed, potentially, the latter could be target for a takeover by some institution of counterexamples

However, at least in codimension 2, Conjecture 5.1 is reinforced by the following facts:

First of all, one has

Theorem 5.3 (Ref. 44). *Let Y and $\mathbf{P}_N$ be as in Conjecture 5.2. Assume that $n = N - 2$ and $N \geq 6$. Then there exist a rank 2 bundle v on $\mathbf{P}_N$ and a section $\sigma \in \Gamma(\mathbf{P}_N.v)$ such that. $Y \cong \{\sigma = 0\}$. Furthermore Y is a complete intersection iff v is a direct sum of line bundles*

In Ref. 13 it was shown that rank 2 bundles on $\mathbf{P}_N$ are decomposable for large N. On the other hand, experimental results in Refs. 48,49 confirmed that Y is a complete intersection if its degree is small compare with N. Although, the jury is still out on this issue, one would undoubtedly, hold the breath for a final verdict, regardless which way will swing the pendulum.

In sharp contrast with the compact case, such a dual suspense does not occur in $\mathbf{C}^N$. Indeed that tension was short-circuited by the following

Theorem 5.4 (Refs. 6,66). *Let Y be a locally complete intersection C-analytic subvariety of pure codimension q in $\mathbf{C}^N$. Then $X := \mathbf{C}^N \backslash Y$ is q-complete.*

5.2. *Pluribus unum*

Even before the appearance of Ref. 1, there are many notions of q-convexity introduced in the literature. We would like to mention and compare few of them

Definition 5.2 (Refs. 24,57,74). *A domain $D \subset \mathbf{C}^n$ is called locally Rothstein q-convex (with $1 \leq q \leq n-1$) if any $x \in \partial D$, has a neighborhood N_x such that for any $U \subset \cap N_x$, there exists a relative compact subset $U' \subset N_x \cap D$ with $U \subset U'$ having the following properties.*

For all $z' \in U'$ one can find $n-q$ holomorphic functions $\{f_1, \ldots, f_{n-q}\}$ on U' such that $|f_i(z')| > 1$ for $1 \leq i \leq n-q$ and for all $z \in U$, $\min |f_i(z)| < 1$.

In the special case where $q = n - 1$, one recovers the notion of "holomorph-convexity" in Ref. 18.

Example 5.2. Let $Y \subset \mathbf{C}^n$ be an irreducible C-analytic subvariety of C-dim$= q$. Then $X := \mathbf{C}^n \backslash Y$ is locally Rothstein q-convex.

In Ref. 24, were stated the following results

Theorem 5.5 (Ref. 24, Theorem 1, p.66; Ref. 75). *Let D be a bounded domain in $\mathbf{C}^n$ with C^∞ boundary. Then D is locally Rothstein $(n-q)$ convex iff D is locally weakly q-convex.*

Theorem 5.6. *(Ref. 24, Theorem 2, p.67) Let D be as in Theorem 5.5. Then D is locally weakly q-convex iff $H^i(D, O_D) = 0$ for all $i \geq q$.*

Now let us quote the following remark from Ref. 24, p.67: "...Theorem 2 is much easier to prove than Theorem 1, we'll assume Theorem 2 to prove Theorem 1".

In the hindsight, one realizes that Ref. 24 was way ahead of its times; indeed Theorem 5.6 is true (in view of Theorem 4.5, but it was not easy!).

Notice that Theorem 5.5 is false (Ref. 57,§12) if one replaces locally Rothstein $(n-q)$-convexity by a global one , (see e.g. Ref. 57 §3)

Initially, the convexity of domains in $\mathbf{C}^n$ is defined by functions which are not even continuous.

Definition 5.3 (Ref. 50). *A function $u : D \to R \cup \{-\infty\}$ defined in an open set $D \subset \mathbf{C}$ is called subharmonic if*

(a) u is upper semicontinuous, i.e. $\{z|u(z) < r\}$ is open for any $r \in \mathbb{R}$,
(b) For every compact set K and every continuous function h on K which is harmonic in the interior of K and is $\geq u$ on ∂K, we have $h \geq u$ in K.

Notice that a function h defined in an open set $D \subset \mathbf{C}$ is said to be harmonic if the laplacian $\Delta h := 4\partial^2 h/\partial z \partial z = 0$ in D. For any holomorphic function f on D, $|f|$ is subharmonic.

Definition 5.4 (Ref. 50). *A function $u : D \to R \cup \{-\infty\}$ defined in an open set $D \subset \mathbf{C}^n$ is called plurisubharmonic if*

(a) u is upper semi-continuous,
(b) For arbitrary z and w, the function $\tau \to u(\tau z + w)$ is subharmonic, for any $\tau \in \mathbf{C}$ such that $\tau z + w \in D$.

For any holomorphic function f, log $|f|$ is plurisubharmonic.

Remark 5.1. Let $\phi : D \to \mathbf{R}$ be a continuous function. Then one could define ϕ to be *pseudoconvex* if for each $x \in D$, there exist a neighborhood U of x in D, and finitely many plurisubharmonic functions $\{f_1, \ldots, f_k\}$ such that $\phi|U = \max\{f_1, \ldots, f_k\}$.

However, as pointed out in Ref. 59(2.2) this concept is no more general than the initial one.

Definition 5.5. (see e.g. Ref. 50) A domain $D \subset \mathbf{C}^n$ is said to be *pseudoconvex* if each point $x \in \partial D$ admits a neighborhood $U \subset \mathbf{C}^n$ and a plurisubharmonic function ϕ on U such that $U \cap D = \{x \in U | \phi(x) < 0\}$.

The upshot of this story resides in the following fundamental result of Oka.

Theorem 5.7. *(see e.g. Ref. 50) Any pseudoconvex domain $D \subset \mathbf{C}^n$ is Stein.*

In order to generalize such a concept, following the tradition of Oka, the Japanese school (Nishino, Fujita, Takadoro, among others) introduced the following:

Definition 5.6 (Ref. 30). *Let $D \subset \mathbf{C}^n$ be a domain and let q be an integer with $1 \leq q \leq n$. A function $\phi : D \to R \cup \{-\infty\}$ is said to be subpluriharmonic if*

(i) ϕ is upper semi continuous,
(ii) Let $G \subset D$ be a relative compact domain, let G' be some neighborhood of G in D and let $h : G' \to D$ be a pluriharmonic function (i.e. locally the real part of some holomorphic function). If $\phi \leq h$ on ∂G then $\phi \leq h$ on G.

Definition 5.7. Let $D \subset \mathbf{C}^n$ be a domain and let q be an integer with $1 \leq q \leq n$. Then $\phi : D \to R \cup \{-\infty\}$ is said to be *pseudoconvex of order n-q* on D if

(i) ϕ is upper semi continuous, and
(ii) For any domain $G \subset \mathbf{C}^q$ and all holomorphic application $f : G \to D$ the composite $\phi \circ f$ is subpluriharmonic in G.

Remark 5.2. A function ϕ is pseudoconvex of order $n-1$ iff it is plurisubharmonic.

Proposition 5.1 (Ref. 30, Prop. 8). *For any integer q with $1 \leq q \leq n$, a C^2 function $\phi : D \to \mathbb{R}$ is weakly q-convex iff ϕ is pseudoconvex of order $n - q$.*

Definition 5.8 (Ref. 54). *Let M be a connected paracompact C-analytic manifold with C-dim $M = n$ and let q be an integer with $1 \leq q \leq n$. Then* *a domain $D \subset M$ is said to be pseudoconvex of order $n-q$ in M, if the complement $M \backslash D$ satisfies "the Hartogs continuity principle of dimension $n-q$" (i.e. pseudoconcave of order n-q in the sense of Ref. 88).*

Remark 5.3.

(i) Any open subset in M is by definition pseudoconvex of order 0,
(ii) A domain $D \subset M$ is pseudoconvex of order $n-1$ iff it is pseudoconvex,
(iii) A domain $D \subset M$ is pseudoconvex of order $n-q$ iff every point $x \in \partial D$ admits a neighborhood U such that $U \cap D$ is pseudoconvex of order $n-q$.

The connection between the above Definitions 5.7 and 5.8, is lying in the following

Theorem 5.8 (Ref. 30, Th. 2). *For any domain $D \subset \mathbf{C}^n$, the following conditions are equivalent:*

(i) D is pseudoconvex of order $n-q$ in $\mathbf{C}^n$,
(ii) D admits an exhaustion function which is pseudoconvex of order $n-q$ on D.

Example 5.3 (Ref. 54). *Let Y be a C-analytic subvariety in M and let $k :=$ minimum of C-dimension of irreducible components of Y. Then $X := M \backslash Y$ is pseudoconvex of order $n-q$ iff $k \geq n-q$.*

Example 5.4. A domain $D \subset M$ is pseudoconvex of order $n-q$ in M if for all $x \in \partial D$, there exists a C-analytic subvariety S of pure C-dim $n-q$, defined near x such that

$$x \in S \text{ and } S \subset M \setminus D.$$

Example 5.5 (Counterexample, Ref. 30). *Let $n := 2q, q \geq 2$ and let $\mathbf{C}^n$ with coordinates $z := (z_1, \ldots, z_q, z_{q+1}, \ldots, z_n)$. Let*

$$Y_1 := \{z | z_1 = \cdots = z_q = 0\}, \quad Y_2 := \{z | z_{q+1} = \cdots = z_n = 0\}$$

and let $X := \mathbf{C}^n \backslash Y$ where $Y := Y_1 \cup Y_2$.

Then one can check Ref. 87 that

$$H^{n-2}(X, \Omega^n) \neq 0, \tag{22}$$

Then it follows from Example 5.3 that X is pseudoconvex of order $n-q$ in $\mathbf{C}^n$. But in view of (22) and Theorem 4.4, X is not even cohomologically q-convex, since $n-2 \geq q$.

To remedy this defect, one has

Theorem 5.9. *Let $D \subset M$ be a domain with $1 \leq q < n$. Assume that:*

(i) ∂D is C^2,
(ii) M is Stein,

Then the following conditions are equivalent:

(a) D is pseudoconvex of order $n - q$ in M,
(b) D is locally weakly q-convex,
(c) D is q-complete.

Remark 5.4. Few words of cautions are in order here:

(i) For the equivalence (a)$\leftrightarrow$ (b), the hypothesis (ii) is superfluous (see Ref. 54, Ex. 2.3).
(ii) As far as the equivalence (c) $\leftrightarrow$ (b) (the proof of it appeared in Ref. 93 is concerned, the hypothesis (i) is redundant when $q = 1$, in view of Oka Theorem.
(iii) The proof of the equivalence (c) $\leftrightarrow$ (a) can be found also in Ref. 54, Theorem 7.6.
(iv) Theorem 5.9 is trivial if $q = n$ in view of Ref. 35.
(v) Example 5.5 shows that the assumption (i) is crucial here.

In contrast with Remark 5.1, let us consider the following

Definition 5.9 (Ref. 64). *A continuous function $\phi \to \mathbb{R}$ is said to be q-convex with corners, if any point $x \in X$ has an open neighborhood U and if there are q-convex functions $\{f_1, \ldots, f_k\}$ defined on U such that*

$$\phi|U = \max\{f_1, \ldots, f_k\}.$$

Obviously any q-convex function is q-convex with corners but not vice versa.

Definition 5.10. A C-analytic space X is said to be *q-complete with corners* if there exists an exhaustion function $\phi : X \to \mathbb{R}$ which is q-convex with corners on X.

Example 5.6 (Ref. 64). *Let $M := C^n$ or $\mathbf{P}_n$, let $Y \subset M$ be a closed C-analytic subvariety and let $q :=$ maximum of codimension of irreducible components of Y. Then $X := M \backslash Y$ is q-complete with corners .*

Definition 5.11. A domain $D \subset M$ is said to be *locally q-complete with corners* if any point $x \in \partial D$ has an open neighborhood U such that $U \cap D$ is q-complete with corners.

In this spirit we have

Theorem 5.10 (Ref. 54, Pro. 2.2). *Let $D \subset M$ be a domain.*

Then D is pseudoconvex of order $n-q$ iff D is locally q-complete with corners

6. Epilogue

Now let us step back and reassess our resources. First of all let us look at an important relative case

Definition 6.1. Refs. 53,81 Let $\pi : X \to Y$ be a morphism of C-analytic spaces. Then π is said to be a *q-convex morphism* if there exist, a constant $\gamma \in \mathbb{R}$

(a) a function $\phi : X \to \mathbb{R}$ such that $\phi|\{x|\phi(x) > \gamma\}$ is q-convex and
(b) $\pi|\{x|\phi(x) \leq c\}$ is proper for any $c \in \mathbb{R}$.

In this situation, one has the following important result:

Theorem 6.1 (Ref. 81). *Let $\pi : X \to Y$ be a q-convex morphism of C-analytic spaces. Then, for any $\mathcal{F} \in Coh(X)$ and any $k \geq q$*

(a) $\mathcal{R}^k\pi * \mathcal{F} \in Coh(Y)$,
(b) $H^k(\pi^{-1}(S'), \mathcal{F})$ has a Hausdorff topology for any Stein open set $S' \subset Y$ and
(c) $H^k(\pi^{-1}(S'), \mathcal{F}) \cong H^0(S', R^k\pi * \mathcal{F})$.

Notice that when $Y =$ one point (resp. when π is proper), one recovers the result in Ref. 1 (resp. Ref. 33). This is a very fine tuned endeavor, since only special cases of it, were known earlier Refs. 53,84. This technique is quite innovative and deserves to be investigated further.

The seminal paper Ref. 66 with sophisticated and difficult techniques showed strong promise and should be pursued further.

Notice that q-concave spaces introduced in Ref. 1 are not mentioned at all here; this is due mainly to the lack of author's expertise in this direction. This theory has a profound impact in other branches of mathematics: Arithmetic groups Refs. 5,15, CR structures Refs. 72,73,79,etc.

Since Stein spaces are holomorphically convex and holomorphically spreadable, the investigations carried out in Chapters 2 and 3 naturally adapted to such a philosophy. On the other hand, Stein manifolds are Kahlerian; consequently in depth investigations of q-convex manifolds, within the framework of Kahlerian geometry have been, in the past, and are still, currently, active research topics by differential geometers.. In this respect we strongly recommend the excellent expository survey Ref. 105.

For any C-algebraic variety X, its underlying topological space X acquires a structure of a C-analytic space; similarly one can associate to each algebraic coherent sheaf $\mathcal{F}$, an $\mathcal{F} \in Coh(X)$ Refs. 39,80. Consequently, one obtains a natural morphism of cohomology groups

$$H^k(X, F) \to H^k(X, \mathcal{F})$$

for any $k \geq 0$. Thus the notion of cohomologically q-convexity could be transplanted within the algebraic context and some comparison theorems could be established Ref. 43. Such an approach has been initiated in Refs. 7, 101,106.

For a study of q-convex manifolds from the classical constructive method of integral representations, the excellent monograph Ref. 47 is strongly recommended.

Definitely, fresh ideas and astucious strategies are needed in order to produce some road map for new frontiers. However, as was experienced in the past Ref. 12, and quite recently in Refs. 103,104, some concrete and constructive examples could go a long way to unveil certain unsolved mysteries.

Acknowledgements

This project began vigorously some 30 years ago, when the author was in exile in Italy: University of Firenze (1976), University of Calabria (Spring 77) and University of Ferrara (Summer 77). Unfortunately, over the years, it was waning off, due to the lack of: personal courage, technical know-how and professional means. The author would like to thank Prof. Nguyen Minh Chuong for his encouragement which helps this journey, at long last, to catch a glimpse of the finishing line.

References

1. A. Andreottiand H. Grauert, *Bull. Soc. Math. France*, **90**, 193 (1962).
2. A. Andreotti and R. Narasimhan, *Trans. Amer. Math. Soc.*, **111**, 499 (1966).
3. A. Andreotti and F. Norguet, *Ann. Scuo. Pisa*, **20**, 197 (1966).
4. A. Andreotti and F. Norguet, *Ann. Scuo. Pisa*, **21** 31 (1967).
5. A. Andreotti and E. Vesentini, *Acta Math.*, bf 112, 249 (1964).
6. E. Ballico, *Atti Accad. Dei Lincei*, **71**, 60 (1981).
7. E. Ballico, *Rendi. Mate. Roma*, **3** 381 (1983).

8. C. Banica and O. Stanasila, *Methodes algebriques dans la theorie globale des espaces complexes* (Gauthier-Villars, 1977).
9. D. Barlet, D.*Lect. Notes in Math.*, **807**, 102 (1977/79).
10. D. Barlet and V. Vajaitu, *Mich. Math. J.*, **50**, 57 (2002).
11. W. Barth,*Inv. Math.*, **9**, 135 (1969-70).
12. W. Barth,*Math. Ann.*, **187**, 150 (1970).
13. W. Barth and A. Van de Ven, *Invent. Math.*, **25**, 91 (1974).
14. R. Benedetti, *Math. Ann.*, **139**, 135 (1977).
15. A. Borel, *Pseudoconcavite et groupes arithmetiques, Essay on Topology and related topics, Memoires dedies a G. de Rham* (Springer, 1970) pp. 70–84.
16. M. Buchner, K. Fritzsche, and T. Sakai, *J. Reine Angew. Math.*, **323**, 1 (1981).
17. H. Cartan, *Quotients of complex analytic spaces, in Contribution to Function Theory* (International Colloqium, Tata Inst. of Fund.Research, Bombay, 1960) pp. 1–15.
18. H. Cartan and P. Thullen, *Math. Ann.*, **106**, 617 (1932).
19. M. Coltoiu, M. *J. reine und angew. Math.*, **412**, 108 (1990).
20. M. Coltoiu, *Nagoya Math.J.1*, **45**, 137 (1997).
21. Del Pezzo *Rendi. Cir. Mate. Palermo*, **1** (1886).
22. J. P. Demailly, *Math. Z.*, **204**, 283 (1990).
23. M. G. Eastwood and G. Vigna Suria, *Comment. Math. Helv.*, **55** 413 (1980).
24. L. Ehrenpreis, *Some applications of the theory of distributions to several complex variables* in *Seminar of complex spaces*, Vol. I. (Princeton, 1957) pp. 65–79.
25. W. Fischer, *Math. Ann.*, **184**, 297 (1969).
26. J. Fornaess, *Math. Ann.*, **234**, 275 (1978).
27. O. Forster, *Math. Z.*, **97**, 376 (1967).
28. K. Fritzsche, *Math. Ann.*, **221**, 251 (1976).
29. K. Fritzsche, *Math. Ann.* **230**, 107 (1977).
30. O. Fujita, *J. Math. Kyoto Univ.*, **30**, 634 (1990).
31. H. Grauert, *Math. Ann.*, **129**, 233 (1955).
32. H. Grauert, *Bull.Soc.Math.France*, **87**, 341 (1959).
33. H. Grauert,*Publi. Math, IHES, Paris*, **5**(1960).
34. H. Grauert, *Math. Ann*, **146**, 331 (1962).
35. R. Greene and H. H. Wu, *Ann. Inst. Fourier*, **25**, 215 (1975).
36. P. Griffiths, *Univ. Tokyo Press*, 185 (1969).
37. P. Griffiths and J. Harris, *Principles of Algebraic geometry* (Wiley-Interscience, New York, 1978).
38. P. A. Griffiths, and W. Schmid, *Acta Math.*, **123**, 253 (1970).
39. A. Grothendieck, *Sur les faisceaux algebriques et faisceaux analytiques coherents*, Expose 2, Seminaire, H. Cartan, Vol. 9 (1956/57).
40. R. Gunning and H. Rossi, *Analytic functions of several complex variables* (Prentice Hall series in Modern Analysis, 1965).
41. H. Hamm, *J. Reine Angew. Math.*, **364**, 1 (1986).
42. R. Hartshorne, *Ann. Math.*, **88**, 403 (1968).
43. R. Hartshorne, *Lect. Notes in Math.*, Vol. 156 (Springer Verlag, Berlin, 1970).

44. R. Hartshorne, *Proc. Symp. Pure Math.*, **29** (1973) *Amer. Math. Soc.* (Providence, RI) pp. 129–164.
45. R. Hartshorne, *Bull. Amer. Math. Soc.*, **80**, 1017 (1974).
46. R. Hartshorne, *Agebraic Geometry* (Berlin, Heidelberg, New York, Springer Verlag, Graduate Texts in Mathematics, 1977).
47. G. Henkin and J. Leiterer, *Andreotti- Grauert theory by integral representations*, Vol. 74 (Birkhauser, Progress in Mathematics, 1988).
48. A. Holme, A. *Manus. Math.*, **65**, 205 (1989).
49. A. Holme and M. Schneider, *J. fur Reine und Ange. Math.*, **357**, 205 (1985).
50. L. Hormander, *An introduction to Complex Analysis in Several complex variables* (North Holland Publishing Co, Amsterdam, 1973).
51. G. Horrocks and D. Mumford, *Topology*, **12**, 63 (1973).
52. M. Jurchescu, *Math. Ann.*, **138**, 332 (1959).
53. K. Knorr and M. Schneider, *Math. Ann.*, **193**, 238 (1971).
54. K. Matsumoto, *J.Math. Soc. Japan*, **48**, 85 (1996).
55. M. E. Larsen, *Inven. Math.*, **19**, 251 (1971).
56. J. Le Potier, *Bull.Soc. Mathe. France*, **98**, 319 (1970).
57. V. Lindenau, *Schriftenreihe Math. Inst. Munster*, **31** (1964).
58. R. Narasimhan, *Math. Ann.*, **142**, 355 (1961).
59. R. Narasimhan, *Math. Ann.*, **146**, 195 (1961).
60. F. Norguet, *Theoremes de finitude pour la cohomologie des espaces complexes*, d'apres, A. Andreotti and H. Grauert, *Seminaire de Bourbaki*, 1961/62, Vol. 7, Expose 234, (Soc. Math. France 1995) pp. 191–205.
61. T. Ohsawa, *Publ. R.I.M.S.*, **20**, 683 (1984).
62. C. Okonek, *Le. Notes in Math,* Vol. 1194 (Springer Verlag, 1986) pp. 104–126.
63. C. Peskine and L. Szpiro, *Publ. IHES.*, **42**, 49 (1973).
64. M. Peternell, *Inv. Math.*, **85**, 249 (1986).
65. M. Peternell, *Math. Z.*, **195**, 443 (1987).
66. M. Peternell, *Math. Z.*, **200**, 547 (1989).
67. D. Prill, *Duke Math. J.*, **34**, 375 (1967).
68. P. Ramis, G. Ruget, et J. Verdier, *Inv. Math.*, **13**, 261 (1971).
69. H. J. Reiffen, *Math. Ann.*, **164**, 271 (1966).
70. R. Remmert, *Reduction of complex spaces, in Seminar of complex spaces* Vol. I (Princeton, 1957) pp. 190–205.
71. H. Rossi, *Math. Ann.*, **146**, 129 (1962).
72. H. Rossi, *Proc. Conf on Complex Analysis* (Minneapolis, Springer, 1964) pp. 242–256.
73. H. Rossi, *Proc. Conference in Complex Analysis, 1972, Rice Univ.*, **59**, **1**, 131 (1973).
74. W. Rothstein, *Math. Ann.*, **129**, 96 (1955).
75. W. Rothstein and H. Sperling, *Einsetzen analytische Flachenstucken in Zyklen auf komplexen Raumen* (Festchrift K. Weiertrass, 1965) pp. 531–554.
76. W. Schmid, *Proc. Nat. Acad. Sci. USA*, **59**, 56 (1968).
77. W. Schmidt, *Shrift. Math. Insti. Munster*, **17** (1979).
78. M. Schneider, *Math.Ann.*, **201**, 221 (1973).
79. W. Schwarz, *Math. Z.*, **210**, 259 (1992).

80. J. P. Serre, *Ann. Inst. Fourier*, **6**, 1 (1956).
81. P. Siegfried, *Comm. Math. Helve.*, **49**, 417 (1974).
82. Y. T. Siu, *Trans. Amer. Math. Soc.*, **142**, 77 (1969).
83. Y. T. Siu, *Ann. Scuo. Pisa*, **26**, 649 (1972).
84. Y. T. Siu, *Inv. Math.*, **38**, 89 (1976).
85. A. Sommese, *Proc. in Pure Math. AMS*, **40**, 253 (1983).
86. G. Sorani, *Ann. Scuola. Norm. Sup. Pisa* **16**, 299 (1962).
87. G. Sorani and V. Villani, *Trans. Amer. Math. Soc.*, **125**, 432 (1966).
88. M. Tadokoro, *J. Math. Soc. Japan*, **17**, 281 (1968).
89. V. Vaijaitu, *Math. Zeit.*, **217**, 215 (1994).
90. V. Vaijaitu, *Math. Z.*, **221**, 217 (1996).
91. L. Verdi, *Boll. Unione Mat. Italia*, **6**, 47 (1986).
92. G. Vigna Suria, *Enseigne. Math.*, **29**, 329 (1983).
93. G. Vigna Suria, *Comp. Math.*, **53**, 105 (1984).
94. T. Vo Van, *On the classification of q-convex complex spaces by their compact analytic subvarieties*, Ph.D. Thesis (Brandeis University, 1974).
95. T. Vo Van, *Atti Accad. Naz. dei Lincei*, **60**, 345 (1976).
96. T. Vo Van, *Proc. Sympo. In Pure Math. AMS*, **30**, 53 (1977).
97. T. Vo Van, *Springer Verlag Lec. Notes in Math.*, Vol. 578, (1977) pp. 71-92.
98. T. Vo Van, On the geometry of holomorph-convex spaces, unpublished manuscript.
99. T. Vo Van, *Commen. Math. Helve.*, **58**, 678 (1983).
100. T. Vo Van, *Bull. Sci. Math., Paris*, **108**, 275 (1984).
101. T. Vo Van, *Expo. Math.*, **10**, 353 (1992).
102. T. Vo Van, *Inter. J. Math.*, **7**, 265 (1996).
103. T. Vo Van, *Bull. Sci. Math. Paris*, **129**, 501 (2005).
104. T. Vo Van, *On the analytic Picard group of proper modifications of certain affine varieties* (to appear).
105. H. H. Wu, *Proc. Sympo. In Pure Math. AMS*, **41**, 253 (1984).
106. F. L. Zak, *Izv. Akad. Nauk SSR*, **39**, 27 (1975).

PART B

P-ADIC AND STOCHASTIC ANALYSIS

Chapter III

OVER P-ADIC FIELD

Harmonic, Wavelet and p-Adic Analysis
Eds. N. M. Chuong *et al.* (pp. 271–290)

§12. HARMONIC ANALYSIS OVER P-ADIC FIELD I. SOME EQUATIONS AND SINGULAR INTEGRAL OPERATORS

NGUYEN MINH CHUONG[‡], NGUYEN VAN CO, AND LE QUANG THUAN

Institute of Mathematics, VAST,
18 Hoang Quoc Viet Road, Cau Giay District, 10307, Hanoi, Vietnam
[‡]*Email: nmchuong@math.ac.vn*

In this paper a solution of a Cauchy problem for a class of pseudodifferential equations over the p-adic field is given. Furthermore the boundedness of the p-adic Hilbert transform in some p-adic spaces, such as $L^q(B^n_{\gamma_0}), L^q(\mathbb{Q}^n_p), L^q(B^n_{\gamma_0}, l_k)$, bounded mean oscillation $BMO(\mathbb{Q}^n_p)$ and vanishing mean oscillation $VMO(Q^n_p)$, is proved.

1. Introduction

In 1994, in Ref. 29, theory of p-adic L- Schwartz distributions was introduced by V.S. Vladinirov, I.V. Volovich, E.I. Zelenov. Most interest in p-adic physics is the p-adic string theory. In 1988-1990 p-adic quantum mechanics and field theory were studied. In 1984, V.S. Vladimirov and I.V. Volovich applied p-adic to superfield theory. Even in theory of probability, probabilities of events can belong not only to the segment [0,1] of the field $\mathbb{R}$, $\mathbb{C}$, but also to some subset of the p-adic field. All the above mentioned results may be refered to Ref. 6, and references therein.

It seems that, modern and future science and technology would work probably not only on the usual $\mathbb{R}$, $\mathbb{C}$ fields but also on the p-adic field, generally non-Archimedean fields, local fields.

The purpose of our joint works is to study harmonic analysis over p-adic field. Some first results on this way were obtained in Refs. 3,4. Here we would present some facts on some differential and singular integral operators over p-adic field in some p-adic spaces, namely we will use the usual Fourier method to solve a Cauchy problem for a class of pseudodifferential equations over p-adic field and we will prove the boundedness of a class of p-adic Hilbert transform in some p-adic spaces.

Note that in most recent years, there has been a growing interest to p-adic pseudodifferential operators, to p-adic wavelets, especially to the very interesting, exciting relation between wavelet analysis and p-adic spectral analysis (see e.g Ref. 12,14–23).

The paper is organized as follows:

2. Preliminaries
3. A p-adic Cauchy Problem
4. The p-adic Hilbert transfom
5. The boundedness in the p-adic space L^q
6. The boundedness in the p-adic space $BMO(\mathbb{Q}_p^n)$
7. The boundedness in the p-adic space $VMO(\mathbb{Q}_p^n)$
8. The boundedness in the p-adic space $L^q(\mathbb{Q}_p^n, \omega(x), l_k)$

References

2. Preliminaries

Denote by p the prime numbers, $p = 2, 3, 5, \ldots$, by Q_p the p-adic field, (see Ref. 29). The space $\mathbb{Q}_p^n$ consists of elements $x = (x_1, \ldots, x_n), x_j \in Q_p$, equipped with p-adic norm $|x|_p = \max\limits_{1\leqslant j\leqslant n} |x_j|$. If $b \in Q_p$ and $x \in \mathbb{Q}_p^n$, then $bx = (bx_1, \ldots, bx_n) \in \mathbb{Q}_p^n$. For each $a \in \mathbb{Q}_p^n$ and $\gamma \in Z$, let us denote by

$$B_\gamma^n(a) = \{x \in \mathbb{Q}_p^n : |x - a|_p \leqslant p^\gamma\},$$
$$S_\gamma^n(a) = \{x \in \mathbb{Q}_p^n : |x - a|_p = p^\gamma\},$$

respectively a ball, a sphere of radius p^γ at center a. We use also the notations $B_\gamma^n = B_\gamma^n(0), S_\gamma^n = S_\gamma^n(0), B_\gamma = B_\gamma^1, S_\gamma = S_\gamma^1, \mathbb{Z}_p^* = S_0 = \{x \in Q_p : |x|_p = 1\}$. Some following simple properties are obvious.

1. $B_\gamma^n(a) = B_{\gamma-1}^n(a) \bigcup S_\gamma^n(a)$
2. $B_\gamma^n(a), S_\gamma^n(a)$ are open and closed compact sets
3. $\mathbb{Q}_p^n = \bigcup\limits_{\gamma\in\mathbb{Z}} B_\gamma^n(a) = \bigcup\limits_{\gamma=-\infty}^{+\infty} S_\gamma^n(a)$
4. $\bigcap\limits_{\gamma\in\mathbb{Z}} B_\gamma^n(a) = \{a\}$
5. $\mathbb{Q}_p^n$ is an additive local compact group and a disconnected space.

There is a Haar measure on $\mathbb{Q}_p^n$ normalized by $\int\limits_{B_0^n} dx = 1$ Ref. 29. If $S \subset \mathbb{Q}_p^n$ is a compact set, $O \subset \mathbb{Q}_p^n$ is an open set, then let us denote by $C[S]$ the space of all continuous complex-valued functions, with the norm $||f||_{C[S]} = \sup\limits_{x\in S} |f(x)|$, $L^l[O]$ the space of all measurable complex-valued functions such that

$$||f||_{L^l[O]} = \Big(\int\limits_O |f(x)|^l dx\Big)^{\frac{1}{l}} < +\infty,$$

$$L^l_{loc}[O] = \{f : O \to C, f \in L^l[S], \forall S \subset O\}.$$

We use also the notations:

$$L^l_{loc} = L^l_{loc}[\mathbb{Q}_p^n], L^l = L^l[\mathbb{Q}_p^n].$$

Let us now consider the space BMO $[\mathbb{Q}_p^n]$ consisting of all $f \in L^l_{loc}$ such that

$$||f||_{BMO(\mathbb{Q}_p^n)} = \sup_{\substack{z\in\mathbb{Q}_p^n \\ \gamma\in\mathbb{Z}}} \frac{1}{|B_\gamma^n(z)|} \int\limits_{B_\gamma^n(z)} |f(y) - f_{B_\gamma^n(z)}|dy < +\infty,$$

where $|B_\gamma^n| = p^{n\gamma}$ and $f_{B_\gamma^n(z)} = \frac{1}{|B_\gamma^n(z)|} \int\limits_{B_\gamma^n(z)} f(x)dx$. The space BMO $(\mathbb{R}^n)$ is first introduced by F. John and L. Nirenberg.[11] It is obvious that, if we identify two functions $f, g \in$ BMO when $f(x) - g(x) = constant$ almost everywhere, then (BMO,$||.||_{BMO(\mathbb{Q}_p^n)}$) is a normed space.

3. A *p*-adic Cauchy Problem

3.1. *The problem and some spectral properties*

Let us solve the following Cauchy problem for a class of p-adic equations :

$$Lu(x,t) \equiv \frac{\partial^2 u(x,t)}{\partial t^2} + a^2 D_x^\alpha u(x,t) = F(x,t), \tag{1}$$

with initial conditions

$$u(x,0) = f(x), \tag{2}$$

$$u_t'(x,0) = g(x), \tag{3}$$

where a and α are positive real numbers, $x \in Q_p, t \in [0,b] \subset \mathbb{R}$, $b > 0; F(x,t), f(x), g(x)$ are complex-valued functions satisfying some certai conditions..

Here $\frac{\partial^2 u(x,t)}{\partial t^2}$ means the classical derivative of order 2 with respect to the real variable t of a function $u(x,t)$ usual in t , and $D_x^\alpha u(x,t)$ mean the p-adic distributional derivatives of order α with respect to p-adic variable x of the p-adic function $u(x,t)$.

We denote by $\chi_p(x)$ the normal additive characteristic function on the p-adic field Q_p. The Fourier transform of a basic function $\varphi \in \mathcal{D}(Q_p)$ is defined by

$$\widetilde{\varphi}(\xi) = \int_{Q_p} \varphi(x)\chi_p(\xi x)dx.$$

If $\varphi \in L^2(Q_p), \widetilde{\varphi}(\xi) \in L^2(Q_p)$ is definet by

$$\tilde{\varphi}(\xi) = \int_{Q_p} \varphi(x)\chi_p(\xi x)dx = \lim_{\gamma\to+\infty} \int_{B_\gamma} \varphi(x)\chi_p(\xi x)dx$$

in $L^2(Q_p)$ and we have

$$\varphi(\xi) = \int_{Q_p} \widetilde{\varphi}(x)\chi_p(-\xi x)dx.$$

The derivative operator D^α is defined by

$$(D^\alpha\varphi)(x) = \int_{Q_p} |\xi|_p^\alpha \widetilde{\varphi}(\xi)\chi_p(-\xi x)d\xi,$$

where α is a positive real number and $\varphi \in \mathcal{D}(Q_p)$ (see Ref. 29).

More general, if the distribution $\varphi \in \mathcal{D}'(Q_p)$ such that the convolution $f_{-\alpha} * \varphi$ exists, $\alpha \neq -1$, α is a real number, then D^α is defined by $D^\alpha\varphi = f_{-\alpha} * \varphi$, where

$$f_\alpha(x) = \frac{|x|_p^{\alpha-1}}{\Gamma_p(\alpha)}, \; \Gamma_p(\alpha) = \frac{1-p^{\alpha-1}}{1-p^{-a}}.$$

Denote the domain of D^α by

$$\mathcal{M}(D^\alpha) = \{\varphi \in L^2(Q_p) : |\xi|_p^\alpha\widetilde{\varphi}(\xi) \in L^2(Q_p)\}.$$

The problem to find eigenfunctions $\varphi \in \mathcal{M}(\mathcal{D}^\alpha)$, with eigenvalues $\lambda \in \mathbb{R}$, that is such that $D^\alpha\varphi = \lambda\varphi$, is solved in Ref. 29. We recall here some main results.

The eigenvalues $\lambda \neq 0$ have the form $\lambda = p^{\alpha n}, n$ is an integer.

Case 1. If $p \neq 2$, p is a prime number, $\lambda_n = p^{\alpha n}$, n is an integer, then we have orthonormal eigenfunctions of type I, II as follows:

I. $\varphi^l_{n,k,\varepsilon_l}(x) = p^{\frac{n+1-l}{2}}.\delta(|x|_p - p^{l-n})\chi_p(\varepsilon_l p^{l-2n}x^2)$

for $l = 2,3,...;\; k = 1,2,...,p-1;\; \varepsilon_l = \varepsilon_0 + \varepsilon_1 p + \cdots + \varepsilon_{l-2}p^{l-2}$, where

$$\delta(t) = \begin{cases} 1, & t=0 \\ 0, & t\neq 0,\; t\in\mathbb{R}. \end{cases}$$

II. For $l = 1,\; \varepsilon = 0$

$$\varphi^l_{n,k,0}(x) = p^{\frac{n-1}{2}}.\Omega(p^{n-1}.|x|_p).\chi_p(kp^{-n}x);\; k = 1,2,...,p-1,$$

where

$$\Omega(t) = \begin{cases} 1, & 0 \leqslant t \leqslant 1 \\ 0, & t > 1. \end{cases}$$

Case 2. If $p = 2$, $\lambda_n = 2^{\alpha n}$, $n \in Z$, then we have orthonormal eigenfunctions of type I, II as follows:
I. $\varphi_{n,k,\varepsilon_l}(x) = 2^{n-1}.\delta(|x|_2 - 2^{l+1-n})\chi_2(\varepsilon_l 2^{l-2n}x^2 + 2^{l-n-k}x)$,
for $l = 2, 3, ...$; $k = 0, 1$; $\varepsilon_l = 1 + \varepsilon_l 2 + \cdots + \varepsilon_{l-2}2^{l-2}$.
II. for $l = 1$; $\varepsilon = 0$; $k = 0, 1$.

$$\varphi^l_{n,k,0}(x) = 2^{\frac{n-1}{2}}.\{\Omega(2^n.|x - k2^{n-2}|_2) - \delta(|x - k2^{n-2}|_2 - 2^{1-n}\}.$$

The following properties are also needed in the sequel.

Property 1.

$$D^\alpha \varphi^l_{n,k,\varepsilon_l} = p^{\alpha n}\varphi^l_{n,k,\varepsilon_l}, \tag{4}$$

That is $\varphi^l_{n,k,\varepsilon_l}(x)$ is a solution of the equation $D^\alpha X = p^{\alpha n}X$

Property 2. For each fixed prime number p the system of all eigenfunctions $\{\varphi^l_{n,k,\varepsilon_l}\}$ is a complete orthonormal one in $L^2(Q_p)$, which means that for any $f \in L^2(Q_p)$ we have a unique expansion of Fourier series

$$f(x) = \sum_{n=-\infty}^{+\infty} \sum_{l=1}^{+\infty} \sum_k \sum_{\varepsilon_l} f^l_{n,k,\varepsilon_l}.\varphi^l_{n,k,\varepsilon_l}(x), \tag{5}$$

with

$$f^l_{n,k,\varepsilon_l} = \int_{Q_p} f(x).\overline{\varphi^l_{n,k,\varepsilon_l}(x)}dx. \tag{6}$$

The series (4) converges in $L^2(Q_p)$ and

$$\|f\|^2_{L^2(Q_p)} = \sum_{n,l,k,\varepsilon_l} \left|f^l_{n,k,\varepsilon_l}\right|^2 < \infty. \tag{7}$$

Property 3. If the series $\sum\limits_n \sum\limits_l \sum\limits_k \sum\limits_{\varepsilon_l} a^l_{n,k,\varepsilon_l}.\varphi^l_{n,k,\varepsilon_l}(x)$ converges in $L^2(Q_p)$ to function $f(x)$, $x \in Q_p$ then

$$a^l_{n,k,\varepsilon_l} = \int_{Q_p} f(x).\overline{\varphi^l_{n,k,\varepsilon_l}(x)}dx \tag{8}$$

Let us now solve a class of p-adic homogeneous and non-homogeneous equations.

Definition 3.1. Let $C^n\left([a,b]\,; \mathcal{M}\left(\mathcal{D}^\beta\right)\right)$ be the space of complex - valued $u(x,t)$ such that $\frac{\partial^m u(x,t)}{\partial t^m}$ and $D^\beta_x\left(\frac{\partial^m u(x,t)}{\partial t^m}\right)$ are continuous and belong to $L^2(Q_p)$ in x for all $t \in [a,b]$, where n is a non - negative integer and $m = 0, 1, 2, ..., n$, β is a non - negative real number, $x \in Q_p$, a and b are positive reals, $\frac{\partial^m u(x,t)}{\partial t^m}$ are the classical derivatives in t and D_x^β are the Vladimirov p-adic derivatives.

3.2. *Some results*

Theorem 3.1. *Let $f \in \mathcal{M}(D^\alpha), g \in \mathcal{M}(D^{\frac{\alpha}{2}})$, then for the homogeneous equation*

$$\frac{\partial^2 u(x,t)}{\partial t^2} + a^2 D_x^\alpha u(x,t) = 0 \tag{9}$$

with the inital conditions

$$u(x,0) = f(x) \tag{10}$$

$$u_t^{'}(x,0) = g(x) \tag{11}$$

there exists solution being the series of orthonormal eigenvectors $\left\{\varphi_{n,k,\varepsilon_l}^l(x)\right\}$in $L^2(Q_p)$ of the operator of differentiation D^α introduced in the above cases 1 and 2, that is

$$u(x,t) = \sum u_{n,k,\varepsilon_l}^l(t)\varphi_{n,k,\varepsilon_l}^l(x) \tag{12}$$

Here instead of
$\sum_n \sum_l \sum_k \sum_{\varepsilon_l}$, *we use* $\sum$.
At the sane time, the function u belongs to

$$u \in \mathcal{U} = C\left([0,b]; \mathcal{M}(D^\alpha)\right) \cap C^2\left([0,b]; L^2(Q_p)\right)$$

Proof. By $f \in \mathcal{M}(D^\alpha), g \in \mathcal{M}(D^{\frac{\alpha}{2}})$ we have

$$\begin{cases} f(x) = \sum f_{n,k,\varepsilon_l}^l \varphi_{n,k,\varepsilon_l}^l(x) \text{ in } L^2(Q_p) \\ f_{n,k,\varepsilon}^l = \int\limits_{Q_p} f(x)\overline{\varphi_{n,k,\varepsilon_l}^l(x)}dx \text{ and } \sum \left|f_{n,k,\varepsilon_l}^l\right|^2 < +\infty \end{cases} \tag{13}$$

$$\begin{cases} (D^\alpha f)(x) = \sum p^{\alpha n} f_{n,k,\varepsilon_l}^l \varphi_{n,k,\varepsilon_l}^l(x) \text{ in } L^2(Q_p) \text{ with} \\ \sum p^{2\alpha n}\left|f_{n,k,\varepsilon_l}^l\right|^2 < +\infty \end{cases} \tag{14}$$

$$\begin{cases} g(x) = & \sum g^l_{n,k,\varepsilon_l}\varphi^l_{n,k,\varepsilon_l}(x) \text{ in } L^2(Q_p) \text{ with} \\ g^l_{n,k,\varepsilon} = \int\limits_{Q_p} g(x)\overline{\varphi^l_{n,k,\varepsilon_l}(x)}dx & \text{and } \sum \left|g^l_{n,k,\varepsilon_l}\right|^2 < +\infty \end{cases} \tag{15}$$

$$\begin{cases} \left(D^{\frac{\alpha}{2}}g\right)(x) = \sum p^{\frac{\alpha n}{2}} g^l_{n,k,\varepsilon_l}\varphi^l_{n,k,\varepsilon_l}(x) \text{ in } L^2(Q_p) \text{ and} \\ \qquad \sum p^{\alpha n}\left|g^l_{n,k,\varepsilon_l}\right|^2 < +\infty \end{cases} \tag{16}$$

The necessity. Let the series (12) belong to the class $\mathcal{U}$. So we can obtain $\frac{\partial^2}{\partial t^2}$ and D^α_x in terms and

$$\frac{\partial^2 u(x,t)}{\partial t^2} = \sum \left(u^l_{n,k,\varepsilon_l}\right)''(t)\varphi^l_{n,k,\varepsilon_l}(x)$$

and

$$\left(D^a_x u\right)(x,t) = \sum \left(u^l_{n,k,\varepsilon_l}\right)(t)p^{\alpha n}\varphi^l_{n,k,\varepsilon_l}(x)$$

Substituting in Eq. (9), identifying the coefficient of $\varphi^l_{n,k,\varepsilon_l}(x)$, we get

$$\left(u^l_{n,k,\varepsilon_l}\right)''(t) + a^2p^{\alpha n}u^l_{n,k,\varepsilon_l}(t) = 0$$

for all (n,l,k,ε_l).
implying

$$u^l_{n,k,\varepsilon_l}(t) = A^l_{n,k,\varepsilon_l}\cos\left(ap^{\frac{\alpha n}{2}}t\right) + B^l_{n,k,\varepsilon_l}\sin\left(ap^{\frac{\alpha n}{2}}t\right)$$

Substituting these functions in (12), using (10), (11), (13), (15), we obtain

$$A^l_{n,k,\varepsilon_l} = f^l_{n,k,\varepsilon_l} \text{ and } B^l_{n,k,\varepsilon_l} = \frac{1}{a}p^{-\frac{\alpha n}{2}}g^l_{n,k,\varepsilon_l}$$

The series (12) has the form

$$u(x,t) = \sum \left[f^l_{n,k,\varepsilon_l}\cos\left(ap^{\frac{\alpha n}{2}}t\right) + \frac{1}{a}p^{\frac{-\alpha n}{2}}g^l_{n,k,\varepsilon_l}\sin\left(ap^{\frac{\alpha n}{2}}t\right)\right]\varphi^l_{n,k,\varepsilon_l}(x) \tag{17}$$

The sufficiency. The function $u(x,t)$ in (17) belong to the class $\mathcal{U}$ and is the solution of the Cauchy problem of Theorem 1. It will be proved that $u(x,t)$ statisfies

a)

$$u \in \mathcal{U} = C\left([0,b]\,;\mathcal{M}\left(D^\alpha\right)\right) \cap C^2\left([0,b]\,;L^2(Q_p)\right)$$

The series (17) converge in $L^2(Q_p)$ in x, uniformly in $t \in [0,b]$, because

$$\sum \left|u^l_{n,k,\varepsilon_l}(t)\right|^2 \leq 2\sum\left(\left|f^l_{n,k,\varepsilon_l}\right|^2 + \left(\frac{1}{a}p^{-\frac{\alpha n}{2}}\left|g^l_{n,k,\varepsilon_l}\right|\left|\sin\left(ap^{\frac{\alpha n}{2}}t\right)\right|\right)^2\right)$$
$$\leq 2\sum\left(\left|f^l_{n,k,\varepsilon_l}\right|^2 + b^2\left|g^l_{n,k,\varepsilon_l}\right|^2\right) < +\infty; \text{ by (13), (15).}$$

Here we have used $\left|\sin\left(ap^{\frac{\alpha n}{2}}t\right)\right| \leq ap^{\frac{\alpha n}{2}}t \leq ap^{\frac{\alpha n}{2}}b$ because $t \in [0, b]$ Hence the sum (17) equals to a continuous functions $u(x,t)$ in x, t and

$$\left(D^\alpha_x u\right)(x,t) = \sum\left[p^{\alpha n} f^l_{n,k,\varepsilon_l}\cos\left(ap^{\frac{\alpha n}{2}}t\right) + \frac{1}{a}p^{\frac{\alpha n}{2}} g^l_{n,k,\varepsilon_l}\sin\left(ap^{\frac{\alpha n}{2}}t\right)\right]\varphi^l_{n,k,\varepsilon_l}(x) \tag{18}$$

By the assumptions it is obvious that $u \in C\left([0,b]; \mathcal{M}\left(\mathcal{D}^\alpha\right)\right)$

b) By the assumptions too, we get

$$\frac{\partial u(x,t)}{\partial t} = \sum\left[-ap^{\frac{\alpha n}{2}} f^l_{n,k,\varepsilon_l}\sin\left(ap^{\frac{\alpha n}{2}}t\right) + g^l_{n,k,\varepsilon_l}\cos\left(ap^{\frac{\alpha n}{2}}t\right)\right]\varphi^l_{n,k,\varepsilon_l}(x)$$

because

$$\sum\left[-ap^{\frac{\alpha n}{2}} f^l_{n,k,\varepsilon_l}\sin\left(ap^{\frac{\alpha n}{2}}t\right) + f^l_{n,k,\varepsilon_l}\cos\left(ap^{\frac{\alpha n}{2}}t\right)\right]^2$$
$$\leq \sum 2\left(a^4 p^{2\alpha n} b^2\left|f^l_{n,k,\varepsilon_l}\right|^2 + \left|g^l_{n,k,\varepsilon_l}\right|^2\right) < +\infty$$

(by (14), (15), the right hand side of $\frac{\partial u}{\partial t}(x,t)$ converges in $L^2(Q_p)$ uniformly in $t \in [0,b]$.

And

$$\frac{\partial^2 u(x,t)}{\partial t^2} = \sum\left[-a^2 p^{\alpha n} f^l_{n,k,\varepsilon_l}\cos\left(ap^{\frac{\alpha n}{2}}t\right) - ap^{\frac{\alpha n}{2}} g^l_{n,k,\varepsilon_l}\sin\left(ap^{\frac{\alpha n}{2}}t\right)\right]\varphi^l_{n,k,\varepsilon_l}(x) \tag{19}$$

converging in $L^2(Q_p)$ uniformly in $t \in [0,b]$, since

$$\sum\left[-ap^{\frac{\alpha n}{2}} f^l_{n,k,\varepsilon_l}\sin\left(ap^{\frac{\alpha n}{2}}t\right) + g^l_{n,k,\varepsilon_l}\cos\left(ap^{\frac{\alpha n}{2}}t\right)\right]^2$$
$$\leq \sum 2\left(a^4 p^{2\alpha n}\left|f^l_{n,k,\varepsilon_l}\right|^2 + a^2 p^{\alpha n}\left|g^l_{n,k,\varepsilon_l}\right|^2\right) < +\infty$$

and by (14), (16), $u \in C^2([0,b]; L^2(Q_p))$.

From (18) and (19) it follows that $u(x,t)$ satisfies equation (9) and $u \in \mathcal{U}$.

c) The solution (17) satisfies obviously the initial values (10), (11). □

Theorem 3.2. *Assume that $f \in \mathcal{M}\left(D^\alpha\right), g \in \mathcal{M}\left(D^{\frac{\alpha}{2}}\right)$ and $F(x,t) \in C\left([0,b]; \mathcal{M}\left(D^{\frac{\alpha}{2}}\right)\right)$. Then for the non-homogeneous*

$$\frac{\partial^2 u(x,t)}{\partial t^2} + a^2 D^\alpha_x u(x,t) = F(x,t) \tag{20}$$

with the initial conditions

$$u(x,0) = f(x) \tag{21}$$

$$u_t^{'}(x,0) = g(x) \tag{22}$$

there exists a solution having the form of the series (12) and belonging to the class $\mathcal{U}$

Proof.

Similarly to the proof of Theorem 3.1 we have

$$u(x,t) = \sum \left[f_{n,k,\varepsilon_l}^l \cos\left(ap^{\frac{\alpha n}{2}} t\right) + \frac{1}{a} p^{-\frac{\alpha n}{2}} g_{n,k,\varepsilon_l}^l \sin\left(ap^{\frac{\alpha n}{2}} t\right) + \right.$$

$$\left. \frac{1}{a} p^{-\frac{\alpha n}{2}} \int_0^t F_{n,k,\varepsilon_l}^l(\tau) \sin\left(ap^{\frac{\alpha n}{2}}(t-\tau)\right) d\tau\right] \varphi_{n,k,\varepsilon_l}^l(x) \tag{23}$$

where

$$F_{n,k,\varepsilon_l}^l(t) = \int_{Q_p} F(x,t)\overline{\varphi_{n,k,\varepsilon_l}^l(x)}dx$$

□

Example 3.1. For the Eq. (9) with the initial conditions (10), (11), in Theorem 3.1 we take

$$f(x) = g(x) = \begin{cases} 1, \ x \in S_0, x_0 = 1 & (x = 1 + x_1 p + \cdots + x_n p^n + \cdots) \\ 0 & \text{otherwise} \end{cases}$$

a, $p \neq 2$

Then

$$f_{n,k,\varepsilon_l}^l = g_{n,k,\varepsilon_l}^l = \int_{Q_p} f(x)\overline{\varphi_{n,k,\varepsilon_l}^l(x)}dx$$

$$= \begin{cases} p^{\frac{n-3}{2}} \chi_p\left(-kp^{-n}\right), \ n \leq 1, k = \overline{1, p-1}, l = 1, \varepsilon_1 = 0 \\ 0 \qquad \text{otherwise} \end{cases}$$

The solution (18) is

$$u(x,t) = \sum_{n \leq 1} \sum_{k=1}^{p-1} p^{\frac{n-3}{2}} \chi_p\left(-kp^{-n}\right) \left[\cos\left(ap^{\frac{\alpha n}{2}} t\right) + \frac{1}{a} p^{-\frac{\alpha n}{2}} \sin\left(ap^{\frac{\alpha n}{2}} t\right)\right] \varphi_{n,k,0}^1(x) \tag{24}$$

b) $p = 2$

$$f^l_{n,k,\varepsilon_l} = g^l_{n,k,\varepsilon_l} = f^1_{n,0,0} = \begin{cases} -\frac{1}{2} & n = 1, k = 0, l = 1, \varepsilon_l = 0 \\ 2^{\frac{n-3}{2}} & n \leq 1, k = 0, l = 1, \varepsilon_l = 0 \\ 0 & \textit{otherwise} \end{cases}$$

The solution (18) has the form

$$u(x,t) = -\frac{1}{2}\left[\cos\left(a2^{\frac{\alpha}{2}}t\right) + \frac{1}{a}2^{-\frac{\alpha}{2}}\sin\left(ap^{\frac{\alpha}{2}}t\right)\right]\varphi^1_{1,0,0}(x)$$

$$+\sum_{n\leq 0} 2^{\frac{n-3}{2}}\left[\cos\left(a2^{\frac{\alpha n}{2}}t\right) + \frac{1}{a}2^{-\frac{\alpha n}{2}}\sin\left(a2^{\frac{\alpha n}{2}}t\right)\right]\varphi^1_{n,0,0}(x). \tag{25}$$

Obviously the function (25) satisfies (9), (10), (11).

Example 3.2.

Consider an example for the problem (20), (21), (22). In Theorem 3.2, with

$f(x) = g(x)$, as in Example 3.1, then
$f^l_{n,k,\varepsilon_l} = g^l_{n,k,\varepsilon_l}$ and $F(x,t) = tf(x)$,
It is clear that

$$F^l_{n,k,\varepsilon_l}(t) = \int_{Q_p} tf(x)\overline{\varphi^l_{n,k,\varepsilon_l}(x)}dx = tf^l_{n,k,\varepsilon_l}$$

a) $p \neq 2$

The solution (23) is

$$u(x,t) = \sum_{n\leq 1}\sum_{k=1}^{p-1} p^{\frac{n-3}{2}}\chi_p\left(-kp^{-n}\right)\left[\cos\left(ap^{\frac{\alpha n}{2}}t\right) + \frac{1}{a}p^{-\frac{\alpha n}{2}}\sin\left(ap^{\frac{\alpha n}{2}}t\right)\right.$$
$$\left. + \frac{1}{a^2}p^{-\alpha n}\left(t - \frac{1}{a}p^{-\frac{\alpha n}{2}}\sin ap^{\frac{\alpha n}{2}}t\right)\right]\varphi^1_{n,k,0}(x) \tag{26}$$

b) $p = 2$

The solution is

$$u(x,t) = \sum_{n\leq 1} f^l_{n,0,0}\left[\cos\left(a2^{\frac{\alpha n}{2}}t\right) + \frac{1}{a}2^{-\frac{\alpha n}{2}}\sin\left(a2^{\frac{\alpha n}{2}}t\right)\right.$$
$$\left. + \frac{1}{a}2^{\alpha n}\left(t - \frac{1}{a}2^{-\frac{\alpha n}{2}}\sin a2^{\frac{\alpha n}{2}}t\right)\right]\varphi^1_{n,0,0}(x) \tag{27}$$

where

$$f^1_{n,0,0} = \begin{cases} -\frac{1}{2} & \text{for } n = 1 \\ 2^{\frac{n-3}{2}} & \text{for } n \leq 0 \\ 0 & \text{for } n \geq 2 \end{cases}$$

The solution (26), (27) satisfy all assumptions of Theorem 3.2.

4. The p-adic Hilbert Transform

Let $s \in \mathbb{R}, \gamma_0 \in \mathbb{Z}$ and f be a locally integrable function on $\mathbb{Q}_p^n$. A p-adic Hilbert transform of f is defined by

$$H_{\gamma_0} f(x) = \lim_{\varepsilon \to -\infty} \int\limits_{p^\varepsilon \leqslant |y|_p} \frac{f(x-y)}{|y|_p^s} \Omega(p^{\gamma_0} y) dy, \ x \in \mathbb{Q}_p^n, \tag{28}$$

if the integral in the right hand side exists, where $\Omega(x)$ is the characteristic function of the p-adic unit ball

$$\Omega(x) = \begin{cases} 1 & , x \in B_0^n \\ 0 & , x \notin B_0^n \end{cases}$$

5. The Boundedness in p-adic Space L^q

Theorem 5.1. *Let $1 < q < +\infty$, q' be the conjugate of q, $\dfrac{1}{q} + \dfrac{1}{q'} = 1$, and $s \in \mathbb{R}$ satisfying $sq' < n$. Then the p-adic Hilbert transform (28) is a bounded operator from $L^q(B^n_{\gamma_0})$ to $L^k(B^n_{\gamma_0})$ with $1 \leqslant k < +\infty$, that is, there exists a constant $A_{s,n,k,q}$ independent of f such that*

$$||H_{\gamma_0} f||_{L^k} \leqslant A_{s,n,k,q} ||f||_{L^q}, \ \forall f \in L^q(B^n_{\gamma_0}),$$

Proof.

With any $f \in L^q(B^n_{\gamma_0})$ we have

$$||f||_{L^q(B^n_{\gamma_0})} = \Big(\int\limits_{B^n_{\gamma_0}} |f(x)|^q dx \Big)^{\frac{1}{q}} < +\infty.$$

With any $x \in B^n_{\gamma_0}$ we obtain

$$(H_{\gamma_0} f)(x) = \int\limits_{\mathbb{Q}_p^n} \frac{f(x-y)}{|y|_p^s} \Omega(p^{\gamma_0} y) dy = \int\limits_{||y||_p \leqslant p^{\gamma_0}} \frac{f(x-y)}{|y|_p^s} dy$$

By Holder inequality

$$\Big|\int\limits_{|y|_p\leqslant p^{\gamma_0}} \frac{f(x-y)}{|y|_p^s}dy\Big| \leqslant \int\limits_{|y|_p\leqslant p^{\gamma_0}} \frac{|f(x-y)|}{|y|_p^s}dy$$

$$\leqslant \Big(\int\limits_{|y|_p\leqslant p^{\gamma_0}} |f(x-y)|^q dy\Big)^{\frac{1}{q}}.\Big(\int\limits_{|y|_p\leqslant p^{\gamma_0}} \frac{1}{|y|_p^{sq'}}dy\Big)^{\frac{1}{q'}}$$

$$\leqslant \Big(\int\limits_{|x|_p\leqslant p^{\gamma_0}} |f(x)|^q dx\Big)^{\frac{1}{q}}.\Big(\int\limits_{|y|_p\leqslant p^{\gamma_0}} \frac{1}{|y|_p^{sq'}}dy\Big)^{\frac{1}{q'}}$$

$$\leqslant ||f||_{L^q(B^n_{\gamma_0})}.\Big(\int\limits_{|y|_p\leqslant p^{\gamma_0}} \frac{1}{|y|_p^{sq'}}dy\Big)^{\frac{1}{q'}}$$

For $sq' < n$ we get

$$\int\limits_{|y|_p\leqslant p^{\gamma_0}} \frac{1}{|y|_p^{sq'}}dy = (1-\frac{1}{p^n})\frac{p^{\gamma_0(n-sq')}}{1-p^{(sq'-n)}}$$

Indeed by $sq' < n$

$$\int\limits_{|y|_p\leqslant p^{\gamma_0}} \frac{dy}{|y|_p^{sq'}} = \sum_{\gamma=-\infty}^{\gamma_0}\int\limits_{S_\gamma}\frac{dy}{p^{sq'\gamma}} = \sum_{\gamma=-\infty}^{\gamma_0}\frac{1}{p^{sq'\gamma}}\int\limits_{S_\gamma}dy = \sum_{\gamma=-\infty}^{\gamma_0}\frac{1}{p^{sq'\gamma}}(1-\frac{1}{p^n})p^{n\gamma}$$

$$= (1-\frac{1}{p^n})\sum_{\gamma=-\infty}^{\gamma_0} p^{\gamma(n-sq')} = (1-\frac{1}{p^n})\sum_{\gamma=-\gamma_0}^{+\infty} p^{-\gamma(n-sq')}$$

$$= (1-\frac{1}{p^n})\sum_{\gamma=-\gamma_0}^{+\infty}\frac{1}{[p^{n-sq'}]^\gamma} = (1-\frac{1}{p^n})\frac{1}{[p^{n-sq'}]^{-\gamma_0}}\frac{1}{1-p^{sq'-n}}$$

$$= (1-\frac{1}{p^n})\frac{p^{\gamma_0(n-sq')}}{1-p^{(sq'-n)}},$$

consequently

$$\Big|\int\limits_{|y|_p\leqslant p^{\gamma_0}} \frac{f(x-y)}{|y|_p^s}dy\Big| \leqslant ||f||_{L^q}\Big[(1-\frac{1}{p^n})\frac{p^{\gamma_0(n-sq')}}{1-p^{(sq'-n)}}\Big]^{\frac{1}{q'}}.$$

Therefore

$$|H_{\gamma_0}f(x)| \leqslant ||f||_{L^q}\Big[(1-\frac{1}{p^n})\frac{p^{\gamma_0(n-sq')}}{1-p^{(sq'-n)}}\Big]^{\frac{1}{q'}}, \forall x\in B^n_{\gamma_0}.$$

Set

$$B_{s,n,q} = \Big[(1-\frac{1}{p^n})\frac{p^{\gamma_0(n-sq')}}{1-p^{(sq'-n)}}\Big]^{\frac{1}{q'}}$$

So

$$\Big(\int_{B^n_{\gamma_0}} |Hf(x)|^k dx\Big)^{\frac{1}{k}} \leqslant \Big(\int_{B^n_{\gamma_0}} [||f||_{L^q}.B_{s,n,q}]^k dx\Big)^{\frac{1}{k}} \leqslant ||f||_{L^q}B_{s,n,q}.\Big(\int_{B^n_{\gamma_0}} dx\Big)^{\frac{1}{k}}$$

$$\leqslant ||f||_{L^q}B_{s,n,q}.(p^{n\gamma_0})^{\frac{1}{k}} = ||f||_{L^q}.\ B_{s,n,q}.\ p^{\frac{n\gamma_0}{k}}.$$

Set

$$A_{s,n,q,k} = B_{s,n,q}.p^{\frac{n\gamma_0}{k}}.$$

It follows that

$$||H_{\gamma_0}f||_{L^k(B^n_{\gamma_0})} \leqslant A_{s,n,q,k}||f||_{L^q},\ \forall f \in L^q(B^n_{\gamma_0}).$$

Thus H_{γ_0} is a bounded operator from $L^q(B^n_{\gamma_0})$ to $L^k(B^n_{\gamma_0})$ and

$$||H_{\gamma_0}||_{L^q \to L^k} \leqslant A_{s,n,q,k},$$

with

$$A_{s,n,q,k} = \Big[(1-\frac{1}{p^n})\frac{p^{\gamma_0(n-sq')}}{1-p^{(sq'-n)}}\Big]^{\frac{1}{q'}} p^{\frac{n\gamma_0}{k}}. \qquad \square$$

Theorem 5.2. *Let $1 \leqslant q < \infty$, q' the conjugate of q and $s \in \mathbb{R}$ satisfying $s < n$. Then the p-adic Hilbert transform (28) is a bounded operator from $L^q(\mathbb{Q}^n_p)$ to $L^q(\mathbb{Q}^n_p)$ and the following estimate hods true*

$$||H_{\gamma_0}f||_{L^q} \leqslant A_{s,n,p,q}||f||_{L^q},\ \forall f \in L^q(\mathbb{Q}^n_p), \tag{29}$$

where $A_{s,n,p,q}$ is a constant independent of f.

Proof. First we consider the case $1 < q < \infty$. For any $f \in L^q(\mathbb{Q}^n_p)$ with $x \in \mathbb{Q}^n_p$ we have

$$|(H_{\gamma_0}f(x)| \leqslant |\int_{\mathbb{Q}^n_p} \frac{f(x-y)}{|y|^s_p}\Omega(p^{\gamma_0}y)dy| \leqslant \int_{\mathbb{Q}^n_p} \frac{|f(x-y)|}{|y|^s_p}\Omega(p^{\gamma_0}y)dy$$

By the Hölder inequality we get

$$|(H_{\gamma_0}f(x)| \leqslant \Big(\int_{\mathbb{Q}^n_p} \frac{|f(x-y)|^q}{|y|^s_p}\Omega(p^{\gamma_0}y)dy\Big)^{\frac{1}{q}}\Big(\int_{\mathbb{Q}^n_p} \frac{\Omega(p^{\gamma_0}y)}{|y|^s_p}dy\Big)^{\frac{1}{q'}}. \tag{30}$$

Since

$$\int_{\mathbb{Q}^n_p} \frac{\Omega(p^{\gamma_0}y)}{|y|^s_p}dy = \int_{|y|_p \leqslant p^{\gamma_0}} \frac{dy}{|y|^s_p} = (1-\frac{1}{p^n})\frac{p^{\gamma_0(n-s)}}{1-p^{(n-s)}},$$

by substituting in (30) we obtain

$$|(H_{\gamma_0}f(x)| \leqslant B_{s,n,\gamma_0,q}\Big(\int\limits_{\mathbb{Q}_p^n}\frac{|f(x-y)|^q}{|y|_p^s}\Omega(p^{\gamma_0}y)dy\Big)^{\frac{1}{q}}. \tag{31}$$

With

$$B_{s,n,\gamma_0,q} = [(1-\frac{1}{p^n})\frac{p^{\gamma_0(n-s)}}{1-p^{n-s}}]^{\frac{1}{q'}}$$

from (30) and using Fubini theorem we get

$$\begin{aligned}\Big(\int\limits_{\mathbb{Q}_p^n}|(H_{\gamma_0}f(x)|^q dx\Big)^{\frac{1}{q}} &\leqslant B_{s,n,\gamma_0,q}\Big[\int\limits_{\mathbb{Q}_p^n}[\int\limits_{\mathbb{Q}_p^n}\frac{|f(x-y)|^q}{|y|_p^s}\Omega(p^{\gamma_0}y)dy]dx\Big]^{\frac{1}{q}}\\ &= B_{s,n,\gamma_0,q}\Big[\int\limits_{\mathbb{Q}_p^n}\frac{\Omega(p^{\gamma_0}y)}{|y|_p^s}dy[\int\limits_{\mathbb{Q}_p^n}|f(x-y)|^q dx]\Big]^{\frac{1}{q}}\\ &= B_{s,n,\gamma_0,q}||f||_{L^q(\mathbb{Q}_p^n)}(\int\limits_{\mathbb{Q}_p^n}\frac{\Omega(p^{\gamma_0}y)}{|y|_p^s}dy)^{\frac{1}{q}}\\ &= B_{s,n,\gamma_0,q}||f||_{L^q(\mathbb{Q}_p^n)}(\int\limits_{B_{\gamma_0}^n}\frac{dy}{|y|_p^s})^{\frac{1}{q}}\\ &= (B_{s,n,\gamma_0,q})^{1+\frac{q'}{q}}||f||_{L^q(\mathbb{Q}_p^n)}.\end{aligned}$$

Setting

$$A_{s,n,\gamma_0,q} = (B_{s,n,\gamma_0})^{1+\frac{q'}{q}}$$

we obtain the estimate

$$||H_{\gamma_0}f||_{L^q(\mathbb{Q}_p^n)} \leqslant A_{s,n,\gamma_0,q}||f||_{L^q(\mathbb{Q}_p^n)},\ \forall f \in L^q(\mathbb{Q}_p^n).$$

Let us consider the case $q = 1$. For any $x \in \mathbb{Q}_p^n$

$$|H_{\gamma_0}f(x)| \leqslant \int\limits_{\mathbb{Q}_p^n}\frac{|f(x-y)|}{|y|_p^s}\Omega(p^{\gamma_0}y)dy,$$

hence

$$\int_{\mathbb{Q}_p^n} |(H_{\gamma_0}f)(x)|dx \leqslant \int_{\mathbb{Q}_p^n} \Big\{ \int_{\mathbb{Q}_p^n} \frac{|f(x-y)|}{|y|_p^s} \Omega(p^{\gamma_0}y)dyBig\Big\}dx$$

$$\leqslant \int_{\mathbb{Q}_p^n} \frac{\Omega(p^{\gamma_0}y)}{|y|_p^s} \Big(\int_{\mathbb{Q}_p^n} |f(x-y)|dx \Big) dy$$

$$= ||f||_{L^1(\mathbb{Q}_p^n)} \cdot \int_{\mathbb{Q}_p^n} \frac{\Omega(p^{\gamma_0}y)}{|y|_p^s} dy.$$

By computing we have

$$||H_{\gamma_0}f||_{L^1(\mathbb{Q}_p^n)} \leqslant (1 - \frac{1}{p^n}) \frac{p^{\gamma_0(n-s)}}{1-p^{n-s}} ||f||_{L^1(\mathbb{Q}_p^n)}, \ \forall f \in L^1(\mathbb{Q}_p^n)$$

$$\leqslant A_{s,n,\gamma_0,1} ||f||_{L^1(\mathbb{Q}_p^n)}, \ \forall f \in L^1(\mathbb{Q}_p^n)$$

with

$$A_{s,n,\gamma_0,1} = (1 - \frac{1}{p^n}) \frac{p^{\gamma_0(n-s)}}{1-p^{n-s}}.$$

So H_{γ_0} a bounded operator. □

6. The Boundedness the p-adic Space BMO $(\mathbb{Q}_p^n)$

Theorem 6.1. *Let $s \in \mathbb{R}$ satisfying $s < n$. Then the p-adic Hilbert transform (28) is a bounded operator from the space $BMO(\mathbb{Q}_p^n)$ to the space $BMO(\mathbb{Q}_p^n)$.*

Proof. Let $f \in BMO(\mathbb{Q}_p^n)$. Then for any ball $B_\gamma^n \subset \mathbb{Q}_p^n$ we have

$$\int_{B_\gamma^n} |H_{\gamma_0}f(x)|dx = \int_{B_\gamma^n} \Big| \int_{\mathbb{Q}_p^n} \frac{f(x-y)}{|y|_p^s} \Omega(p^{\gamma_0}y)dy \Big| dx$$

$$\leqslant \int_{\mathbb{Q}_p^n} \frac{\Omega(p^{\gamma_0}y)}{|y|_p^s} \Big(\int_{B_\gamma^n} |f(x-y)|dx \Big) dy$$

$$\leqslant \int_{\mathbb{Q}_p^n} \frac{\Omega(p^{\gamma_0}y)}{|y|_p^s} \Big(\int_{B_{\gamma_1}^n} |f(x)|dx \Big) dy$$

$$\leqslant (1 - \frac{1}{p^n}) \frac{p^{\gamma_0(n-s)}}{1-p^{n-s}} \int_{B_{\gamma_1}^n} |f(x)|dx,$$

where $\gamma_1 = \max\{\gamma, \gamma_0\}$.
By $f \in L^1_{loc}$ we get $\int\limits_{B^n_{\gamma_1}} |f(x)|dx < \infty$. It follows that $\int\limits_{B^n_\gamma} |H_{\gamma_0} f(x)|dx < \infty$, for every $\gamma \in \mathbb{Z}$.
So $H_{\gamma_0} f \in L^1_{loc}(\mathbb{Q}_p^n)$. Moreover, for any $z \in \mathbb{Q}_p^n$ we obtain

$$\begin{aligned}(H_{\gamma_0} f)_{B^n_\gamma(z)} &= \frac{1}{|B^n_\gamma(z)|} \int\limits_{B^n_\gamma(z)} (H_{\gamma_0} f)(x)dx \\ &= \frac{1}{|B^n_\gamma(z)|} \int\limits_{B^n_\gamma(z)} \Big(\int\limits_{\mathbb{Q}_p^n} \frac{f(x-y)}{|y|_p^s} \Omega(p^{\gamma_0} y)dy\Big) dx\end{aligned}$$

By Fubini theorem

$$(H_{\gamma_0} f)_{B^n_\gamma(z)} = \int\limits_{\mathbb{Q}_p^n} \frac{\Omega(p^{\gamma_0} y)}{|y|_p^s} \Big(\frac{1}{|B^n_\gamma(z)|} \int\limits_{B^n_\gamma(z)} f(x-y)dx\Big) dy,$$

thus for any $x \in B^n_\gamma(z)$ we get

$$\begin{aligned}&|(H_{\gamma_0} f)(x) - (H_{\gamma_0} f)_{B^n_\gamma(z)}| = \\ &|\int\limits_{\mathbb{Q}_p^n} \frac{\Omega(p^{\gamma_0} y)}{|y|_p^s} [f(x-y) - \frac{1}{|B^n_\gamma(z)|} \int\limits_{B^n_\gamma(z)} f(x-y)dx]dy| \\ &\leqslant \int\limits_{\mathbb{Q}_p^n} \frac{\Omega(p^{\gamma_0} y)}{|y|_p^s} |f(x-y) - \frac{1}{|B^n_\gamma(z)|} \int\limits_{B^n_\gamma(z)} f(x-y)dx|dy\end{aligned}$$

Hence for any $z \in \mathbb{Q}_p^n$ and $\gamma \in \mathbb{Z}$ we have

$$\begin{aligned}&\frac{1}{|B^n_\gamma(z)|} \int\limits_{B^n_\gamma(z)} |(H_{\gamma_0} f)(x) - (H_{\gamma_0} f)_{B^n_\gamma(z)}|dx \\ &\leqslant \frac{1}{|B^n_\gamma(z)|} \int\limits_{B^n_\gamma(z)} \Big\{\int\limits_{\mathbb{Q}_p^n} \frac{\Omega(p^{\gamma_0} y)}{|y|_p^s} |f(x-y) - \frac{1}{|B^n_\gamma(z)|} \int\limits_{B^n_\gamma(z)} f(x-y)dx|dy\Big\} dx \\ &\leqslant \int\limits_{\mathbb{Q}_p^n} \frac{\Omega(p^{\gamma_0} y)}{|y|_p^s} \Big\{\frac{1}{|B^n_\gamma(z)|} \int\limits_{B^n_\gamma(z)} |f(x-y) - \frac{1}{|B^n_\gamma(z)|} \int\limits_{B^n_\gamma(z)} f(x-y)dx|dx\Big\} dy \\ &\leqslant ||f||_{BMO(\mathbb{Q}_p^n)} \int\limits_{\mathbb{Q}_p^n} \frac{\Omega(p^{\gamma_0} y)}{|y|_p^s} dy \leqslant (1 - \frac{1}{p^n}) \frac{p^{\gamma_0(n-s)}}{1-p^{n-s}} ||f||_{BMO(\mathbb{Q}_p^n)} < \infty\end{aligned}$$

Therefore $\quad H_{\gamma_0} f \in BMO(\mathbb{Q}_p^n)$ and

$$||H_{\gamma_0} f||_{BMO(\mathbb{Q}_p^n)} \leqslant (1 - \frac{1}{p^n}) \frac{p^{\gamma_0(n-s)}}{1 - p^{n-s}} ||f||_{BMO(\mathbb{Q}_p^n)}, \ \forall f \in BMO(\mathbb{Q}_p^n).$$

This proves that H_{γ_0} is a bounded operator from $BMO(\mathbb{Q}_p^n)$ to $BMO(\mathbb{Q}_p^n)$ and

$$||H_{\gamma_0}||_{BMO \to BMO} \leqslant (1 - \frac{1}{p^n}) \frac{p^{\gamma_0(n-s)}}{1 - p^{n-s}}. \qquad \square$$

7. The Boundedness in the p-adic Space VMO($\mathbb{Q}_p^n$)

Theorem 7.1. *Let s be a real number satisfying $s < n$. Then the p- adic Hilbert transform* (28) *is a bounded operator from the space $VMO(\mathbb{Q}_p^n)$ to the space $VMO(\mathbb{Q}_p^n)$.*

Proof. Let $f \in VMO(\mathbb{Q}_p^n)$. Then $f \in BMO(\mathbb{Q}_p^n)$. Using the proof of Theorem 5 we have $H_{\gamma_0} f \in BMO(\mathbb{Q}_p^n)$ and

$$\frac{1}{|B_\gamma^n(z)|} \int_{B_\gamma^n(z)} |(H_{\gamma_0} f)(x) - (H_{\gamma_0} f)_{B_\gamma^n(z)}| dx \leqslant$$

$$\leqslant \Big(\int_{\mathbb{Q}_p^n} \frac{\Omega(p^{\gamma_0} y)}{|y|_p^s} dy \Big) \cdot \frac{1}{|B_\gamma^n(z)|} \int_{B_\gamma^n(z)} |f(x) - f_{B_\gamma^n(z)}| dx$$

$$\leqslant (1 - \frac{1}{p^n}) \frac{p^{\gamma_0(n-s)}}{1 - p^{n-s}} \frac{1}{|B_\gamma^n(z)|} \int_{B_\gamma^n(z)} |f(x) - f_{B_\gamma^n(z)}| dx,$$

for every $z \in \mathbb{Q}_p^n$ and every $\gamma \in \mathbb{Z}$. In the above inequality letting $\gamma \to -\infty$ and using the assumption that $f \in VMO(\mathbb{Q}_p^n)$ we obtain

$$\lim_{\gamma \to -\infty} \frac{1}{|B_\gamma^n(z)|} \int_{B_\gamma^n(z)} |H_{\gamma_0} f(x) - (H_{\gamma_0} f)_{B_\gamma^n(z)}| dx = 0$$

for any $z \in \mathbb{Q}_p^n$.
So $H_{\gamma_0} f \in VMO(\mathbb{Q}_p^n)$ and

$$\begin{aligned} ||H_{\gamma_0} f||_{VMO(\mathbb{Q}_p^n)} &= ||H_{\gamma_0} f||_{BMO(\mathbb{Q}_p^n)} \\ &\leqslant (1 - \frac{1}{p^n}) \frac{p^{\gamma_0(n-s)}}{1 - p^{n-s}} ||f||_{BMO(\mathbb{Q}_p^n)} \\ &\leqslant (1 - \frac{1}{p^n}) \frac{p^{\gamma_0(n-s)}}{1 - p^{n-s}} ||f||_{VMO(\mathbb{Q}_p^n)}. \end{aligned}$$

Thus H_{γ_0} is a bounded operator from $VMO(\mathbb{Q}_p^n)$ to $VMO(\mathbb{Q}_p^n)$ and we get

$$||H_{\gamma_0}||_{VMO\to VMO} \leqslant (1-\frac{1}{p^n})\frac{p^{\gamma_0(n-s)}}{1-p^{n-s}}. \qquad \square$$

8. The Boundedness in the p-adic Space $L^q(\mathbb{Q}_p^n, \omega(x), l_k)$

Let $0 < q < +\infty$, $0 < k < +\infty$ and $\omega(x) > 0$ be a measurable function in $\mathbb{Q}_p^n$. Then the space $L^q(\mathbb{Q}_p^n, \omega(x), l_k)$ is defined as a linear space of the functions

$$f:\ \mathbb{Q}_p^n \to l_k,$$
$$x \mapsto f(x) = (f_1(x), ..., f_n(x), ...)$$

such that

(i) $f_j \in L^q(\mathbb{Q}_p^n), \forall j = 1, 2, ...;$

(ii) $||f||_{L^q(\mathbb{Q}_p^n,\omega(x),l_k)} := \{\int\limits_{\mathbb{Q}_p^n} \omega^q(x)(\sum_{j=1}^\infty |f_j(x)|^k)^{\frac{q}{k}} dx\}^{\frac{1}{q}} < \infty$

In particular case when $\omega(x) = 1, \forall x \in \mathbb{Q}_p^n$, instead of $L^q(\mathbb{Q}_p^n, 1, l_k)$ we write $L^q(\mathbb{Q}_p^n, l_k)$ For $s \in \mathbb{R}$ $\gamma_0 \in \mathbb{Z}$ the p-adic Hilbert transform of the function$f \in L^q(\mathbb{Q}_p^n, \omega(x), l_k)$ is a function defined by

$$(\mathbb{H}_{\gamma_0} f)(x) = (\sum_{j=1}^{\infty}[(H_{\gamma_0} f_j)(x)]^k)^{\frac{1}{k}} \tag{32}$$

where H_{γ_0} is the p-adic Hilbert transform defined by (28). The formula (32) can be rewritten in the form

$$(\mathbb{H}_{\gamma_0} f)(x) = (\sum_{j=1}^{\infty}[\int\limits_{\mathbb{Q}_p^n} \frac{f_j(x-y)}{|y|_p^s}\Omega(p^{\gamma_0}y)dy]^k)^{\frac{1}{k}} \tag{33}$$

Theorem 8.1. *Let $1 \leqslant q < +\infty$, $1 \leqslant k < +\infty$ and $s \in \mathbb{R}$ satisfying $s < n$. Then the p-adic Hilbert transform* (33) *is a bounded operator from the space $L^q(\mathbb{Q}_p^n, l_k)$ to the space $L^q(\mathbb{Q}_p^n)$ and the following estimate holds true*

$$||\mathbb{H}_{\gamma_0} f||_{L^q(\mathbb{Q}_p^n)} \leqslant (1-\frac{1}{p^n})\frac{p^{\gamma_0(n-s)}}{1-p^{n-s}}||f||_{L^q(\mathbb{Q}_p^n,l_k)}, \forall\ f \in L^q(\mathbb{Q}_p^n, l_k)$$

Proof. For any $f \in L^q(\mathbb{Q}_p^n, l_k)$ we have

$$f(x) = (f_1(x), f_2(x), ..., f_n(x), ...),\ x \in \mathbb{Q}_p^n$$

By the definition of $L^q(\mathbb{Q}_p^n, l_k)$ we obtain

$$f_j \in L^q(\mathbb{Q}_p^n), \quad \forall j = 1, 2, ...$$

It follows that

$$|\mathbb{H}_{\gamma_0} f(x)| \leqslant \{\sum_{j=1}^{\infty} |\int_{\mathbb{Q}_p^n} \frac{f_j(x-y)}{|y|_p^s} \Omega(p^{\gamma_0} y) dy|^k\}^{\frac{1}{k}}$$

By the Minkowski inequality

$$|\mathbb{H}_{\gamma_0} f(x)| \leqslant \int_{\mathbb{Q}_p^n} \frac{\Omega(p^{\gamma_0} y)}{|y|_p^s} \{\sum_{j=1}^{\infty} |f_j(x-y)|^k\}^{\frac{1}{k}} dy.$$

It follows that

$$\begin{aligned}
(\int_{\mathbb{Q}_p^n} |\mathbb{H}_{\gamma_0} f(x)|^q dx)^{\frac{1}{q}} &\leqslant [\int_{\mathbb{Q}_p^n} \{\int_{\mathbb{Q}_p^n} \frac{\Omega(p^{\gamma_0} y)}{|y|_p^s} (\sum_{j=1}^{\infty} |f_j(x-y)|^k)^{\frac{1}{k}} dy\}^q dx]^{\frac{1}{q}} \\
&\leqslant \int_{\mathbb{Q}_p^n} \frac{\Omega(p^{\gamma_0} y)}{|y|_p^s} \{\int_{\mathbb{Q}_p^n} (\sum_{j=1}^{\infty} |f_j(x-y)|^k)^{\frac{q}{k}} dx\}^{\frac{1}{q}} dy \\
&\leqslant ||f||_{L^q(\mathbb{Q}_p^n, l_k)} \cdot \int_{\mathbb{Q}_p^n} \frac{\Omega(p^{\gamma_0} y)}{|y|_p^s} dy \\
&= (1 - \frac{1}{p^n}) \frac{p^{\gamma_0 (s-n)}}{1 - p^{s-n}} ||f||_{L^q(\mathbb{Q}_p^n, l_k)}.
\end{aligned}$$

Therefore

$$||\mathbb{H}_{\gamma_0} f||_{L^q(\mathbb{Q}_p^n)} \leqslant (1 - \frac{1}{p^n}) \frac{p^{\gamma_0 (s-n)}}{1 - p^{s-n}} ||f||_{L^q(\mathbb{Q}_p^n, l_k)}, \forall\ f \in L^q(\mathbb{Q}_p^n, l_k). \qquad \square$$

Acknowledgment

The authors are grateful to the referee for helpful comments.

References

1. A.Kh. Bikulov, *Teoret. Mat. Fiz.* **87**, 376 (1991).
2. Nguyen Minh Chuong (Author in-chief), Ha Tien Ngoan, Nguyen Minh Tri, Le Quang Trung, *Partial differential equations* (Publishing House Education, Hanoi, 2000).
3. Nguyen Minh Chuong and Nguyen Van Co, *Proc. Americ. Math. Soc.* **3**, 685 (1999)

4. Nguyen Minh Chuong and Ha Duy Hung, *Preprint* 03/22, (Inst. of Math, Hanoi, 2003).
5. R.R. Coifman and C. Fefferman, *Studia Mathematica,* **11**, 211 (1874).
6. B.M. Dwork, *Duke Math. J.* **62**, 689 (1991).
7. C. Fefferman, *Bulletin AMS,* **77**, 587 (1971).
8. Shai Haran, *Ann. Inst. Fourier, Grenpble,* **43**, 905 (1993).
9. Kazuo Ikeda, Taekyun Kim, Katsumi Shiratani, *Kyushu University, Ser. A* **46**, 341 (1992).
10. Lee-Chae-Jamg, Min-Soo Kim, Jin-Woo-Son, Taekyun Kim, and Seog-Hoon Rim, *JMAA,* **264**, 21 (2001).
11. F.John and L. Nirenberg, *Comm. Pure Appl. Math.* **14**, 415 (1961).
12. A. Yu Khrennikov, *p-adic valued distributions in mathematical physics* (Kluwer Academic Publisher, Dordrecht-Boston-London, 1994).
13. A. Yu Khrennikov, S.V. Kozyrev, *Apll.Comput. Harmon. Anal.* **19**, 61 (2005).
14. A. Khrennikov and M. Nilsson, *J. of Number Theory,* **90**, 255 (2001).
15. A. N. Kochubei, *Pseudodifferential equations and stochastics over non-archimedean fields* (Marcel Dekker, Inc. New York–Basel, 2001).
16. A. N. Kochubei, *Potential Anal.,* **6**, 105 (1997).
17. A.N. Kochubei, *Potential Anal.* **10**, 305 (1999).
18. A.N. Kochubei, *Pseudo-differrental equations and Stochastics over Non-Archimedean Fields* (Marcel Dekker, 2001).
19. S.V. Kozyrev, A. Yu Khrennikov, *Izv. Ross. Akad. Nauk Ser. Mat.,* **69**, 133 (2005).
20. S.V. Kozyrev, *Izv. Ross. Akad. Nauk Ser. Mat.,* **66**, 149 (2002).
21. S.V. Kozyrev, *Teoret. Mat. Fiz.,* **138**, 322 (2004).
22. Shanzhen Lu and Dachun Yang, *Math. Nachr,* **191**, 229 (1998).
23. Tao Mei, *C. R. Acad. Sci. Paris, Ser. I,* **336**, 1003 (2003).
24. C.W. Onneweer and T.S. Quek, *J. Australia Math. Soc. Series A,* **65**, 370 (1999).
25. C.W. Onneweer and Su Weyil, *Studia Mathematica,* **93**, 17 (1989).
26. Keith Phillips and Mitchell Taibleson, *Pacif. J. Math.,* **30**, 200 (1969).
27. Elias M. Stein, *Harmonic Analyis, real variable methods, orthogonality and Oscillatory integrals* (Princeton Universty Press, 1993).
28. Otto Vejvoda, *Partial Differential Equations : time-periodic solutions* (Martinus Nijhoff Publishers, The Hague/Boston/London, 1982).
29. V.S.Vladimirov, I.V. Volovich, E.I. Zelenov, *p-adic analysis and mathematical physics* (Russian), (Nauka, Moscow 1994).

Harmonic, Wavelet and p-Adic Analysis
Eds. N. M. Chuong *et al.* (pp. 291–309)

§13. *P*-ADIC AND GROUP VALUED PROBABILITIES

ANDREI KHRENNIKOV

International Center for Mathematical Modeling in Physics and Cognitive Sciences,MSI, University of Växjö, S-35195, Sweden
Andrei.Khrennikov@vxu.se

We develop an analogue of probability theory for probabilities taking values in topological groups. We also present a review of non-Kolmogorovian models with negative, complex, and p-adic valued probabilities. We discuss applications of non-Kolmogorovian models in physics and cognitive sciences.

1. Historical Introduction

Since the creation of the modern probabilistic axiomatics by A. N. Kolmogorov in 1933 Ref. 1, probability theory was merely reduced to the theory of normalized σ-additive measures taking values in the segment [0,1] of the field of real numbers **R**. In particular, the main competitor of Kolmogorov's measure-theoretic approach, von Mises' frequency approach to probability,[2,3] practically totally disappeared from the probabilistic arena. On one hand, this was a consequence of difficulties with von Mises' definition of randomness (via place selections), see e.g., Refs.4–6.* On the other hand, von Mises' approach (as many others) could not compete with precisely and simply formulated Kolmogorov's theory.

We mentioned von Mises' approach not only, because its attraction for applications, but also because von Mises' model with frequency probabilities played the important role in the process of formulation of the conventional axiomatics of probability theory. If one opens Kolmogorov's book Ref.1, he will see numerous remarks about von Mises' theory. Andrei Nikolaevich Kolmogorov used properties of the frequency probability to justify his choice of the axioms for probability. In particular, Kolmogorov's probability belongs to the segment [0,1] of the real line **R**, because the same takes place for

*However, see also Ref.7, where von Mises' approach was simplified, generalized, and then fruitfully applied to theoretical physics.

von Mises' frequency probability (relative frequencies $\nu_N = n/N$ as well as their limits - probabilities – always belong to the segment [0,1] of the real line **R**). In the same way Kolmogorov's probability is additive, because the frequency probability is additive: the limit of the sum of two frequencies equals to the sum of limits. And so on... Thus by using *THEOREMS* of von Mises' frequency theory Kolmogorov justified AXIOMATIZATION of probability as a normalized finite-additive measure taking values in [0,1]. Finally, he added the condition of σ-additivity.

We would like to mention that Kolmogorov's (as well as von Mises') assumptions were also based on a fundamental, but hidden, assumption: *Limiting behavior of relative frequencies is considered with respect to one fixed topology on the field of rational numbers* **Q**, *namely, the real topology.* In particular, the consideration of this asymptotic behavior implies that probabilities belong to the field real numbers **R**.

However, it is possible to study asymptotic behavior of relative frequencies (which are always rational numbers) in other topologies on field of rational numbers **Q**. In this way we derive another probability-like structure that recently appeared in theoretical physics. This is so called *p-adic probability.*

We recall that p-adic numbers are applied intensively in different domains of physics – quantum logic, string theory, cosmology, quantum mechanics, quantum foundations, see, e.g., Refs.8–12, dynamical systems Refs.11,13,14, biological and cognitive models Ref.11,14–16.

In this paper we shall concentrate our study to probabilistic models that could be obtained through changing the range of values of probabilities. Thus our "generalized probabilities" do not more belong to the segment [0,1] of **R**, cf. with Refs.17–22: there were considered negative and complex "probabilities."

We consider natural generalizations of properties of probability that are obtained through the transition from **R** to an arbitrary topological group. We consider **R** as a topological group (with respect to addition) and extract the main properties of Kolmogorov's measure-theoretic or von Mises' frequency probability corresponding to the group structure (algebraic and topological) on **R**. Then we use generalizations of these properties to define generalized probabilities that take values in an arbitrary topological group G.

Before developing the general axiomatics, we will pay more attention to the p-adic valued probabilities Ref.7,23–25. In fact, it was the first example of the mathematically rigorous formalism for probabilities that take values in a topological group G which is different from $\mathbf{R}$.

2. p-adic Lessons

2.1. *p-adic numbers*

The field of real numbers $\mathbf{R}$ is constructed as the completion of the field of rational numbers $\mathbf{Q}$ with respect to the metric $p(x,y) = |x-y|$, where $|\cdot|$ is the usual valuation given by the absolute value. The fields of p-adic numbers $\mathbf{Q}_p$ are constructed in a corresponding way, but by using other valuations. For a prime number p the p-adic valuation $|\cdot|_p$ is defined in the following way. First we define it for natural numbers. Every natural number n can be represented as the product of prime numbers, $n = 2^{r_2}3^{r_3}\dots p^{r_p}\dots$, and we define $|n|_p = p^{-r_p}$, writing $|0|_p = 0$ and $|-n|_p = |n|_p$. We then extend the definition of the p-adic valuation $|\cdot|_p$ to all rational numbers by setting $|n/m|_p = |n|_p/|m|_p$ for $m \neq 0$. The completion of $\mathbf{Q}$ with respect to the metric $\rho_p(x,y) = |x-y|_p$ is the locally compact field of p-adic numbers $\mathbf{Q}_p$.

The number fields $\mathbf{R}$ and $\mathbf{Q}_p$ are unique in a sense, since by Ostrovsky's theorem, see e.g., Ref.26, $|\cdot|$ and $|\cdot|_p$ are the only possible valuations on $\mathbf{Q}$, but have quite distinctive properties. The field of real numbers $\mathbf{R}$ with its usual valuation satisfies $|n| = n \to \infty$ for valuations of natural numbers n and is said to be *Archimedian*. By a well known theorem of number theory[26] the only complete Archimedian fields are those of the real and the complex numbers. In contrast, the fields of p-adic numbers, which satisfy $|n|_p \leq 1$ for all $n \in N$, are examples of *non-Archimedian* fields.

Unlike the absolute value distance $|\cdot|$, the p-adic valuation satisfies the strong tringle inequality:

$$|x+y|_p \leq \max[|x|_p, |y|_p], x, y \in \mathbf{Q}_p.$$

Consequently the p-adic metric satisfies the strong triangle inequality $\rho_p(x,y) \leq \max[\rho_p(x,z), \rho_p(z,y)], x, y, z \in \mathbf{Q}_p$, which means that the metric ρ_p is an *ultrametric*,[26] Write $U_r(a) = \{x \in \mathbf{Q}_p : |x-a|_p \leq r\}$, where $r = p^n$ and $n = 0, \pm 1, \pm 2, \dots$ These are the "closed" balls in $\mathbf{Q}_p$ while the sets $\mathbf{S_r(a)} = \{\mathbf{x} \in \mathbf{Q_p} : |\mathbf{x} - \mathbf{a}|_\mathbf{p} = \mathbf{r}\}$ are the spheres in $\mathbf{Q}_p$ of such radii r. These sets (balls and spheres) have a somewhat strange topological structure from the viewpoint of our usual Euclidian intuition: they are both open and closed at the same time, and as such are called *clopen* sets. Finally, any p-adic ball $U_r(0)$ is an additive subgroup of $\mathbf{Q}_p$, while the ball $U_1(0)$ is also a ring, which is called the *ring of p-adic integers* and is denoted by $\mathbf{Z}_p$.

The p-adic *exponential function* $e^x = \sum_{n=0}^{\infty} \frac{x^n}{n!}$. The series converges in $\mathbf{Q}_p$ if $|x|_p \leq r_p$, where $r_p = 1/p, p \neq 2$ and $r_2 = 1/4$. p-adic trigonometric functions $\sin x$ and $\cos x$ are defined by the standard power series. These series have the same radius of convergence r_p as the exponential series.

2.2. *p-adic frequency probability model*

As in the ordinary probability theory Refs.2,3, the first p-adic probability model was the frequency one, Refs.7,9,10,23–25. This model was based on the simple remark that relative frequencies $\nu_N = \frac{n}{N}$ always belong to the field of rational numbers $\mathbf{Q}$. And $\mathbf{Q}$ can be considered as a (dense) subfield of $\mathbf{R}$ as well as $\mathbf{Q}_p$ (for each prime number p). Therefore behaviour of sequences $\{\nu_N\}$ of (rational) relative frequencies can be studied not only with respect to the real topology on Q, but also with respect to any p-adic topology on Q. Roughly speaking a p-adic probability (as real von Mises' probability) is defined as:

$$\mathbf{P}(\alpha) = \lim_N \nu_N(\alpha). \tag{1}$$

Here α is some label denoting a result of a statistical experiment. Denote the set of all such labels by the symbol Ω. In the simplest case $\Omega = \{0, 1\}$. Here $\nu_N(\alpha)$ is the relative frequency of realization of the label α in the first N trials. The $\mathbf{P}(\alpha)$ is the frequency probability of the label α.

The main p-adic lesson is that it is impossible to consider, as we did in the real case, limits of the relative frequencies ν_N when the $N \to \infty$. Here the point "∞" belongs, in fact, to the real compactification of the set of natural numbers. So $|N| \to \infty$, where $|\cdot|$ is the real absolute value. The set of natural numbers $\mathbf{N}$ is bounded in $\mathbf{Q}_p$ and it is densely embedded into the ring of p-adic integers $\mathbf{Z}_p$ (the unit ball of $\mathbf{Q}_p$). Therefore sequences $\{N_k\}_{k=1}^{\infty}$ of natural numbers can have various limits $m = \lim_{k\to\infty} N_k \in \mathbf{Z}_p$.

In the p-adic frequency probability theory we proceed in the following way to provide the rigorous mathematical meaning for the procedure (1), see Refs. 31,32. We fix a p-adic integer $m \in \mathbf{Z}_p$ and consider the class, L_m, of sequences of natural numbers $s = \{N_k\}$ such that $\lim_{k\to\infty} N_k = m$ in $\mathbf{Q}_p$.

Let us consider the fixed sequence of natural numbers $s \in L_m$. We define a p-adic s-probability

$$\mathbf{P}(\alpha) = \lim_{k\to\infty} \nu_{N_k}(\alpha), s = \{N_k\}.$$

This is the limit of relative frequencies with respect to the fixed sequence $s = \{N_k\}$ of natural numbers. For any subset A of the set of labels Ω, we define its s-probability as

$$\mathbf{P}(A) = \lim_{k\to\infty} \nu_{N_k}(A), s = \{N_k\},$$

where $\nu_{N_k}(A)$ is the relative frequency of realization of labels α belonging to the set A in the first N trials. As $\mathbf{Q}_p$ is an additive topological semigroup (as well as $\mathbf{R}$), we obtain that the p-adic probability is additive:

Theorem 2.1.

$$\mathbf{P}(A_1 \cup A_2) = \mathbf{P}(A_1) + \mathbf{P}(A_2), A_1 \cap A_2 = \emptyset. \tag{2}$$

As $\mathbf{Q}_p$ is even an additive topological group (as well as $\mathbf{R}$), we get that

Theorem 2.2.

$$\mathbf{P}(A_1 \setminus A_2) = \mathbf{P}(A_1) - \mathbf{P}(A_1 \cap A_2). \tag{3}$$

Trivially, for any sequence $s = \{N_k\}$, $\mathbf{P}(\Omega) = \lim_{k\to\infty} \nu_{N_k}(\Omega) = 1$, as $\nu_N(\Omega) = \frac{N}{N} = 1$ for any N. As $\mathbf{Q}_p$ is a multiplicative topological group (as well as $\mathbf{R}$) , we get (see von Mises Refs.2,3 for the real case and Ref.7 for the p-adic case) Bayes' formula for conditional probabilities:

Theorem 2.3.

$$\mathbf{P}(A|B) = \lim_{k\to\infty} \frac{\nu_{N_k}(A \cap B)}{\nu_{N_k}(A)} = \frac{\mathbf{P}(A \cap B)}{\mathbf{P}(A)}, \mathbf{P}(A) \neq 0. \tag{4}$$

As we know, frequency probability played the crucial role in conventional probability theory in determination of the range of values (namely, the segment [0,1]) of a probabilistic measure, see remarks on von Mises' theory in Kolmogorov's book Ref.1. Frequencies always lie between zero and one. Thus their limits (with respect to the real topology) belong to the same range.

In the p-adic case we can proceed in the same way. Let $r \equiv r_m = \frac{1}{|m|_p}$ (where $r = \infty$ for $m = 0$). We can easily get, see Refs.23,24, that for the p-adic frequency s-probability, $s \in L_m$, the values of $\mathbf{P}$ always belong to the p-adic ball $U_r(0) = \{x \in \mathbf{Q}_p : |x|_p \leq r\}$. In the p-adic probabilistic model such a ball $U_r(0)$ plays the role of the segment [0,1] in the real probabilistic model.

2.3. *Measure-theoretic approach*

As in the real case, the structure of an additive topological group of $\mathbf{Q}_p$ induces the main properties of probability that can be used for the axiomatization in the spirit of Kolmogorov,[1] Let us fix $r = p^{\pm l}, l = 0, 1, \ldots,$ or $r = \infty$.

Axiomatics 1. *Let Ω be an arbirary set (a sample space) and let F be a field of subsets of Ω (events). Finally, let $\mathbf{P} : F \to U_r(0)$ be an additive function (measure) such that $\mathbf{P}(\Omega) = 1$. Then the triple $(\Omega, F, \mathbf{P})$ is said to be a p-adic r-probabilistic space and* $\mathbf{P}$ *p-adic r-probability .*

Following to Kolmogorov we should find some technical mathematical restriction on $\mathbf{P}$ that would induce fruitful integration theory and give the possibility to define averages. Kolmogorov (by following Borel, Lebesque, Lusin, and Egorov) proposed to consider the σ-additivity of measures and the σ-structure of the field of events. Unfortunately, in the p-adic case the situation is not so simple as in the real one. One could not just copy Kolmogorov's approach and consider the condition of σ-additivity. There is, in fact, a No-Go theorem, see, e.g., Ref.27:

Theorem 2.4. *All σ-additive p-adic valued measures defined on σ-fields are discrete.*

Here the difficulty is not induced by the condition of σ-additivity, but by an attempt to extend a measure from the field F to the σ-field generated by F. Roughly speaking there exist σ-additive "continuous" $\mathbf{Q}_p$-valued measures, but they could not be extended from the field F to the σ-field generated by F. Therefore it is impossible to choose the σ-additivity as the basic integration condition in the p-adic probability theory.

The first important condition (that was already invented in the first theory of non-Archimedian integration of Monna and Springer[28]) is *boundedness:* $\|A\|_{\mathbf{P}} = \sup\{|\mathbf{P}(A)|_p : A \in F\} < \infty$.

Of course, if $\mathbf{P}$ is a p-adic r-probability with $r < \infty$, then this condition is fulfilled automatically. It is nontrivial only if the range of values of a p-adic probability is unbounded in $\mathbf{Q}_p$.† We pay attention to one important particular case in that the condition of boundedness alone implies fruitful integration theory. Let Ω be a compact zero-dimensional topological space.‡ Then the integral

$$E\xi = \int_{\Omega} \xi(\omega)\mathbf{P}(d\omega)$$

is well defined for any continuous function $\xi : \Omega \to \mathbf{Q}_p$. For example, this theory works well for the following choice: Ω is the ring of q-adic integers

†In the frequency formalism this corresponds to considering of p-adic (frequency) s-probabilities for $s \in L_0$; e.g., $s = \{N_k = p^k\}$. In this case $m = \lim_{k\to\infty} p^k = 0$.
‡There exists a basis of neighborhoods that are open and closed at the same time.

Z_q, and $\mathbf{P}$ is a bounded p-adic r-probability, $r < \infty$. The integral is defined as the limit of Riemannian sums Refs.27,28.

But in general boundedness alone does not imply fruitful integration theory. We should consider another condition, namely *continuity* of $\mathbf{P}$. The most general continuity condition was proposed by A. van Rooij Ref.27.[§]

Definition 2.1. A p-adic valued measure that is bounded, continuous, and normalized is called p-adic probability measure.

Everywhere below we consider p-adic probability spaces endowed with p-adic probability measures.

Let $(\Omega, F, \mathbf{P})$ be a p-adic probabilistic space. *Random variables* $\xi : \Omega \to \mathbf{Q}_p$ are defined as $\mathbf{P}$-integrable functions.

As the frequency p-adic probability theory induces, see Ref.7, (as a Theorem) Bayes' formula for conditional probability, we can use (4) as the definition of conditional probability in the p-adic axiomatic approach (as it was done by Kolmogorov in the real case).

Example 2.1. (p-adic valued uniform distribution on the space of q-adic sequences). Let p and q be two prime numbers. We set $X_q = \{0, 1, \ldots, q-1\}, \Omega_q^n = \{x = (x_1, \ldots, x_n) : x_j \in X_q\}, \Omega_q^\star = \bigcup_n \Omega_q^n$ (the space of finite sequences), and

$$\Omega_q = \{\omega = (\omega_1, \ldots, \omega_n, \ldots) : \omega_j \in X_q\}$$

(the space of infinite sequences). For $x \in \Omega_q^n$, we set $l(x) = n$. For $x \in \Omega_q^\star, l(x) = n$, we define a cylinder U_x with the basis x by $U_x = \{\omega \in \Omega_q : \omega_1 = x_1, \ldots, \omega_n = x_n\}$. We denote by the symbol $F_{\rm cyl}$ the field of subsets of Ω_q generated by all cylinders. In fact, the $F_{\rm cyl}$ is the collection of all finite unions of cylinders.

First we define the uniform distribution on cylinders by setting $\mu(U_x) = 1/q^{l(x)}, x \in \Omega_q^\star$. Then we extend μ by additivity to the field $F_{\rm cyl}$. Thus $\mu : F_{\rm cyl} \to \mathbf{Q}$. The set of rational numbers can be considered as a subset of any $\mathbf{Q}_p$ as well as a subset of $\mathbf{R}$. Thus μ can be considered as a p-adic valued (for any prime number p) as well as the real valued measure. We use symbols $\mathbf{P}_p$ and $\mathbf{P}_\infty$ to denote these measures. The probability space for the uniform p-adic measure is defined as the triple

$$\mathcal{P} = (\Omega, F, \mathbf{P}), \text{ where } \Omega = \Omega_q, F = F_{\rm cyl} \text{ and } \mathbf{P} = \mathbf{P}_p.$$

[§] We remark that in many cases continuity coincides with σ-additivity.

The $\mathbf{P}_p$ is called a *uniform p-adic probability distribution.*

The uniform p-adic probability distribution is a probabilistic measure iff $p \neq q$. The range of its values is a subset of the unit p-adic ball.

Remark 2.1. Values of $\mathbf{P}_p$ on cylinders coincide with values of the standard (real-valued) uniform probability distribution (Bernoulli measure) $\mathbf{P}_\infty$. Let us consider, the map $j_\infty(\omega) = \sum_{j=0}^\infty \frac{\omega_j}{2^{j+1}}$. The j_∞ maps the space Ω_q onto the segment [0, 1] of the real line $\mathbf{R}$ (however, j_∞ is not one to one correspondence). The j_∞-image of the Bernoulli measure is the standard Lebesque measure on the segment [0,1] (the uniform probability distribution on the segment [0,1]).

Remark 2.2. The map $j_q : \Omega_q \to \mathbf{Z}_q, j_q(\omega) = \sum_{j=0}^\infty \omega_j q^j$, gives (one to one!) correspondence between the space of all q-adic sequences Ω_q and the ring of q-adic integers Z_q. The field F_{cyl} of cylindrical subsets of Ω_q coincides with the field $B(\mathbf{Z}_q)$ of all clopen (closed and open at the same time) subsets of $\mathbf{Z}_q$. If Ω_q is realized as $\mathbf{Z}_q$ and F_{cyl} as $B(Z_q)$, then μ_p is the p-adic valued Haar measure on $\mathbf{Z}_q$. The use of the topological structure of $\mathbf{Z}_q$ is very fruitful in the integration theory (for $p \neq q$). In fact, the space of integrable functions $f : \mathbf{Z}_q \to \mathbf{Q}_p$ coincides with the space of continuous functions (random variables) $C(\mathbf{Z}_q, \mathbf{Q}_p)$, see Refs.7,26–28.

3. *p*-adic Limit Theorems

3.1. *p-adic asymptotics of bernoulli probabilities*

Everywhere in this section p is a prime number distinct from 2. We start with considering the classical Bernoulli scheme (in the conventional probabilistic framework) for random variables $\xi_j(\omega) = 0, 1$ with probabilities $1/2, j = 1, 2, \ldots$. First we consider a finite number n of random variables: $\xi_1(\omega), \ldots, \xi_n(\omega)$. A sample space corresponding to these random variables can be chosen as the space $\Omega_2^n = \{0, 1\}^n$. The probability of an event A is defined as

$$\mathbf{P}^{(n)} = \frac{|A|}{|\Omega_2^n|} = \frac{|A|}{2^n},$$

where the symbol $|B|$ denotes the number of elements in a set B. The typical problem of ordinary probability theory is to find the asymptotic behavior of the probabilities $\mathbf{P}^{(n)}(A), n \to \infty$. It was the starting point of the theory of limit theorems in conventional probability theory.

But the probabilities $\mathbf{P}^{(n)}(A)$ belong to the field of rational numbers $\mathbf{Q}$. We may study behavior of $\mathbf{P}^{(n)}(A)$, not only with respect to the usual real metric $\rho_\infty(x,y)$ on $\mathbf{Q}$, but also with respect to an arbitrary metric $\rho(x,y)$ on $\mathbf{Q}$. We have studied the case of the p-adic metric on $\mathbf{Q}$, see Refs.29,30. We remark that $\mathbf{P}^{(n)}(A) = \sum_{x\in A}\mu(U_x)$, where μ is the uniform distribution on Ω_2. By realizing μ as the (real valued) probability distribution $\mathbf{P}_\infty$ we use the formalism of conventional probability theory. By realizing μ as the p-adic valued probability distribution $\mathbf{P}_p$ we use the formalism of p-adic probability theory.

What kinds of events A are naturally coupled to the p-adic metric? Of course, such events must depend on the prime number p. As usual, we consider the sums

$$S_n(\omega) = \sum_{k=1}^{n} \xi_n(\omega).$$

We are interested in the following question. Does p divide the sum $S_n(\omega)$ or not? Set $A(p,n) = \{\omega \in \Omega_2^n : p$ divides the sum$S_n(\omega)\}$. Then $\mathbf{P}^{(n)}(A(p,n)) = L(p,n)/2^n$, where $L(p,n)$ is the number of vectors $\omega \in \Omega_2^n$ such that p divides $|\omega| = \sum_{j=1}^{n}\omega_j$. As usual, denote by $\bar{A}$ the complement of a set A. Thus $\bar{A}(p,n)$ is the set of all $\omega \in \Omega_q^n$ such that p does not divide the sum $S_n(\omega)$. We shall see that the sets $A(p,n)$ and $\bar{A}(p,n)$ are asymptotically symmetric from the p-adic point of view:

$$\mathbf{P}^{(n)}(A(p,n)) \to \frac{1}{2} \text{ and } \mathbf{P}^{(n)}(\bar{A}(p,n)) \to \frac{1}{2} \tag{5}$$

in the p-adic metric when $n \to 1$ in the same metric. Already in this simplest case we shall see that the behavior of sums $S_n(\omega)$ depends crucially on the choice of a sequence $s = \{N_k\}_{k=1}^{\infty}$ of natural numbers. A limit distribution of the sequence of random variables $S_n(\omega)$, when $n \to \infty$ in the ordinary sense, does not exist. We have to describe all limiting distributions for different sequences s converging in the p-adic topology.

Let $(\Omega, F, \mathbf{P})$ be a p-adic probabilistic space and $\xi_n : \Omega \to \mathbf{Q}_p (n = 1,2,\ldots)$ be a sequence of equally distributed independent random variables, $\xi_n = 0,1$ with probability $1/2$.¶ We start with the following result that can be obtained through purely combinatorial considerations (behavior of binomial coefficients C_m^r in the p-adic topology).

Theorem 3.1. *Let* $m = 0,1,\ldots,p^s-1 (s = 1,2,\ldots), r = 0,\ldots,m,$ *and*

¶Here $1/2$ is considered as a p-adic number. In the conventional theory $1/2$ is considered as a real number.

$l \geq s$. Then

$$\lim_{n \to m} \mathbf{P}(\omega : S_n(\omega) \in U_{1/p^l}(r)) = \frac{C_m^r}{2^m}.$$

Formally this theorem can be reformulated as the following result for the convergence of probabilistic distributions: *The limiting distribution on $\mathbf{Q}_p$ of the sequence of the sums $S_n(\omega)$, where $n \to m$ in $\mathbf{Q}_p$, is the discrete measure $\kappa_{1/2,m} = 2^{-m} \sum_{r=0}^{m} C_m^r \delta_m$.*

We consider the event $A(p,n,r) = \{\omega : S_n(\omega) = pi + r\}$ for $r = 0, 1, \ldots, p-1$. This event consists of all ω such that the residue of $S_n(\omega)$ mod p equals to r. Note that the set $A(p,n,r)$ coincides with the set $\{\omega : S_n(\omega) \in U_{1/p}(r)\}$.

Corollary 3.1. *Let $n \to m$ in $\mathbf{Q}_p$, where $m = 0, 1, \ldots, p-1$. Then the probabilities $\mathbf{P}^{(n)}(A(p,n,r))$ approach $C_m^r/2^m$ for all residues $r = 0, \ldots, m$.*

In particular, as $A(p,n) \equiv A(p,n,0)$, we get (5). What happens in the case $m \geq p$? We have only the following particular result:

Theorem 3.2. *Let $n \to p$ in $\mathbf{Q}_p$ and $r = 0, 1, 2, \ldots, p$. Then*

$$\lim_{n \to p} \mathbf{P}(\omega : S_n(\omega) \in U_{1/p^l}(r)) = \frac{C_p^r}{2^p},$$

where $s \geq 2$ for $r = 0, p$ and $s \geq 1$ for $r = 1, \ldots, p-1$.

Remark 3.1. (Bernard-Letac asymtotics) In Ref.31 J. Bernard and G. Letac have studied p-adic asymptotic of multi binomial coefficients. Although they did not consider the p-adic probabilistic terminology (at that moment there were no physical motivations to consider the p-adic generalization of probability), their results may be interpreted as a kind of a limit theorem for p-adic probability.

3.2. *Laws of large numbers*

We now study the general case of dichotomic equally distributed independent random variables: $\xi_n(\omega) = 0, 1$ with probabilities q and $q' = 1 - q, q \in \mathbf{Z}_p$. We shall study the weak convergence of the probability distributions $\mathbf{P}_{S_{N_k}}$ for the sums $S_{N_k}(\omega)$. We consider the space $C(\mathbf{Z}_p, \mathbf{Q}_p)$ of continuous functions $f : \mathbf{Z}_p \to \mathbf{Q}_p$. We will be interested in convergence of integrals

$$\int_{\mathbf{Z}_p} f(x) d\mathbf{P}_{S_{N_k}}(x) \to \int_{\mathbf{Z}_p} f(x) d\mathbf{P}_S(x), f \in C(\mathbf{Z}_p, \mathbf{Q}_p),$$

where $\mathbf{P}_S$ is the limiting probability distribution (depending on the sequence $s = \{N_k\}$). To find the limiting distribution $\mathbf{P}_S$, we use the method of characteristic functions. We have for characteristic functions

$$\phi_{N_k}(z, q, a) = \int_\Omega \exp\{zS_{N_k}(\omega)\}d\mathbf{P}(\omega) = (1 + q'(e^z - 1))^{N_k}.$$

Here z belong to a sufficiently small neighborhood of zero in the $\mathbf{Q}_p$; see Ref.10 for detail about the p-adic method of characteristic functions. Let a be an arbitrary number from $\mathbf{Z}_p$. Let $s = \{N_k\}_{k=1}^\infty$ be a sequence of natural numbers converging to a in the $\mathbf{Q}_p$. Set $\phi(z, q, a) = (1 + q'(e^z - 1))^a$. This function is analytic for small z. It is easy to see that the sequence of characteristic functions $\{\phi_{N_k}(z, q, a)\}$ converges (uniformly on every ball of a sufficiently small radius) to the function $\phi(z, q, a)$. Unfortunately, we could not prove (or disprove) a p-adic analogue of Levy's theorem. Therefore in the general case the convergence of characteristic functions does not give us anything. However, we shall see that we have Levy's situation in the particular case under consideration: There exists a bounded probability measure distribution, denoted by $\kappa_{q,a}$, having the characteristic function $\phi(z, q, a)$ and, moreover, $\mathbf{P}_{S_{N_k}} \to \mathbf{P}_S = \kappa_{q,a}, N_k \to a$.

We start with the first part of the above statement. Here we shall use Mahlers integration theory on the ring of p-adic integers, see e.g., Refs.9, 10,26,27. We introduce a system of binomial polynomials: $C(x, k) = C_x^k = \frac{x(x-1)\ldots(x-k+1)}{k!}$ (that are considered as functions from $\mathbf{Z}_p$ to $\mathbf{Q}_p$). Every function $f \in C(\mathbf{Z}_p, \mathbf{Q}_p)$ is expanded into a series (a Mahler expansion, see Ref.?) $f(x) = \sum_{k=0}^\infty a_k C(x, k)$. It converges uniformly on $\mathbf{Z}_p$. If μ is a bounded measure on $\mathbf{Z}_p$, then

$$\int_{\mathbf{Z}_p} f(x)\mu(dx) = \sum a_k \int_{\mathbf{Z}_p} C(x, n)\mu(dx).$$

Therefore to define a p-adic valued measure on $\mathbf{Z}_p$ it suffices to define coefficients $\int_{\mathbf{Z}_p} C(x, n)\mu(dx)$. A measure is bounded iff these coefficients are bounded. Using the Mahler expansion of the function $\phi(z, q, a)$, we obtain

$$\langle_m(q, a) = \int_{\mathbf{Z}_p} C(x, m)\kappa_{q,a}(dx) = (1 - q)^m C(a, m).$$

As $|C(a, m)|_p \leq 1$ for $a \in \mathbf{Z}_p$, we get that the distribution $\kappa_{q,a}$ (corresponding to $\phi(z, q, a)$) is bounded measure on $\mathbf{Z}_p$. Set $\langle_{mn}(q, a) = \int_\Omega C(S_n(\omega), m)dP(\omega)$. We find

$$\langle_{mN_k}(q, a) = (1 - q)^m C_{N_k}^m.$$

Thus $\langle_{mN_k}(q,a) \to \langle_m(q,a), N_k \to a$. This implies the following limit theorem.

Theorem 3.3. *(p-adic Law of Large Numbers.) The sequence of probability distributions* $\{\mathbf{P}_{S_{N_k}}\}$ *converges weakly to* $\mathbf{P}_S = \kappa_{q,a}$, *when* $N_k \to a$ *in* $\mathbf{Q}_p$.

3.3. *The central limit theorem*

Here we restrict our considerations to the case of symmetric random variables $\xi_n(\omega) = 0, 1$ with probabilities $1/2$. We study the p-adic asymptotic of the normalized sums

$$G_n(\omega) = \frac{S_n(\omega) - ES_n(\omega)}{\sqrt{DS_n(\omega)},} \tag{6}$$

Here $ES_n = n/2, D\xi_n = E\xi^2 - (E\xi)^2 = 1/4$ and $DS_n = n/4$. Hence

$$G_n(\omega) = \frac{S_n(\omega) - n/2}{\sqrt{n}/2} = \sum_{j=1}^{n} \frac{2\xi_n}{\sqrt{n}} - \sqrt{n}.$$

By applying the method of characteristic functions we can find the characteristic function of the limiting distribution. Let us compute the characteristic function of random variables $G_n(\omega)$:

$$\psi_n(z) = (cosh\{z/\sqrt{n}\})^n.$$

Set $\psi(z,a) = (cosh\{z/\sqrt{a}\})^a, a \in \mathbf{Z}_p, a \neq 0$. This function belongs to the space of locally analytic functions. There exists the p-adic analytic generalized function, see Ref.10 for detail, γ_a with the Borel-Laplace transform $\psi(z,a)$. Unfortunately, we do not know so much about this distribution (an analogue of Gaussian distribution?). We only proved the following theorem:

Theorem 3.4. *The* γ_1 *is the bounded measure on* $\mathbf{Z}_p$.

Open Problems:

1). Boundedness of γ_a for $a \neq 1$.

2). Weak convergence of $\mathbf{P}_{G_n}$ to $\mathbf{P}_G = \gamma_a$ (at least for $a = 1$).

4. Axiomatics for Probability Valued in a Topological Group

Let G be a commutative (additive) topological group. In general, it can be nonlocally compact.[‖] Let us choose a fixed subset Δ of the group G.

[‖]In principle, we could proceed in the same way in the non-commutative case.

Axiomatics 2. *Let Ω and F be as in Axiomatics 1. Let $\mathbf{P} : F \to \Delta$ be an additive function (measure). The triple $(\Omega, F, \mathbf{P})$ is said to be a G-probabilistic space (with the Δ-range of probability).*

We also have to add an integration condition. Such a condition depends on the topological structure of G. It seems to be impossible to propose a general condition providing fruitful integration theory.

The reader might say that our definition of a G-probabilistic measure is too general. Moreover, our real probabilistic intuition would protest against disappearance of the *unit probability* from consideration. We shall discuss this problem in Sec. 5.

We now consider a modification of the above axiomatics that includes a kind of 'unit probability'. Let $E = \mathbf{P}(\Omega)$ be a nonzero element in G. Let G be metrizable (with the metric ρ). The additional "unit-probability axiom" should be of the following form:

$$\sup_{A \in F} \rho(0, \mathbf{P}(A)) = \rho(0, E). \tag{7}$$

A G-probabilistic space in that (7) holds is called a G-probabilistic space with *unit probability axiom.* Of course, the consideration of such probabilistic spaces seems to be more natural from the standard probabilistic viewpoint. Therefore it would be natural to start with consideration of such models. However, for many important G-probabilistic spaces the unit probability axiom does not hold true. At the moment we know a few examples of G-probabilities having applications:

1). $G = \mathbf{R}$ and $\Delta = [0, 1]$(the conventional probability theory);

2). $G = \mathbf{R}$ and $\Delta = \mathbf{R}$ ("negative probabilities", see, e.g., Refs.17–20, they are realized as signed measures, charges).

3). $G = \mathbf{C}$ and $\Delta = \mathbf{C}$ ("complex probabilities", see, e.g., Refs.21,22, they are realized as $\mathbf{C}$-valued measures).

4). $G = \mathbf{Q}_p$ and Δ is a ball in $\mathbf{Q}_p$ ("p-adic probabilities", Refs.7,9,10,23,24, they are realized as $\mathbf{Q}_p$-valued measures).

The p-adic model can be essentially (and rather easily) generalized. Let K be an arbitrary complete non-Archimedian field with the valuation (absolute value) $|\cdot|$. We can define K-valued probabilistic measures by using the same integration conditions as in the p-adic case, namely boundedness and continuity.

We note that in all considered examples the additive group G has the additional algebraic structure, namely the field structure. The presence of such a field structure gives the possibility to develop an essentially richer probabilistic calculus than in the general case. Here we can introduce conditional probability by using Bayes' formula and define the notion of independence of events.

The following slight generalization gives the possibility to consider a few new examples. Let G be a non-Archimedian normed ring. To simplify considerations, we again consider the commutative case. Here:
(1) $\|x\| \geq 0, \|x\| = 0 \leftrightarrow x = 0$;
(2) $\|x\| \, \|y\| \leq \|x\| \, \|y\|$ and $\|x + y\| \leq \max(\|x\|, \|y\|)$.

We set, for $A \in F$, $\|A\|_{\mathbf{P}} = \sup\{\|\mathbf{P}(B)\| : B \in F, B \subset A\}$. We define a G-probabilistic measure as a normalized G-valued measure satisfying to the conditions of boundedness and continuity. Corresponding integration theory is developed in the same way as in the case of a non-Archimedian field. One of the most important examples of non-Archimedian normed rings is a ring of m-adic numbers $\mathbf{Q}_m$, where $m \neq p^k, p - prime$. It is a locally compact ring. We can present numerous examples of non-Archimedian normed rings by considering various functional spaces of $\mathbf{Q}_p$ (or $\mathbf{Q}_m$)-valued functions.

For a ring G, we can define averages for G-valued random variables, $\xi : \Omega \rightarrow G$. In particular, we can represent the probability distribution of the sum $\eta = \xi_1 + \xi_2$ of two G-valued random variables as the convolution of corresponding probability distributions. Here we define the convolution of two G-valued measures on G as:

$$\int_G f(x) M_1 \star M_2(dx) = \int_{G \times G} f(x_1 + x_2) M_1(dx_1) M_2(dx_2),$$

where $f : G \rightarrow G$ is a "sufficiently good" function. If G is a ring and $A \in F$ is such that $\mathbf{P}(A)$ is invertible, then we can define conditional probabilities by using Bayes' formula.

We obtain a large class of new mathematical problems related to G-probabilistic models. We emphasize that, despite a rather common opinion, probability theory is not just a part of functional analysis (measure theory). Probability theory has also its own ideology. The probabilistic ideology induces its own problems. Such problems would be impossible to formulate in the framework of functional analysis (of course, methods of functional analysis can be essentially used for the investigation of these problems).

One of the most important problems is to find analogues of limit theorems, compare, e.g., with Ref.32.

Open problem:

Let $s = \{N_k\}$ be a sequence of natural numbers and let

$$M_{11}, M_{12}, M_{1n_1};$$

$$M_{21}, M_{22}, \ldots, M_{2n_2};$$

$$M_{k1}, M_{k2}, \ldots, M_{kN_k};$$

be G-probabilistic measures. As usual, we have to study behavior of convolutions:

$\alpha_k = M_{k1} \star M_{k2} \star \ldots \star M_{kN_k}$

to find analogues of limit theorems. For example, an analogue of the law of large numbers could be formulated in the following way. Let d be a nonzero element of a topological additive group G. Let $s = \{N_k\}_{k=1}^{\infty}$ be a sequence of natural numbers. Suppose that the corresponding sequence $\{N_k d\}_{k=1}^{\infty}$ of elements of G converges to some element $a \in G$ or to $a = \infty$. The latter has the standard meaning: for each neighborhood U of zero in G there exists t such that $N_k d \notin U$ for all $k \geq t$.

Let $(\Omega, F, \mathbf{P})$ be a G-probabilistic space. Let $\xi_n(\omega) = 0, d$, with G-probabilities $q, q' = E - q$ where $E = \mathbf{P}(\Omega)$, be a sequence of independent random variables. Let $S_n(\omega)$ be the sum of n first variables.

Open Problem:

Does the sequence of probability distributions $\mathbf{P}_{S_{N_k}}$ converge weakly to some probability distribution $\mathbf{P}_S$, when $k \to \infty$?

The simplest variant of this problem is to generalize Theorem 3.1: to find (if it exists) $\lim_{k\to\infty} \mathbf{P}(S_{N_k}(\omega) \in U_r(0))$. In the case of a metrizable group G, $U_r(0) = \{g \in G : \rho(g, 0) \leq r\}, r > 0$, is a ball in G. In the case when G is a field we can consider normalized sums (6) and try to get an analogue of the central limit theorem.

5. Interpretation of Probabilities with Values in a Topological Group, Statistics

In fact, Kolmogorov's probability theory has two (more or less independent) counterparts: (a) axiomatics (a mathematical representation); (i) interpretation (rules for application). The first part is the measure-theoretic formalism. The second part is a mixture of frequency and ensemble interpretations: "... we may assume that to an event A which has the following characteristics: (a) one can be practically certain that if the complex of conditions

$\sum$ is repeated a large number of times, N, then if n be the number of occurrences of event A, the ratio n/N will differ very slightly from $\mathbf{P}(A)$; (b) if $\mathbf{P}(A)$ is very small, one can be practically certain that when conditions $\sum$ are realized only once the event A would not occur at all", Ref.1.

As we have already noticed, (a) and (i) are more or less independent. Therefore Kolmogorov's measure-theoretic formalism, (a), is used successfully, for example, in the subjective probability theory..

In practice we apply Kolmogorov's (conventional) interpretation, (i), in the following way. First of all we have to fix $0 < \epsilon < 1$, *significance level.* If the probability $\mathbf{P}(A)$ of some events A is less than ϵ, this event is considered as practically impossible. We now generalize the conventional interpretation of probability to the case of G-valued probabilities. First of all we have to fix some neighborhood of zero, V, *significance neighborhood.*

If the probability $\mathbf{P}(A)$ of some event A belongs to V, this event is considered as practically impossible.

If a group G is metrizable, then the situation is even more similar to the standard (real) probability. We choose $\epsilon > 0$ and consider the ball $V_\epsilon = \{x \in G : \rho(0, x) < \epsilon\}$. If $\rho(0, \mathbf{P}(A)) < \epsilon$, then the event A is considered as practically impossible.

Let us borrow some ideas from statistics. We are given a certain sample space Ω with an associated distribution $\mathbf{P}$. Given an element $\omega \in \Omega$, we want to test the hypothesis "ω belongs to some reasonable majority." A reasonable majority $\mathcal{M}$ can be described by presenting *critical regions* $\Omega^{(\epsilon)}(\in F)$ of the significance level $\epsilon, 0 < \epsilon < 1 : \mathbf{P}(\Omega^{(\epsilon)}) < \epsilon$. The complement $\bar{\Omega}^{(\epsilon)}$ of a critical region $\Omega^{(\epsilon)}$ is called $(1 - \epsilon)$ confidence interval. If $\omega \in \Omega^{(\epsilon)}$, then the hypothesis '$\omega$ belongs to majority $\mathcal{M}$' is rejected with the significance level ϵ. We can say that ω fails the test to belong to $\mathcal{M}$ at the level of critical region $\Omega^{(\epsilon)}$.

G-statistical machinery works in the same way. The only difference is that, instead of significance levels ϵ, given by real numbers, we consider significance levels V given by neighborhoods of zero in G. Thus we consider critical regions $\Omega^{(V)}(\in F)$:

$$\mathbf{P}(\Omega^{(V)}) \in V.$$

If $\omega \in \Omega^{(V)}$, then the hypothesis "ω belongs to majority $\mathcal{M}$" (represented by the statistical test $\{\Omega^{(V)}\}$) is rejected with the significance level V. If G is metrizable, then we have even more similarity with the standard (real) statistics. Here $V = V_\epsilon, \epsilon > 0$.

Of course, the strict mathematical description of the above statistical considerations can be presented in the framework of Martin-Löf Refs.4,6, 7 statistical tests. We remark that such a p-adic framework was already developed in Ref.7. In the p-adic case (as in the real case) it is possible to enumerate effectively all p-adic tests for randomness. However, a universal p-adic test for randomness does not exist.[7] If the group G is metrizable we can proceed in the same way as in the real and p-adic case[11] and define G-random sequences, namely sequences $\omega = (\omega_1, \ldots, \omega_N, \ldots), \omega_j = 0, 1$, that are random with respect to a G-valued probability distribution. However, if G is not metrizable, then the notion of a recursively enumerable set would not be more the appropriative basis for such a theory. In any case we have an interesting

Open problem:

Development of randomness theory for an arbitrary topological group.

The general scheme of the application of G-valued probabilities is the same as in the ordinary case: 1) we find initial probabilities; 2) then we perform calculations by using calculus of G-valued probabilities; 3) finally, we apply the above interpretation to resulting probabilities.

The main question is "How can we find initial probabilities?" Here the situation is more or less similar to the situation in the ordinary probability theory. One of possibilities is to apply the frequency arguments (as R. von Mises). We have already discussed such an approach for p-adic probabilities. Another possibility is to use subjective approach to probability. I think that everybody agrees that there is nothing special in segment $[0, 1]$ as the set of labels for the measure of belief in the occurrence of some event. In the same way we can use, for example, the segment [-1,1] (signed probability) or the unit complex disk (complex probability) or the set of p-adic integers $\mathbf{Z}_p$ (p-adic probability). If G is a field we can apply the machinery of Bayesian probabilities and, finally, use our interpretation of probabilities to make a statistical decision. The third possibility is to use symmetry arguments, Laplacian approach. For example, by such arguments we can choose (in some situations) the uniform $\mathbf{Q}_p$-valued distribution.

We now turn back to the role of the unit probability and, in particular, the axiom (7). In fact, by considering the interpretation of probability based on the notion of the significance level we need not pay the special attention to the probability $E = \mathbf{P}(\Omega)$. It is enough to consider V-impossible events, $V = V(0)$. If V is quite large and $\mathbf{P}(A) \not\in V$, then an event A can be considered as practically definite.

Example 5.1. (A p-adic statistical test) Theorem 3.1. implies that, for each p-adic sphere $\mathbf{S}_{1/p^l}(r)$, where l, r, m were done in Theorem 3.1:

$$\lim_{k\to\infty} \mathbf{P}(\{\omega \in \Omega_2 : S_{N_k}(\omega) \in \mathbf{S}_{1/p^l}(r)\}) = 0,$$

for each sequence $s = \{N_k\}, N_k \to m, k \to \infty$. We can construct a statistical test on the basis of this limit theorem (as well as any other limit theorem). Let $s = \{N_k\}, N_k \to m$, be a fixed sequence of natural numbers. For any $\epsilon > 0$, there exists k_ϵ such that, for all $k \geq k_\epsilon$,

$$|\mathbf{P}(\{\omega \in \Omega_2 : S_{N_k}(\omega) \in \mathbf{S}_{1/p^l}(r)\})|_p < \epsilon.$$

We set $\Omega^{(\epsilon)} = \bigcup_{k\geq k_\epsilon}\{\omega \in \Omega_2 : S_{N_k}(\omega) \in \mathbf{S}_{1/p^l}(r)\}$. We remark that

$$|\mathbf{P}(\Omega^{(\epsilon)})|_p < \epsilon.$$

We now define reasonable majority of outcomes as sequences that do not belong to the sphere $\mathbf{S}_{\frac{1}{p^l}}(r)$, "nonspherical majority." Here the set $\Omega^{(\epsilon)}$ is the critical region on the significance level ϵ.

Suppose that a sequence ω belongs to the set $\Omega^{(\epsilon)}$. Then the hypothesis "ω belongs nonspherical majority" must be rejected with the significance level ϵ. In particular, such a sequence ω is not random with respect to the uniform p-adic distribution on Ω_2. If, for some sequence of 0 and 1, $\omega = (\omega_j)$ we have $\omega_1 + \ldots + \omega_{N_k} - r = \alpha \mod p^l, \alpha = 1, \ldots, p-1$, for all $k \geq k_\epsilon$, then it is rejected.

The simplest test is given by $m = 1, r = 0, N_k = 1 + p^k$ and $\omega_1 + \ldots + \omega_{N_k} = \alpha \mod p, \alpha = 1, \ldots, p-1$.

Acknowledgments

I would like to thank S. Albeverio, A. Bendikov, W. Hazod, H. Heyer, G. Letac, V. Maximov, D. Neuenschwander, A. Shiryaev, Yu. Prohorov, T. Hida, V. Vladimirov and I. Volovich for fruitful discussions.

References

1. A. N. Kolmogorov, *Grundbegriffe der Wahrscheinlichkeitsrechnung*, (Springer Verlag, Berlin, 1933); reprinted: *Foundations of the Probability Theory*, (Chelsey Publ. Comp., New York, 1956).
2. R. Von Mises, *Math. Z.*, **5**, 52 (1919).
3. R. Von Mises, *The Mathematical Theory of Probability and Statistics*, (Academic, London, 1964).

4. M. Li, P. Vitanyi, *An Introduction to Kolmogorov Complexity and its Applications*, (Springer, Berlin-Heidelberg-New York, 1997).
5. A. N. Kolmogorov, *Problems Inform. Transmition*, **1**, 1 (1965).
6. P. Martin-Löf, *Theory of Probability Appl.* **11**, 177 (1966).
7. A. Yu. Khrennikov, *Interpretation of probability*, (VSP Int. Publ., Utrecht, 1999).
8. E. Beltrametti and G. Cassinelli, *Found. of Physics*, **2**, 1 (1972).
   I. V. Volovich, *Class. Quant. Grav.* **4**, 83(1987).
   B. Dragovic, *Mod. Phys. Lett.*, **6**, 2301 (1991).
9. V. S. Vladimirov, I. V. Volovich, and E. I. Zelenov, *p-adic Analysis and Mathematical Physics*, (World Scientific Publ., Singapore, 1993).
10. A. Yu. Khrennikov, *p-adic Valued Distributions in Mathematical Physics*, (Kluwer Academic Publishers, Dordrecht, 1994).
11. A. Yu. Khrennikov, *Dynamical Systems and Biological Models*, (Kluwer Academic Publishers, Dordrecht, 1997).
12. A. Yu. Khrennikov, *Physica A*, **215**, 577 (1995).
13. E. Thiran, D. Verstegen, and J. Weyers, *J. Stat. Phys.*, **54**, 893 (1989).
14. A. Yu. Khrennikov, M. Nilsson, *p-adic Deterministic and Random Dynamical Systems*, (Kluwer Academic Publishers, Dordrecht, 2004).
15. D. Dubischar, V. M. Gundlach, O. Steinkamp, and A. Yu. Khrennikov, *Physica D*, **130**, 1 (1999).
    S. Albeverio, A. Yu. Khrennikov, and P. Kloeden, *Biosystems*, **49**, 105 (1999).
16. A. Yu. Khrennikov, *Information Dynamics in Cognitive, Psychological and Anomalous Phenomena*, (Kluwer Academic Publishers, Dordrecht, 2004).
17. P. A. M. Dirac,*Proc. Roy. Soc. London*, **A 180**, 1 (1942).
18. E. Wigner, *Quantum-mechanical distribution functions revisted*, in *Perspectives in quantum theory*, Eds., Yourgrau, W., and van der Merwe, A., (MIT Press, Cambridge MA, 1971).
19. W. Muckenheim, *Phys. Reports*, **133**, 338 (1986).
20. A. Yu. Khrennikov, *Int. J. Theor. Phys.*, **34**, 2423 (1995).
21. P. A. M. Dirac, *Rev. of Modern Phys.*, **17**, 195 (1945).
22. E. Prugovecki, *Found. of Physics*, **3**, 3 (1973).
23. A. Yu. Khrennikov, *Dokl. Akad. Nauk USSR*, **322**, 1075 (1992).
24. A. Yu. Khrennikov, *Theory of Probability and Appl.*, **40**, 458 (1995).
25. A. Yu. Khrennikov, *Theory of Probability and Appl.* **46**, 311 (2001).
26. W. Schikhov, *Ultrametric Calculus. Cambridge Univ*, (Press, Cambridge, 1984).
27. A. Van Rooij, *Non-Archimedian Functional Analysis*, (Marcel Dekker, Inc., New York, 1978).
28. A. Monna and T. Springer, *Indag. Math.* **25**, 634 (1963).
29. A. Yu. Khrennikov, *Izvestia Akademii Nauk*, **64**, 211 (2000).
30. A. Yu. Khrennikov, *Statistics and Probability Lett.*, **51**, 269 (2001).
31. J. Bernard and G. Letac, *Illinois J. Math.*, **17**, 317 (1997).
32. H. Heyer, *Probability Measures on Locally Compact Groups*, (Springer-Verlag, Berlin-Heidelberg-New York, 1977).

Chapter IV

ARCHIMEDEAN STOCHASTIC ANALYSIS

Harmonic, Wavelet and p-Adic Analysis
Eds. N. M. Chuong *et al.* (pp. 313–330)

§14. INFINITE DIMENSIONAL HARMONIC ANALYSIS FROM THE VIEWPOINT OF WHITE NOISE THEORY

TAKEYUKI HIDA

Meijo University, Nagoya, Japan
E-mail: thida@ccmfs.meijo-u.ac.jp

White noise measure, which is the standard infinite dimensional Gaussian measure, is invarant under the infinite dimensionl rotation group, so that white noise analysis has an aspect of harmonic analysis ariding from the rotation group. Complexification of white noise and that of groups give us tools of the analysis and suggestions on applications to quantum dynamics.

Keywords: AMS 2000 Mathematics Classifications: 60H40.

1. Introduction

We are going to discuss harmonic analysis arising from the infinite dimensional rotation group. Such an analysis is one of the significant chracteristics of white noise analysis. Further, it is noted that many interesting developments of our theory in this direction can be seen in various applications which would stimulate future directions.

These notes involve some reviews of infinite doimensional rotation group and related topics in order to describe systematic approach to an infinite dimensional harmonic analysis and some new results that can suggest applicable areas of white noise theory.

One may wonder why rotation group is involved in white noise analysis. To answer this question, we can state various plausible reasons. Among others,

i) One uses the strong law of large numbers: there are defined countably infinite number of independent standard Gaussian random variables on white noise measure space. At the same time those random variables are viewed as coordinates of the space. It can therefore be seen, by the strong law of large numbers, that the white noise measure μ is a uniform probability measure on an infinite dimwensional sphere.

ii) The characteristic functional of μ is a function of the square of the norm

$\|\xi\|$. This fact immediately suggests, although formally, that the functional, and hence the measure, is invariant under the rotations.
iii) As the third reason, one may recall the interesting and surprising fact in classical functional analysis. In order to define a uniform measure on the unit sphere of a Hilbert space, they tried to approximate by the uniform measure on the finite dimensional ball S^d. The radius of the ball has to be proportional to $\sqrt{d}$. If d is getting larger and larger, the uniform measure on the sphere approximates the white noise measure μ. There the rotation group should play an essential role. Some opther interpretations may be given, however we claim that those facts can be rephrased within the framework of white noise theory.
iv) Various applications stimulate the use of group to describe invariance, symmetry and other abstract properties, sometimes latent traits. For instance, in quantum dynamics, infrormation sociology, molecular biology, etc. we can seeeffective use of trasformation groups. We may rephrase them in white noise theory and see good interplays.

With these facts in mind, we are going to define an infinite dimensional rotation group and to proceed the analysis that can be thought of a harmonic analysis.

2. Infinite Dimensional Rotation Group $O(E)$

The basic nuclear space is usually taken to be either the Schwartz space or the space D_0 that is isomorphic to the space $C^\infty(S^d)$. The latter is convenient to introduce a subgroup isomorphioc to the conformal group. Now let a nuclear space E be fixed.

Definition 2.1. A continuous linear homeomorohism g acting on E is called a *rotation* of E if the following equality holds for every $\xi \in E$;

$$\|g\xi\| = \|\xi\|.$$

Obviously, the collection of all rotations of E forms a group (algebraically) under the product

$$(g_1 g_2)\xi = g_1(g_2\xi),$$

The group is denoted by $O(E)$, or by O_∞ if there is no need to specify the basic nuclear space E.

Since $O(E)$ is a group of transformations acting on the topological space E, it is quite natural to let it be topologized by the compact-open topology. Thus, we are given a topological group, keeping the same notation, $O(E)$.

The adjoint transformation, denoted by g^*, to $g \in O(E)$ is defined by

$$\langle x, g\xi \rangle = \langle g^* x, \xi \rangle, \quad x \in E^*, \xi \in E,$$

where E^* is the dual space of E.

It is easy to see that the g^* is a continuous linear transfrmation acting on the space E^*. The collection $O^*(E^*) = \{g^*; g \in O(E)\}$ also forms a group.

We can prove that the group $O^*(E^*)$ is (algebraically) isomorphic to the rotation group $O(E)$ under the mapping

$$g \longrightarrow (g^*)^{-1}, \quad g \in O(E).$$

In view of this, the group $O^*(E^*)$ can also be topologized so as to be isomorphic to $O(E)$, and the topological group $O^*(E^*)$ is also called *infinite dimensional rotation group.*

There is a fundamental theorem regarding probabilistic role of the infinite dimensional rotation group. Recall that the white noise measure is introduced on the measurable space $(E^*, \mathcal{B})$, where $\mathcal{B}$ is the sigma-field generated by the cylinder subsets of E^*. The complex Hilbert space $L^2(E^*, \mathcal{B})$ is denoted by (L^2) for simplicity.

Theorem 2.1. *The white noise measure μ is invariant under the infinite dimensonal rotation group $O^*(E^*)$.*

Harmonic Analysis

At present the following topics are worth to be mentioned.

1) Irreducible unitary representations of $O(E)$. There are masny possibilities of introducing unitary representations. It is noted that, unlike finite dimensional Lie group, the dimension of the space on which a representation is defined should be infinite dimensional. This is one of the reason why various representations are accepted.

2) Laplacians. The (finite dimensional) spherical Laplacian tends to the infinite dimensional Laplace-Beltrami operator Δ_∞. Other, in fact more significant, Laplacians are defined and they play different roles in our analysis, respectively.

3) Although the complex Hilbert space (L^2) is not quite like $L^2(G, m)$, G being a Lie group and m being the left or right Haar measure, still one can

see some analogy between them. The former is more like an L^2 space over a symmetric space.

4) There is the Fourier-Wiener transform defined on (L^2). It is given by analogy with the ordinary Fourier transform acting on $L^2(R)$, but essential modification is necessary; for one thing the basic measure is not of Lebesgue type but Gaussian.

5) We emphasize the significance of the roles played by whiskers which are one-parameter subgroups coming from diffeomorphisms of the @arameter space.

3. The Lévy Group, the Windmill Subgroup and the Sign-Changing Subgroup of $O(E)$

The group $O(E)$ is quite big, we shall therefore take suitable subgroups successively, from finite dimensional to infinite dimensional and even further essentially to infinite dimensional ones which are far from finite dimensional approximations.

We start with a simple subgroup which is given below. Take a complete orthonormal system $\{\xi_n\}$ such that $\xi_n \in E$ for every n. Take $\xi_j, 1 \leq j \leq n$, and let E_n be the subspace of E spanned by them. Define G_n by

$$G_n = \{g \in O(E);\ g|_{E_n} \in SO(n),\ g|_{E_n^\perp} = I\}.$$

Obviously, it holds that

$$G_n \cong SO(n).$$

Now the subgroup G_∞ is defined as the inductive limit of the G_n:

$$G_\infty = \operatorname{ind\,lim}_{n\to\infty} G_n.$$

The group G_∞ is certainly infinite dimensional, however, by definition, each member of the group is understood to be a transformation that can be approximated by the finite dimensional rotations.

The Lévy group that is going to be defined was first introduced by P. Lévy in 1922 and systematic approach has been done in his book Ref. 10 as a tool from functional analysis. Since we recognize the significance of this group also in white noise analysis, we rephrase the definition a little, and we wish to find the important roles of this group in our stochastic analysis.

Let π be an automorphism of $Z_+ = \{1, 2, \cdots\}$. Fix a complete orthonormal system $\{\xi_n\}$ in $L^2(R)$ such that ξ_n is in E for every n. A transformation

g_π of $\xi \in E$ is defined by

$$g_\pi : \xi = \sum_1^\infty a_n \xi_n \to g_\pi \xi = \sum_1^\infty a_n \xi_{\pi(n)}.$$

Define the *density* $d(\pi)$ of the automorphism π by

$$d(\pi) = \limsup_{N\to\infty} \frac{1}{N} \#\{n \le N; \pi(n) > N\}.$$

Denote the collection

$$\{g_\pi; d(\pi) = 0, g_\pi \in O(E)\}.$$

by $\mathcal{G}$. Obviously the collection $\mathcal{G}$ forms a subgroup of $O(E)$. It is a discrete infinite group.

Definition 3.1. The group $\mathcal{G}$ is called the *Lévy group.*

The *average power* $a.p.(g)$ of a member $g = g_\pi$ of $O(E)$ is introduced for the benefit of seeing what the finite dimensional approximation means and of seeing how an essentially infinite dimensional transformation looks like.

We continue to fix the complete orthonormal system $\{\xi_n\}$. Now define the average power $a.p.(g_\pi)$ by

$$a.p.(g)(x) = \lim \sup_{N\to\infty} \frac{1}{N} \sum_1^N < x, g\xi_n > -\xi_n >^2 .$$

Definition 3.2. If $a.p.(g)(x)$ is positive almost surely (μ), then we call g_π *essentialy infinite dimensional.* Contrary to this case, if $a.p.(g_\pi)(x) = 0$ almost surely, then g_π is said to be *approximated* by the finite dimernsional rotations.

If g is in the Lévy group, say $g = g_\pi$, then

$$a.p.(g_\pi)(x) = \lim \sup_{N\to\infty} \frac{1}{N} \sum_1^N < x, \xi_{\pi(n)} - \xi_n >^2 .$$

We can see that there are many members in the Lévy group (see Example below) that are essentially infinite dimensional.

Example 3.1. An example of a member of the Lévy group $\mathcal{G}$.

Let π be a permutation of positive integers such that

$$\pi(2n-1) = 2n, \quad \pi(2n) = 2n-1, \quad n = 1, 2, \cdots .$$

Fix a complete orthonormal system $\{\xi_n\}$ in $L^2(R)$ such that every ξ_n is in E. For $\xi = \sum a_n \xi_n$ define $g_\pi \xi$ by

$$g_\pi \xi = \sum a_n \xi_{\pi(n)}.$$

This is, as it were, a pairwise permntation of the coordinates. By actual computation, we have $a.p.(g_\pi) = 2$. Hence, g_π is an essentially ionfinite dimensional rotation.

It is interesting to note that there should be an intimate connections between the Lévy group and the Lévy Laplacian, although some are known.

The Windmill subgroup

There is another subgroup of $O(E)$ that contains essentially infinite dimensional transformation. It is a windmill subgroup $\mathcal{W}$, which is defined in the following manner. Take E to be the Schwartz space S and take a sequence $n(k)$ of positive integers satisfying the condition

$$(n(k+1) - n(k))\frac{n(k+1)}{n(k)} \leq K, \ (K > 1).$$

Let $\xi_n, n \geq 0$, be the complete orthonormal system in $L^2(R)$ such that ξ_n is the eigenfunction of A defined before: $A\xi_n = 2(n+1)\xi_n$. Denote by E_k the $(n(k+1) - n(k))$-dimensional subspace of $E = S$ that is spanned by $\{\xi_{n(k)+1}, \xi_{n(k)+2}, \cdots, \xi_{n(k+1)}\}$. Let G_k be the rotation group acting on E_k. Then, $\mathcal{W} = \mathcal{W}(\{n(k)\})$ is defined by

$$\mathcal{W} = \otimes_k G_k.$$

It is easy to prove

Proposition 3.1. *The system $\mathcal{W} = \mathcal{W}(\{n(k)\})$ forms a subgroup of $O(S)$ and contains infinitely many members that are essentially infinite dimensional.*

Proof is given by evaluating the norm $\|g\xi\|_p, g \in \mathcal{W}$ and by using the requirement on the sequence $n(k)$.

Definition 3.3. The subgroup $\mathcal{W}$ is called a *windmill subgroup.*

The sign-changing subgroup $\mathcal{H}$

The group $\mathcal{H}$ is introduced in the following steps:

1) Take $t \in (0, 1]$. Denote the binary expansion of t by

$$t = \sum_1^\infty \eta_n(t) 2^{-n},$$

where $\eta_n(t) = 0$ or 1.

To guarantee the uniqueness of the determination of $\eta_n(t)$, we define

$$\eta_n(1) = 1 \quad \text{for every } n,$$

$$\eta_n(2^{-k}) = 0, \ n \le k; \quad = 1, n > k.$$

2) Set $\epsilon_n(t) = 2\eta_n(t) - 1$. Then, g_t is defined by

$$g_t : \ \xi = \sum_1^\infty a_n \xi_n \quad \to \quad \sum_1^\infty a_n \epsilon_n(t) \xi_n,$$

where $\{\xi_n\}$ is a fixed complete orthonormal system in $L^2(R)$.

3) Every g_t belongs to the group $O(E)$, since it is a linear transformation on E and preserves the L^2-norm. The collection $\mathcal{H} = \{g_t, t \in (0, 1]\}$ forms a group, and hence it is a subgroup of $O(E)$. The product $g_t g_s$ is defined in the usual manner $(g_t g_s)\xi = g_t(g_s \xi)$ and the result is a transformation denoted by $g_{\phi(t,s)}$. Obviously $\mathcal{H}$ is abelian: $\phi(t, s) = \phi(s, t)$.

4) Since g_1 is the identity, we have $\phi(t, 1) = \phi(1, t) = t$. By definition, we have

$$\phi(t, t) = 1, \quad \text{i.e.} \ \ g_t^2 = e(\text{identity}).$$

5) There exists a member $g_t \in \mathcal{H}$ such that the average power of g_t is positive. We may say that the subgroup $\mathcal{H}$ itself is essential infinite dimensional subgroup of $O(E)$.

Definition 3.4. The subgrou $\mathcal{H}$ is called a *sign-changing subgropup* of $O(E)$.

A significance of the sign-changing subgroup is that it has a connection with certain transformations of sample functions (in fact, generalized functions) of white noise. Indeed, it describes some inherent properties of white noise. In other words, the group is interesting from the viewpoint of the generalized harmonic analysis of stochastic processes. One of the typical examples is given in what follows.

The sign changing group describes local (in time) intrincic transfromation of a sample function of white noise. This fact can be seen with the help of Lévy's construction of Brown ian motion by successive interporation.

4. Whiskers

A one-parameter subgroup of $O(E)$ that comes from diffeomorphisms of the parameter space is called a *whisker*. For convenience, we take the basic nuclear space E is taken to be the D_0 which a subspace of $L^2(R^d)$ and is isomorphic to $C^\infty(S^d)$. A whisker, denoted by $g_t, t \in R$, is determined by a one-parameter group of diffeomorphisms $\psi_t(u), t \in R, u \in R^d$, in such a way that

$$g_t\xi(u) = \xi(\psi_t(u))\sqrt{|\psi_t'(u)|}, \ \psi_t'(u) : Jacobian,$$

where

$$\psi_t o \psi_s = \psi_{t+s}.$$

We further assume continuity of the product of the ψ_t in t.

Typical examples of a whisker are listed below.

1) The shift. Let $e_j, 1 \le j \le d$, be an orthonormal base of R^d. Define S_t^j by

$$S_t^j \xi(u) = \xi(u - te_j).$$

Then, obviously $S_t^j, 1 \le j \le d, t \in R$, are whiskers. Each S_t^j is called a *shift.*

2) Isotropic dilation. The isotropic dilation $\tau_t, t \in R$, is defined by

$$\tau_t\xi(u) = \xi(ue^{\lambda t}u)e^{td/2}.$$

This is another whisker.

3) With a special choice of the basic nuclear space to be $O(D_0)$, we can define $\kappa_t^j, 1 \le j \le d, t \in R$, by using the refelection w, in the following manner:

$$\kappa_t^j = wS_t^j w.$$

Put together the above three examples. And we obtain a subgroup of $O(D_0)$ which is isomorphic to $SO(d+1,1)$. It is called a conformal group $C(d)$.

Take finitely many whiskers that form a Lie group like in the above example. Then, applying the well known technique for finite dimensional

Lie groups, we can describe probabilistic properties of white noise and of white noise functionals.

We are now ready to discuss the the topic *harmonic analysis* that is our main aim. The essential part in our analysis is that white noise $\dot{B}(t)$ (or sometimes Poisson noise $\dot{P}(t)$) is taken to be the *variable* of functionals to be discussed.

Unitary representation of the group $O(E)$

As was announced before, we shall discuss unitary representations of the infinite dimensional rotation group $O(E)$. The definition of the unitary representation and its irreduciblity are the same in the ordinary text book on topological groups, although the group that we have is infinite dimensional.

First, the Hilbert space (L^2) is taken to define a representation. For any $\varphi \in (L^2)$ and for $g \in O(E)$ define U_g by

$$(U_g\varphi)(x) = \varphi(g * x).$$

Then U_g is a unitary operator on (L^2), and the collection $\mathbf{U} = \{U_g, g \in O(E)\}$ forms a group. The group is topologized so as to be isomorphic to the topological group $O(E)$.

We can prove (see Ref. 5)

Theorem 4.1.

(i) $\{U_g, g \in O(E); (L^2)\}$ *is a unitary representation of the infinite dimensional rotationgroup* $O(E)$.

(ii) *The unitary representation above is reduced to* $H_n, n \geq 1$, *which is the subspace of* (L^2) *that appears in the Fock space. Then,* $\{U_g, g \in O(E); H_n\}$ *is an irreducible unitary representation.*

Laplacians

We are familiar with three different Laplacians in white noise analysis. They can be characterized in diffrent manner and their roles are also different.

1) The *Laplace-Beltrami operator*

An infinite dimensional version (unfortunately it is not quite a simple generalization, but something like a version) of the spherical Laplacian or the Laplace-Beltrami operator is the following operator:

$$\Delta_\infty = \sum(\frac{\partial^2}{\partial x_i^2} - x_i\frac{\partial}{\partial x_i}),$$

where $x_n = \langle x, \xi_n\rangle$, $\{\xi_n\}$ being a complete orthonormal system in $L^2(R)$ as before.

A direct characterization has been made in the same spirit as in the case of the finite dimensional spherical Laplacian.

We can easily prove

Proposition 4.1. *The subspace H_n in the Fock space is the eigenspace of the Laplace-Beltrami operator Δ_∞ belonging to the eigenvalue $-n$.*

In order to discriminate from other Laplacians that will appear in what follows, we write the Laplace-Beltrami operator by Δ_∞.

2) The *Volterra Laplacian* Δ_V (which is, in reality, the same as the *Gross Laplacian* Δ_G).

We define a second order functional derivative $\tilde{\Delta}_V$ acting on function spaceis by the formula

$$\begin{aligned}&\tilde{\Delta}_V \int\cdots\int F(u_1, u_2, \cdots, u_n)\xi(u_1)\xi(u_2)\cdots\xi(u_n)du^n\\&= n(n-1)\int\cdots\int F(t, t, u_3, \cdots u_n)\xi(u_3)\cdots\xi(u_n)du^{n-2}dt.\end{aligned}$$

The Volterra Laplacian Δ_V is defined by

$$\Delta_V\varphi = S^{-1}\tilde{\Delta}_V S\varphi, \quad \varphi \in H_n^{(n)},$$

where S is the so-called S-transform.

The domain of Volterra Laplacian is taken to be either $\sum H_n^{(n)}$ or the Hilbert space $(S)_2$.

The Δ_V may be expressed in terms of the white noise derivatives $\frac{\partial}{\partial_t}$, namely

$$\Delta_V = \int \partial_t^2 dt.$$

For details together with notations are refered to Ref. 8.

3) The Lévy Laplacian Δ_L

The Lévy Lapalcian was first introduced in his 1992 book "Lecons d'analyse fonctionnelle, Gauthier-Villars, and more discussions have been given in Ref. 10. Now the Lévy Laplacian is given, in our terminology, by the formula

$$\Delta_L = \int \partial_t^2 (dt)^2.$$

Remark 4.1. One may ask what is its adjoint like. It could be expected that the Laplacian is self-adjoint. But not quite in this case. We can give an interpretation from the view point of irreducible unitary representation of infinite Symmetric group $S(\infty)$ (see Ref. 13).

Example 4.1. Integrals of the (renormalized) quadratic polynomials : $\dot{B}(t)^2$: in $\dot{B}(t)$'s which is realized as

$$\varphi_1(x) = \int f(t) : x(t)^2 : dt,$$

where f is integrable. This example is inolved in the domain of the Lévy Laplacian. It is easy to see

$$\Delta_L \varphi_1(x) = 2 \int f(t) dt.$$

Example 4.2. The Gauss kernels with finite time domain like

$$\varphi_2(x) = N \exp[c \int_0^1 : x(t)^2 : dt],$$

is in the domain of Δ_L, and it holds that

$$\Delta_L \varphi_2(x) = \frac{2c}{1-2c} \varphi_2(x).$$

We see that the Gauss kernel is an eigenfunctional of the Lévy Laplacian. The constant c can be taken to be complex. It appears in the formulation of the Feynman path integrals.

Remark 4.2. Laplacians Δ_V and Δ_L share the roles in the second variation of a white noise functional. The former comes from the regular part, whereas the latter is determined by the singular part.

5. Infinite Dimensional Unitary Group

Let E_c and E_c^* be the complexifications of E and E^*, respectively:

$$E_c = E + iE,$$

$$E_c^* = E^* + iE^*.$$

An element ζ of E_c and an element z of E_c^* are written in the form

$$\zeta = \xi + i\eta, \quad \xi, \eta \in E,$$

$$z = x + iy, \quad x, y \in E^*,$$

respectively. The canonical bilinear form $\langle x, \xi \rangle$ extends to a bilinear form $\langle z, \zeta \rangle, z \in E_c^*, \zeta \in E_c$, that connects E_c and E_c^*. We have

$$\langle z, \zeta \rangle = (\langle x, \xi \rangle + \langle y, \eta \rangle) + i(-\langle x, \eta \rangle + \langle y, \xi \rangle).$$

Take white noise measures μ_1 and μ_2 on $E^*(\cong iE^*)$ with variance $\frac{1}{2}$. Let $\mathbf{B}$ be the sgma-field generated by the cylinder subsets of E_c^*, and form a measurable space $(E_c^*, \mathbf{B})$ on which the product measure

$$\nu = \mu_1 \times \mu_2$$

is introduced.

Definition 5.1. The measure space $(E_c^*, \mathbf{B}, \nu)$ is called a *complex white noise.*

A *complex Brownian motion* $Z(t), t \geq 0$, is defined by

$$Z(t) = \langle z, \chi_{[0,t]} \rangle.$$

The complex Hilbert space $(L_c^2) = L^2(E_c^*, \mathbf{B}, \nu)$ is the space of functionals of complex white noise with finite variance.

Triviality. We have

$$(L_c^2) = (L^2)_x \otimes (L^2)_y.$$

The Fock space in this case is expressed as

$$(L_c^2) = \bigoplus H_n, \quad H_n = \bigoplus_{k=0}^{n} H_{(n-k,k)},$$

where $H_{(n-k,k)}$ is spanned by complex Hermite polynomials of degree $(n-k)$ in $\langle z, \zeta \rangle$ and of k in $\langle \bar{z}, \zeta \rangle$.

We are now ready to define the *infinite dimensional unitary group.* Denote by $U(E_c)$ the collection of all linear transformations g on E_c^* that satisfy the conditions:

1. g is a linear homeomorphism of E_c,

2. g preserves the complex $L^2(R)$-norm:

$$\|g\eta\| = \|\eta\|, \quad \zeta \in E_c.$$

Under the usual product the $U(E_c)$ forms a group. We introduce the compact-open topology to $U(E_c)$ so as to be a topological group.

Definition 5.2. The topological group $U(E_c)$ is called the *infinite dimensional unitary group.*

The adjoint g^* of g in $U(E_c)$ is defined in the usual manner, and the collection of g^*'s forms a group, denoted by $U^*(E_c^*)$.

Proposition 5.1. *For every* $g^* \in U^*(E_c^*)$ *we have*

$$g^*\nu = \nu.$$

Hence, we are given a unitary operator U_g defined by

$$U_g\varphi(z) = \varphi(g^*z), \varphi \in (L_c^2).$$

Subgroups of $U(E_c)$.

1) Finite dimensional unitary group $U(n), n \geq 1$

As in the case of the subgroup G_n of $O(E)$, we can define a subgroup $U(n)$ of $U(E_c)$, as well as the inductive limit of the $U(n)$.

2) Conformal group

Take E_c to be the complexification of D_0. Let it be denoted by $D_{0,c}$. Then, naturally follows the complexification $C_c(d)$ of the conformal group $C(d)(\subset U(D_{0,c})$. The generators of whiskers have the same expression as in $O(E)$.

3) Heisenberg group

It is an advantage of complexification to obtain Heisenberg group and gause transformations.

The basic nuclear space is now specified to be the complex Schwarts space $E_c = S_c = S + iS$. One of the reasons of such a choice is for the benefit of Fourieer transform.

3.1) The simplest gauge transform I_t is defined by

$$I_t : \zeta(u) \longrightarrow I_t\zeta(u) = e^{it}\zeta(u).$$

Obviously $\{I_t\}$ forms a continuous periodic one parameter subgroup of $U(E_c)$ with period 2π.

The group $\{I_t, t \in R\}$ is called the *gauge group*. Let the unitary operator U_t be defined by U_{I_t}, which forms a one-parameter unitary group acting (L_c^2). It has only point spectrum. The eigenspace belonging to the eigenvalue $-n+2k$ is $H_{(n-k,k)}$. Hence, the spce H_n, a member of the Fock space is classified, according to the action of I_t, into its subspaces $H_{(n-k,k)}$.

The infinitesimal generator of the gauge group is iI, I being the identity.

3.2) The shift S_t^j is a member of the Heisenberg group. The generators are denoted by $s_j = -\frac{\partial}{\partial u_j}$.

3.3) Multiplication $\pi_t^j, j = 1, 2, \cdots, d$. Let them be defined to be the conjugate to the shifts via the Fourier transform $\mathcal{F}$:

$$\pi_t^j = \mathcal{F} S_t^j \mathcal{F}^{-1}.$$

Actual expressions are

$$(\pi_t^j\zeta)(u) = e^{itu_j}\zeta(u), \;\; u \in R^d.$$

The infinitesimal generator of the multiplication is denoted by π:

$$i\pi = iu\cdot .$$

Definition 5.3. The subgroup of the $U(S_c)$ generated by the gauge group, the shifts and the multiplication is called the *Heisenberg group*.

It should be noted, for $d = 1$, that we have the commutation relation

$$\pi_t S_s = I_{st} S_s \pi_t.$$

In terms of the generators we are given uncertainty principle:

$$[s, i\pi] = iI.$$

3.4) *Abelian gauge field.* Take the (real) Schwartz space S. For any $\alpha \in S$, we define

$$I_\alpha : \zeta(u) \;\rightarrow\; e^{i\alpha(u)}\zeta(u).$$

Obviously, there is defined a continuous bijection:

$$I_\alpha \;\rightarrow\; i\alpha \in S.$$

We therefore have, as it were, an abelian nuclear gauge group $\{I_\alpha, \alpha \in S\}$.

If one wishes, a one-parameter subgroup $I_{\alpha t}, t \in R$, is defined, and the generator is $i\alpha$. And so forth. Thus, we can easily see good relationship with quantum dynamics.

4) The Fourier-Mehler transforms $\mathcal{F}_\theta$

It is possible to consider the fractional power of the ordinary Fourier transform acting on S_c. It is defined by the integral kernel $K_\theta(u,v)$:

$$K_\theta(u,v) = (\pi(1-\exp[2i\theta]))^{-1/2}\exp\left[-\frac{i(u^2+v^2)}{2tan\theta} + \frac{iuv}{\sin\theta}\right].$$

It defines an operator $\mathcal{F}_\theta$ by writing

$$(\mathcal{F}_\theta\zeta)(u) = \int_{-\infty}^{\infty} K_\theta(u,v)\zeta(v)dv,$$

where $\theta \neq \frac{1}{2}k\pi,\ k \in Z$.

Set, by using the Hermite polynomials H_n,

$$\xi_n(u) = (2^n n!\sqrt{\pi})^{-1/2} H_n(u)\exp[-\frac{u^2}{2}].$$

Then, it is proved that

$$\mathcal{F}_\theta\xi_n(u) = e^{in\theta}\xi_n(u), \quad n \geq 0.$$

With this relationship we can prove that $\mathcal{F}_\theta$ is well defined for every θ (by interporation), and further

$$\mathcal{F}_\theta\mathcal{F}_{\theta'} = \mathcal{F}_{\theta+\theta'} = \mathcal{F}_{\theta''}, \quad \theta+\theta' = \theta''\ (mod. 2\pi).$$

$$\mathcal{F}_\theta \to I, \text{as } \theta \to 0.$$

With the understanding that $\mathcal{F}_{\pi/2} = \mathcal{F}$ and $\mathcal{F}_{(3/2)\pi} = \mathcal{F}^{-1}$, we obtain a periodic one-parameter unitary group including the Fourier transform and its inverse.

The infinitesimal generator of $\mathcal{F}_\theta$ is denoted by if:

$$if = -\frac{1}{2}i\Big(\frac{d^2}{du^2} - u^2 + I\Big).$$

Observing the commutation relations of the generators, so as to have a finite dimensional Lie algebra, either real form and complex form, we are given a generator σ' expressed in the form

$$\sigma' = \frac{1}{2}\Big(\frac{d^2}{du^2} + u^2\Big).$$

We are particularly interested in the probabilistic role or roles of this operator (generator) in quantum dynamics.

An easy and formal interpretation of σ' is such that

$$\frac{1}{i}\partial_t \Psi = \sigma' \Psi$$

is the Schrödinger equation for the *repulsive oscillator.*

For our purpose, it is convenient to take $\sigma = \sigma' + \frac{i}{2}I$, namely we have

$$\sigma = \frac{1}{2}\Big(\frac{d^2}{du^2} + u^2 + iI\Big).$$

Lie Algebras of Generators

We have so far many infinitesimal generators. For simplicity we consider the case $d = 1$, i.e. one-dimensional parameter complex white noise.

The Lie algebra $\mathbf{c(1)}$ of the conformal group is generated by

$$s = -\frac{d}{du}, \tau = u\frac{d}{du} + \frac{1}{2}, \kappa = u^2\frac{d}{du} + u.$$

For the Heisenberg group we have the algebra $\mathbf{h(1)}$ generated by

$$iI, s, i\pi = iu,$$

generators related to the Fourier transform are

$$if = -\frac{i}{2}\Big(\frac{d^2}{du^2} - u^2 + I\Big), \sigma = \frac{1}{2}\Big(\frac{d^2}{du^2} + u^2 + iI\Big).$$

We now consider the Lie algebras $\mathbf{a}$ generated by the above operators. Their commutation relations are listed below:

$$[\tau, s] = -s \quad [\tau, \kappa] = \kappa$$

$$[s, \kappa] = -2\tau \quad [\pi, s] = I$$

$$[f, s] = \pi \quad [\sigma, s] = \pi$$

$$[f, \pi] = s \quad [\sigma, \pi] = -s$$

$$[f, \tau] = -2\sigma + iI \quad [\sigma, \tau] = 2f - I$$

the family $\{I_\alpha, \alpha \in S\}$ shall be dealt with separately.

Theorem 5.1.

(i) The vector space spanned by the operators

$$\{I, s, \pi, \tau, f, \sigma\}$$

forms a Lie algebra, let it be denoted by $\mathbf{g}$.

(ii) The algebra $\mathbf{h}(1)$ *is an ideal of* $\mathbf{g}$ *and is the maximum solvable Lie subalgebra. In fact,* $\mathbf{h}(1)$ *is the radical of* $\mathbf{g}$.

Proof is given by the actual and rather easy computations.

It seems necessary to give some interpretation to the fact that the generator κ of the special conformal transformation is outside of $\mathbf{g}$.

1) Heuristically, the generator κ was a good candidate to be introduced among the possible expressions of generators expressed in the the form $a(u)\frac{d}{du} + \frac{1}{2}a'(u)$. If the basic nuclear space E is taken to be D_0, the κ is acceptable. As a result, we have proved that the algebra generated by those possible generators is isomorphic to $sl(2, R)$ and $\mathbf{c(1)}$ describes the projective invariance of Brownian motion.

Similar to s, the κ is transversal to τ, which defines a flow of the Ornstein-Uhlenbeck process (flow).

2) On the other hand, there are crucial reasons why τ should not be involved in the algebra $\mathbf{g}$.

i) From our viewpoint that the Fourier transform is particularly emphasized. So the complex Schwartz space is fitting for the complex analysis. Namely, the Schwartz space S_c, which is invariant under the Fourier transform, is more significant. While, in order to introduce the κ we need another space like D_0, instead of S_c.

ii) Needless to say, Fourier transform is quite important. If f and κ are involved, we are not given finite dimensionsal Lie algebra. As a result we can not give good probabilistic interpretation to the algebra. Note that κ is concerned with the reglection of the parameter space, and it plays important roles in other situations.

Acknowledgements

The author is grateful to Professor Nguyen Minh Chuong who invited him to the conference with very kind help.

References

1. L. Accardi et al eds., *Selected papers of Takeyuki Hida* (World Scientific Pub. Co. Ltd. 2001).
2. L. Accardi, T. Hida and Si Si, *Innovation approach to some stochastic processes* (Volterra Center Notes. N.537, 2002).
3. T. Hida, *Stationary stochastic processes* (Princeton Univ. Press. 1970).
4. T. Hida, *Carleton Univ. Math. Notes*, (1975).
5. T. Hida, *Brownian motion* (Springer-Verlag, 1980).
6. T. Hida et al, *White Noise. An infinite dimensional calculus*, (Kluwer Academic Pub. Co. 1993).
7. T. Hida and Si Si, *Innovation approach to random field: An application of white noise theory* (World Sci. Pub. Co. Ltd. 2004).
8. T. Hida and Si Si, *Lectures on white noise functionals* (World Sci. Pub. Co. Ltd. 2006).
9. P. Lévy, *Processus stochastiques et mouvement brownien*, (Gauthier-Villars, 1948; 2ème éd. 1965).
10. P. Lévy, *Problème concrets d'analyse fonctionnelle* (Gauthier-Villars, 1951).
11. K. Saitô, *Acta Appl. Math.* bf 63, 363 (2000).
12. Si Si, *Quantum Prob. and Related Topics.* **6**, 609 (2003).
13. Si Si, *Symmetric group in Poissonj noise analysis* (to appear).

Harmonic, Wavelet and p-Adic Analysis
Eds. N. M. Chuong *et al.* (pp. 331–342)

§15. STOCHASTIC INTEGRAL EQUATIONS OF FREDHOLM TYPE

SHIGEYOSHI OGAWA

Dept. of Mathematical Sciences,
Ritsumeikan University,
Kusatsu, Shiga 525-8577, Japan

We are to give in this note an introductory review of the theory of stochastic integral equations of Fredholm type. The subject should be discussed naturally in the framework of the noncausal theory of stochastic calculus, on which we have given in another lecture note[4] a unified review with typical example of applications. We will show some basic results on this subject mainly following the author's earlier articles.[1–3] We will limit our discussions in the case of linear equations of Fredholm type and we will show some basic results concerning the question of existence and uniqueness of solutions.

1. Introduction

Let us consider the boundary value problem of a formal stochastic differential equation as follows;

$$\left\{\frac{d^2}{dt^2}+k^2(\omega)\right\}X(t,\omega)=\{a(t,\omega)\frac{dZ}{dt}+b(t,\omega)\}X(t,\omega)+c(t,\omega),$$
$$X(0,\omega)=x_0,\quad X(1,\omega)=x_1\ (x=0,x_1\in R^1). \tag{1}$$

where the $Z(t,\omega), a(t,\omega), b(t,\omega), c(t,\omega)$ $t\in[0,1]$ are real square integrable stochastic processes defined on a probability space $(\Omega,\mathcal{F},P)$, and $k(\omega)$ is a random variable. We do not necessarily suppose that the process Z_t is differentiable in t, so that we need to understand this formal problem via an integral equation form as we did for the SDE (stochastic differential equation) in Itô's theory of stochastic calculus. So by taking the Green's function corresponding to the above problem, we may in a formal way rewrite the problem into the stochastic integral equation of the following form;

$$X(t)=f(t,\omega)+\int_0^1 L(t,s,\omega)X(s)ds+\int_0^1 K(t,s,\omega)X(s)d_\varphi Z_s. \tag{2}$$

Here the term $\int d_\varphi Z_t$ represents the noncausal stochastic integral with respect to a properly chosen orthonormal basis $\{\varphi_n\}$ in the real Hilbert space $L^2(0,1)$. Notice that this procedure of transformation is formal but the integral equation of Fredholm type (2) thus derived has a concrete meaning in the theory of noncausal stochastic calculus introduced by the author (cf. the introductory review given in Ref. 4).

As a natural generalization of such integral equation, we can think of the integral equation for the random fields, namely the stochastic integral equation for the processes with multi-dimensional parameter $t \in J = [0,1]^d \subset \mathbf{R}^d$.

$$X(t,\omega) = f(t,\omega) + \int_J L(t,s,\omega)X(s)ds + \int_J K(t,s,\omega)X(s)d_\varphi Z(s). \quad (3)$$

Notice that in such case the notion of *Causality* looses its sound meaning because there is no natural order of parameters in the multi-dimensional space, but the stochastic integral with respect to the random field $Z(t,\omega)$, $t \in \mathbf{R}^d$ can be treated in the framework of the noncausal theory without any difficulty in causal measurability of the integrands. Following the author's earlier articles,[1–3] we like to show in this note some important results concerning the basic questions of existence and that of uniqueness of solutions, for these two noncausal integral equations (2),(3). To each of these equations we will apply different methods to develop the discussion. In the next paragraph 2 we will discuss the case of the equation (2), and in the paragraph 3 the case of SIEs like (3) for the random fields.

Throughout the discussions, we fix a probability space $(\Omega, \mathcal{F}, P)$ and an underlying driving process with d-dimensional parameter $Z(t,\omega)$ $t \in [0,1]^d \subset \mathbf{R}^d (p \geq 1)$ defined on $(\Omega, \mathcal{F}, P)$, measurable in (t,ω) with respect to the σ field $\mathcal{B}_{R^d} \times \mathcal{F}$. Given these, we understand by the random functions or random fields, those real functions $f(t,\omega), t \in [0,1]^d (d \geq 1)$, measurable in (t,ω) and almost surely square integrable in t over the interval $[0,1]^d$. Throughout the whole discussions, we also assume that all kernels $L(t,s,\omega), K(t,s,\omega)$ $(s,t \in J = [0,1]^d, d \geq 1)$ are almost surely of Hilbert-Schmidt type, that is;

$$P\left[\int_{J\times J} \{L^2(t,s,\omega) + K^2(t,s,\omega)\} dsdt < \infty\right] = 1$$

A random kernel $G(t,s,\omega)$ will introduce the following integral operators acting on the set of random functions $X(t)$;

$$(GX)(t) = \int_J G(t,s,\omega)X(s)ds,$$
$$((GX))(t) = \int_J G(t,s,\omega)X(s)d_\varphi Z(s). \tag{4}$$

2. Uni-dimensional Case

We like to begin our study with the noncausal SIE of one dimensional parameter, that is the SIE for the random functions $X(t), t \in [0,1]$.

$$X(t) = f(t,\omega) + \int_0^1 L(t,s,\omega)X(s)ds + \int_0^1 K(t,s,\omega)X(s)d_\varphi Z_s, \tag{5}$$

Following the articles in Refs. 1,2 we will show in this paragraph some results, especially the fact that by virtue of the nice properties of our noncausal integral the above SIE can be solved in a very elementary way.

2.1. *Assumptions and notations*

As we have already remarked, the integral $\int d_\varphi Z_t$ should be understood in the sense of noncausal integral (i.e. the Ogawa integral) with respect to a properly chosen orthonormal basis $\{\varphi_n\} \in L^2([0,1])$. This means that, first of all the fundamental pair $(Z_t, \{\varphi_n\})$ should be nice enough so that the term

$$(\varphi_n, \dot{Z}_t) = \int_0^1 \varphi_n(t)dZ_t$$

is well defined and the definition of the noncausal integral

$$\int_0^1 f(t)d_\varphi Z_t := \sum_n^\infty (f,\varphi_n)(\varphi_n,\dot{Z}) \tag{6}$$

becomes meaningful.

Remark 2.1. This constraint can be realized in a natural way when we take as the basis $\{\varphi_n\}$ the system of Haar functions or those orthonormal bases with smooth elements.

If this random series in (6) converges in probability, we will say that the function f is integrable with respect to the basis $\{\varphi_n\}$ (or φ-integrable for short). Moreover, if the sequence of random functions $\int_0^t f(s)dZ_n^\varphi(s), 0 \le$

$t \leq 1$ converges (in probability) to the limit $\int_0^t f(s) d_\varphi Z(s)$ in the $L^2(0,1)$-sense,

$$\lim_{n \longrightarrow \infty} \int_0^1 dt \{ \int_0^t f(s) d_\varphi Z(s) - \int_0^t f(s) dZ_n^\varphi(s) \}^2 = 0,$$

we will say that the f is strongly integrable and will denote by $\mathbf{S}$ the totality of all such strongly integrable functions.
For any φ-integrable (in t) random function $f(t,\omega)$ or a kernel $G(t,s,\omega)$, we will use the following convenient notations to denote their noncausal integrals in the parameter t;

$$\tilde{f}(t,\omega) = \int_0^t f(s) d_\varphi Z(s), \ \tilde{G}(t,s,\omega) = \int_0^t G(r,s,\omega) d_\varphi Z(r).$$

About the choice of the pair $(Z_t, \{\varphi_n\})$ and the regularity of the functions f, K, L, we put the following assumptions (2.1);

Assumption 2.1.

(H,1) $\lim_{n \to \infty} Z_n^\varphi(\cdot) = Z(\cdot)$, $\lim_{n \to \infty} Z_n^\varphi(1) = Z(1)$ *in proba., where,*

$$Z_n^\varphi(t) = \sum_{k \leq n} (\dot{Z}, \varphi_k) \int_0^t \varphi_k(s) ds.$$

Once fixed such pair $(Z_t, \{\varphi_n\})$, we will understand by the noncausal integral the noncausal integral with respect to this pair and denote it simply by the notation, $\int d_* Z(t)$.

(H,2) $f(\cdot), K(\cdot,1) \in \mathbf{S}$ *and* $(LX), (KX), (K_s X) \in \mathbf{S}(^\forall X \in L^2)$, *where* $K_s(t,s) = \frac{\partial}{\partial s} K(t,s)$.

(H,3) $P[\tilde{K}(1,1) \neq 1] = 1$.

2.2. *Simple case*

We begin with the following simpler equation,

$$X(t) = f(t) + ((KX))(t), \ ((KX))(t) = \int_0^1 K(t,s) X(s) d_* Z(s) \quad (7)$$

Theorem 2.1. *Under the assumptions (2.1,1)-(2.1,3), there is a one-to-one correspondence between the $S-$solution X of (7) and the L^2-solution Y of the random integral equation (8) below;*

$$Y(t) = (\tilde{A} f)(t) + (\tilde{g} Y)(t), \quad (8)$$

where the term (Af) and the random kernel g are as follows,

$$
\begin{aligned}
(Af)(t) &= f(t) - \frac{\tilde{f}(1)K(t,1)}{\tilde{K}(1,1)-1} \\
g(t,s) &= \frac{\tilde{K}(t,1)\tilde{K}_s(1,s)}{\tilde{K}(1,1)-1} - K_s(t,s).
\end{aligned}
\tag{9}
$$

Proof. This can be easily verified by applying the integration by parts technique to the term $((KX))(t) = \int_0^1 K(t,s)X(s)d_*Z(s)$. In fact, taking the assumption (2.1) into account, we get

$$((KX))(t) = K(t,1)\tilde{X}(1) - \int_0^1 K_s(t,s)\tilde{X}(s)ds,$$

where, $\tilde{X}(t) = \int_0^t X(r)d_*Z(r)$. Substituting this into the equation (7) we have,

$$X(t) = f(t) + K(t,1)\tilde{X}(1) - \int_0^1 K_s(t,s)\tilde{X}(s)ds. \tag{10}$$

Now taking the stochastic integral over $[0,1]$ of the both sides, we get the following equality for the $\tilde{X}(t)$,

$$\tilde{X}(t) = \tilde{f}(t) + \tilde{K}(t,1)\tilde{X}(1) - \int_0^1 K_s(\tilde{t},s)\tilde{X}(s)ds. \tag{11}$$

In particular putting $t = 1$ we get the following expression for $\tilde{X}(1)$,

$$\tilde{X}(1) = \frac{\tilde{f}(1)}{1-\tilde{K}(1,1)} - \int_0^1 \frac{K_s(1,s)}{1-\tilde{K}(1,1)}\tilde{X}(s)ds.$$

Substituting this into the Eq. (11) we find,

$$\tilde{X}(t) = (\tilde{Af})(t) + (\tilde{g}\tilde{X})(t),$$

which shows that the $Y(t) = \tilde{X}(t)$ is a solution of the random integral Eq. (8).

Conversely given the L^2-solution $Y(\cdot)$ of the RIE (8), we put

$$X(t) = f(t) + K(t,1)Y(1) - \int_0^1 K_s(t,s)Y(s)ds.$$

Now taking the stochastic integral of both sides of the equation, and comparing with the Eq. (8) we see that $Y(t) = \tilde{X}(t)$. Hence we find that the X

defined as above satisfies the Eq. (10), and tracing back the integration by parts procedure we confirm that the X is the solution of the SIE (7). □

Since the RIE (8) is a family of the integral equations parametrized by ω and since the kernel $\tilde{g}(t,s,\omega)$ is of Hilbert-Schmidt type for almost all ω, we get the following result by a simple application of the Riesz-Schauder theory.

Corollary 2.1. *The SIE (7) has the unique* **S**-*solution* X, *provided that the homogeneous equation*

$$X(t) = \int_0^1 K(t,s)X(s)d_*Z(s)$$

does not have a nontrivial **S**-*solution.*

2.3. *General case*

Let us go back to the general case (5),

$$X(t) = f(t,\omega) + \int_0^1 L(t,s,\omega)X(s)ds + \int_0^1 K(t,s,\omega)X(s)d_*Z_s.$$

We continue to suppose the same assumptions (2.1). Then by following a similar argument that we have done for the simpler case (7), we see that to find the **S**-solution is equivalent to find the L^2-solution of the next random integral equation,

$$X(t) = (\mathcal{C}f)(t) + (\mathcal{B}X)(t), \tag{12}$$

where

$$\begin{aligned}
&(\mathcal{C}f)(t) = f(t) + \{(I-G)^{-1}(\tilde{A}f)\}(1)K(t,1) - \{K_s(I-G)^{-1}(\tilde{A}f)\}(t),\\
&\mathcal{B}(t,s,\omega) = L(t,s) + K(t,1)\{(I-G)^{-1}(\tilde{A}L(\cdot,s))\}(1) - \{K_s(I-G)^{-1}(\tilde{A}L(\cdot,s))\}\\
&G(t,s,\omega) = \tilde{g}(t,s,\omega).
\end{aligned}$$

Notice that the kernel $\mathcal{B}(t,s,\omega)$ is again of Hilbert-Schmidt type for almost all ω. Now suppose that the operator $(I-\mathcal{B})(\omega)$ is almost surely invertible. Such situation is realized for example when all eigen values of the $\mathcal{B}$ are non-atomic. Then by the discussion above we see that the Eq. (12) has the unique solution which must belong to the class **S**. The converse can be easily verified, hence we confirm the next result,

Proposition 2.1. *Under the assumptions (H), if the operator* $(I-\mathcal{B})$ *is almost surely invertible, the SIE (5) has the unique* **S**-*solution.*

3. Multi Dimensional Case

Let $Z(t,\omega)((t,\omega) \in R^d \times \Omega)$ be such that the derivative, $\dot{Z}(t,\omega) = \dfrac{\partial^d}{\partial t_1 \cdots \partial t_d} Z(t,\omega)$ is well defined as a $L^2(\Omega)$-valued generalized random field on the Schwartz space $\mathcal{S}(R^d)$. We suppose the Z to have nice property such that the application,

$$\mathbf{S} \ni \varphi(x) \Longrightarrow \dot{Z}(\phi) = (\dot{Z}, \varphi) \in L^2(\Omega),$$

becomes continuous with respect to the topology in $L^2(R^p)$. Thus the application can be extended over the $L^2(R^p)$. Now let $\{\varphi_n\}_{n=1}^{\infty}$ be a complete orthonormal basis in the real Hilbert space $L^2(J)$.

Definition 3.1. The stochastic integral $\int f(t,\omega) d_\varphi Z(t)$ of a random field $f(t,\omega)$ with respect to the pair $(Z, \{\varphi_n\})$ is defined as being the limit in probability of the following random series,

$$\sum_{n=1}^{\infty} (f, \varphi_n)(\varphi_n, \dot{Z}).$$

In this paragraph, we are going to study the basic properties of the SIE of Fredholm type (3) for the random fields,

$$X(t,\omega) = f(t,\omega) + \int_J L(t,s,\omega) X(s) ds + \int_J K(t,s,\omega) X(s) d_\varphi Z(s)$$

and show some results mainly following the article in Ref. 3.

For the SIE in one-dimensional parameter case we have solved the equation by applying the integration by parts method, which does not work for such equation of multi-dimensional parameters. Thus to solve the above SIE we need to introduce another technique, a kind of stochastic Fourier transformation $\mathcal{T}_\epsilon$. This can be done when we suppose a kind of smoothness of the stochastic kernels and functions involved in the equation.

3.1. *Stochastic Fourier transformation*

For the simplicity of discussions, we will fix once for all, another orthonormal basis, $\{\psi_n\}$ in an arbitrary way and we set the next assumption (3.1) which concerns a regularity of the random kernels, K, L.

Assumption 3.1. *There exists a positive sequence $\{\epsilon_n\}$ such that,*

(3.1,1) $\{\epsilon_n \epsilon_m \gamma_{m,n}\} \in l^2$ *(P-a.s), where* $\gamma_{m,n} = \int \psi_m(t)\psi_n(t) d_\varphi Z(t)$,

(3.1,2) $\{k'_{m,n}\}, \{l'_{m,n}\} \in l^2$ *(P-a.s.) where* $k'_{m,n} = k_{m,n}/\epsilon_m \epsilon_n$, $l'_{m,n} = l_{m,n}/\epsilon_m$, $k_{m,n} = (K, \psi_m \otimes \psi_n)$, $l_{m,n} = (L, \psi_m \otimes \psi_n)$.

We will call such sequence $\{\epsilon_n\}$ *the admissible weight.*

Notice that if $\{\epsilon_n\}$, $\{\eta_n\}$ are admissible weights then the sequences, $\{(\epsilon \wedge \eta)_n\}$, $\{(\epsilon \vee \eta)_n\}$, given by $(\epsilon \wedge \eta)_n = \min\{\epsilon_n, \eta_n\}$, $(\epsilon \vee \eta)_n = \max\{\epsilon_n, \eta_n\}$, are also admissible.

Example 3.1 (Brownian sheet). *In the case that* Z *is the Brownian sheet and* $\{\psi_n\}$ *is such that all elements are uniformly bounded on* J, *then any positive* l^2-*sequence satisfies the condition (3.1,1).*

Definition 3.2 (ϵ-smoothness). *We will say that a random field* $g(t, \omega)$ *admits a sequence* $\{\epsilon_n\}$ *as the weight (or shortly,* $\{\epsilon_n\}$*-smooth) if there exists an admissible weight* $\{\epsilon_n\}$*, such that ;*

(t,1) The integral $\hat{g}_n = \int g(t, \omega)\psi_n(t) d_\varphi Z(t)$ *exists for all* $n \in \mathbf{N}$ *and* $\{\epsilon_n \hat{g}_n\} \in l^2(P - a.s.)$.

(t,2) The following limit converges in probability,

$$\lim_{m \to \infty} \sum_{n=1}^{\infty} \{\epsilon_n (\hat{g}_n - \int g(t, \omega)\psi_n(t) dZ_m^\varphi(t))\}^2 = 0.$$

We will denote by $\mathbf{S}_2$ *the totality of all such random fields that are* $\{\epsilon\}$*-smooth for some admissible weight* $\{\epsilon_n\}$.

It is easy to check that if a $\mathbf{S}_2$-field $g(t, \omega)$ admits two sequences, $\{\epsilon_n\}, \{\eta_n\}$ as the weights, then it also admits the sequences $\{(\epsilon \wedge \eta)_n\}, \{(\epsilon \vee \eta)_n\}$ as the weights.

Remark 3.1. In the case that $Z =$ the Brownian sheet and the all elements of $\{\psi_n\}$ are uniformly bounded, we see that $\mathbf{S}_2 \supset L^2(J)$ and that every admissible sequence can be the weight for any $g(t, \omega) \in L^2(J)$.

Associated to the notion of $\mathbf{S}_2$-fields, we introduce the linear stochastic transformation, $\mathcal{T}_\epsilon$ acting on $\mathbf{S}_2$, in the following,

Definition 3.3 (Stochastic Fourier transformation). *For a* $g(t, \omega) \in$

$\mathbf{S}_2$ *admitting a* $\{\epsilon_n\}$ *as the weight, we set,*

$$
\begin{aligned}
&(\mathcal{T}_\epsilon g)(t) = \sum_n \epsilon_n \hat{g}_n \psi_n(t), \\
&\textit{where} \quad \hat{g}_n(\omega) = \int_J g(t,\omega)\psi_n(t) d_\varphi Z(t).
\end{aligned}
\tag{13}
$$

We should notice that the transformation $\mathcal{T}_\epsilon$ depends on the weight $\{\epsilon_n\}$ and that for any $\{\epsilon\}$-smooth g, we have $\mathcal{T}_\epsilon g \in L^2(J), (P - a.s.)$.

3.2. *Results*

Theorem 3.1. *For any* $f(t,\omega) \in \mathbf{S}_2$ *the following integral equation*

$$X(t,\omega) = f(t,\omega) + \int_J K(t,s,\omega)X(s)d_\varphi Z(s), \tag{14}$$

has the unique $\mathbf{S}_2$*-solution provided that the next condition (C) holds,*

> *(C); the homogeneous equation,* $X(t,\omega) = \int_J K(t,s,\omega)X(s)d_\varphi Z(s)$, *does not have nontrivial* $\mathbf{S}_2$*-solutions.*

Proof. Let $\{\epsilon_n\}$ be an admissible weight for the random field $f(t,\omega)$. First we are going to show that the condition (C) is sufficient to assure the existence of a $\mathbf{S}_2$-solution X, which is unique among those functions that admit the same weight $\{\epsilon_n\}$.

Let X be an $\{\epsilon\}$-smooth solution of (14). Then, since

$$K(t,s,\omega) = \sum_{m,n} k_{m,n}\psi_m(t)\psi_n(s),$$

we get the following relation (15) by virtue of the condition (t,2),

$$
\begin{aligned}
&X(t) = f(t) + \sum_{m,n} \epsilon_m \epsilon_n k'_{m,n} \psi_m(t) \hat{x}_n, \\
&\text{where} \quad \hat{x}_n(\omega) = \int_J X(t,\omega)\psi_n(t) d_\varphi Z(t).
\end{aligned}
\tag{15}
$$

Multiplying by $\psi_l(t)$ and taking the stochastic integration over J on both sides of the Eq. (15), we obtain, under the assumption (3.1,2) the next relation,

$$\hat{x}_l = \hat{f}_l + \sum_{m,n} \gamma_{l,m} k_{m,n} \hat{x}_n, \ (\forall l \in N) \tag{16}$$

or equivalently,

$$\epsilon_l \hat{x}_l = \epsilon_l \hat{f}_l + \sum_{m,n} \epsilon_l \epsilon_m \gamma_{l,m} k'_{m,n} \epsilon_n \hat{x}_n. \tag{17}$$

So if we set,

$$\bar{K}(t,s,\omega) = \sum_{l,n} \epsilon_l \{ \sum_m \epsilon_m \gamma_{l,m} k'_{m,n} \} \psi_l(t) \psi_n(s),$$

then by virtue of the condition (t,1), we see that the kernel $\bar{K}(\cdot,\cdot,\omega)$ is of Hilbert-Schmidt type for almost all ω and that the field, $Y = (\mathrm{T}_\epsilon X)(t,\omega)$ satisfies the following random integral equation,

$$Y(t) = (\mathcal{T}_\epsilon f)(t) + \int_J \bar{K}(t,s,\omega) Y(s) ds. \tag{18}$$

Conversely if we set $\hat{x}_n = (Y, \psi_n)/\epsilon_n$ for an L^2- solution Y of (18), then we see that the $\{\hat{x}_n\}$ satisfies the Eq. (16) and so the field $X(t)$ defined through the relation (15) becomes an $\mathbf{S}_2$−solution of (14). As is easily seen, this correspondence between the $\{\epsilon\}$-smooth solution of (14) and the L^2−solution of (18) is one-to-one and onto. Thus the question of the existence and the uniqueness of the $\{\epsilon\}$-smooth solution is reduced to the same question about the L^2−solutions of (18). Hence, by a simple application of The Riesz-Schauder Theory, we confirm that the condition (C) is sufficient for the validity of the prescribed result.

Next, we are going to show that this solution X which has the $\{\epsilon_n\}$ as the weight is unique among all $\{\epsilon\}$- smooth fields. So let X' be another $\mathbf{S}_2$−solution of (14) having a different sequence $\{\eta\}$ as the weight. Then it satisfies a similar relation as (15) from which we see the field $f(t,\omega)$ is $\{\eta\}$-smooth. Since all $\mathbf{S}_2$-fields f, X and X' are $\{(\epsilon \wedge \eta)\}$-smooth, the field X' and X must coincide with each other as the unique $\mathbf{S}_2$-solution admitting the same sequence as the weight. □

Corollary 3.1. *If all elements of the c.o.n.s. $\{\psi_n\}$ are continuous and uniformly bounded over J and if almost all sample of the field $f(t,\omega)$ are continuous. Then the $\mathbf{S}_2$-solution of (14) is also almost surely sample-continuous.*

Proof. Evident from the equality (15) and the fact that,

$$\sum_{m,n} |\epsilon_m k'_{m,n} \epsilon_n \hat{x}_n| < +\infty \text{ (P-a.s.)}$$

□

Now we are to discuss the general case (3). First we notice that the condition (A,2) implies ;

$$(LX)(t,\omega) = \int_J L(t,s,\omega)X(s)ds \in \mathbf{S}_2$$

for any random field X. Let $\{\epsilon_n\}$ be a weight for the $f(t,\omega)$. Then following the same discussion as in the proof of Theorem (3.1), we see that any $\mathbf{S}_2$-solution X admitting the $\{\epsilon_n\}$ as weight, if exists, satisfies the following equation,

$$Y(t,\omega) = \{\mathcal{T}_\epsilon(f + LX)\}(t,\omega) + \int_J \tilde{K}(t,s,\omega)Y(s)ds, \tag{19}$$

where $Y(t,\omega) = (\mathcal{T}_\epsilon X)(t,\omega)$.

Since the operator $\tilde{K}(\omega)$, given by ;

$$(L^2(J) \ni)Y \longrightarrow (\tilde{K}Y)(t,\omega) = \int_J \tilde{K}(t,s,\omega)Y(s)ds \ (\in L^2(J)),$$

is compact for almost all ω. So in the case that the $(I - \tilde{K})$ is almost surely invertible, we get by solving (19) in Y, the following expression,

$$(\mathcal{T}_\epsilon X)(t) = f_1(t) + (L'X)(t) \tag{20}$$

where

$$f_1(t) = (I - \tilde{K})^{-1}(\mathcal{T}_\epsilon f)(t)$$

and

$$(L'X)(t) = \{(I - \tilde{K})^{-1}(LX)\}(t).$$

On the other hand we have the next relation which can be derived in a same way as in the derivation of the (15);

$$X(t) = f(t) + (LX)(t) + (K_1)\mathcal{T}_\epsilon X)(t) \tag{21}$$

where

$$(K_1Y)(t) = \int_J K_1(t,s,\omega)Y(s)ds,$$

$$K_1(t,s,\omega) = \sum_{m,n}(k_{m,n}/\epsilon_n)\psi_m(t)\psi_n(s).$$

Substituting the relation (20) into (21), we get the next,

Proposition 3.1. *If $f(t,\omega) \in \mathbf{S}_2$ and if the operator $(I - \tilde{K})$ is almost surely invertible, then the problem of finding the $\mathbf{S}_2$ solution of the Eq. (3) is reduced to the problem of finding the L^2 solution of the following random integral equation;*

$$X(t) = f_2(t) + (L''X)(t) \tag{22}$$

where

$$f_2(t) = f(t) + (K_1 f_1)(t), \qquad \textit{and}$$

$$(L''Y)(t) = \{(L + K_1 L')Y\}(t) \ (Y \in L^2(J)).$$

Notice again that if the operator $(I - L'')$ is almost surely invertible, the Eq. (22) has a unique $\mathbf{S}_2$-solution and it is immediate to see that this solution does not depend on the choice of the weight $\{\epsilon_n\}$ for the $f(t,\omega)$.

References

1. S. Ogawa, On the stochastic integral equation of Fredholm type. in *Waves and Patterns*, (Kinokuniya and North-Holland, 1986), pp.597-605.
2. S. Ogawa, Topics in the theory of noncausal stochastic calculus, in *Diffusion Processes and Related Problems in Analysis, Vol. 1*, (edt. Mark Pinsky, Birkhäuser Boston Inc., 1990), pp.411-420.
3. S. Ogawa, On a stochastic integral equation for the random fields, in *Séminairs de Probabilités - xxv*, (Springer Verlag, Berlin, Heidelberg, 1991), pp.324-339
4. S. Ogawa, Noncausal Stochastic Calculus Revisitted – around the So-called Ogawa integral, *(to appear in)* Lecture Notes of the Quynhon Symposium, (2005).

Harmonic, Wavelet and p-Adic Analysis
Eds. N. M. Chuong *et al.* (pp. 343–361)

§16. BSDES WITH JUMPS AND WITH QUADRATIC GROWTH COEFFICIENTS AND OPTIMAL CONSUMPTION

SITU RONG

Department of Mathematics, Zhongshan University,
Guangzhou 510275, China
mail: mcsstr@zsu.edu.cn

We obtain the existence and uniqueness of solutions of some BSDEs with quadratic growth coefficients and with jumps. Then the results are applied to get the existence of optimal consumptions for some stochastic consumption problem in financial markets.

Backward stochastic differential equation (BSDE) is a powerful tool in studying financial markets.[1,4,7] In some financial markets with contrains on portfolios to price some contingent claims will lead to continuous quadratic BSDEs.[3] For the case that the coefficient of the continuous BSDE has a less than linear growth in y, where y_t will be the solution of the BSDE, and the coefficient has a quadratic growth in q, where q_t is the integrand of the continuous stochastic integral in the BSDE, the existence and uniqueness results on solutions are discussed in Ref. 2. Here we will obtain the existence and uniqueness of solutions of some BSDEs with junps such that the coefficients have a quadratic growth in both q and y, and a non-linear growth in p, where p_t will be the integrand for the jump stochastic integral. Such results are different from Refs. 2,9 and all others. Then the results are applied to get the existence of some optimal consumption for some stochastic consumption problem maybe happened in some financial markets.

1. Conception

Consider the following backward stochastic differential equation (BSDE) with jumps[4] in R^1 :

$$x_t = X + \int_t^T b(s, x_s, q_s, p_s, \omega)ds - \int_t^T q_s dw_s$$
$$- \int_t^T \int_Z p_s(z)\tilde{N}_k(ds, dz), \quad 0 \leqslant t \leq T, \tag{1}$$

where $w_t^T = (w_t^1, \cdots, w_t^{d_1}), 0 \leqslant t$, is a d_1-dimensional standard Brownian motion (BM), w_t^T is the transpose of w_t; and for simplicity, k_t is a 1−dimensional stationary Poisson point process, $\tilde{N}_k(ds, dz)$ is the Poisson martingale measure generated by k_t satisfying

$$\tilde{N}_k(ds, dz) = N_k(ds, dz) - \pi(dz)ds,$$

where $\pi(.)$ is a σ-finite measure on a measurable space $(Z, \mathfrak{B}(Z))$, $N_k(ds, dz)$ is the Poisson counting measure generated by k_t, and $\mathfrak{F}_t$ is the $\sigma-$algebra generated (and completed) by

$$w_s, k_s, s \leqslant t;$$

and the terminal condition is given as $X \in \mathfrak{F}_T$.

For the precise definition of solution of (1) we need the following notation.

$$S^2_{\mathfrak{F}}(R^1) = \left\{ \begin{array}{c} f(t,\omega) : f(t,\omega) \text{ is } \mathfrak{F}_t - \text{adapted}, R^1 - \text{valued such that} \\ E \sup_{t\in[0,T]} |f(t,\omega)|^2 < \infty \end{array} \right\},$$

$$L^2_{\mathfrak{F}}(R^{1\otimes d_1}) = \left\{ \begin{array}{c} f(t,\omega) : f(t,\omega) \text{ is } \mathfrak{F}_t - \text{adapted}, R^{1\otimes d_1} - \text{valued} \\ \text{such that } E\int_0^T |f(t,\omega)|^2 dt < \infty \end{array} \right\},$$

and

$$\mathcal{F}^2_{\mathfrak{F}}(R^1) = \left\{ \begin{array}{c} f(t,z,\omega) : f(t,z,\omega) \text{ is } R^1 - \text{valued}, \mathfrak{F}_t\text{-predictable} \\ \text{such that } E\int_0^T \int_Z |f(t,z,\omega)|^2 \pi(dz)dt < \infty \end{array} \right\}.$$

Now we can give the definition for the solution.[4,8]

Definition 1.1. . (x_t, q_t, p_t) is said to be a solution of (1), iff $(x_t, q_t, p_t) \in S^2_{\mathfrak{F}}(R^1) \times L^2_{\mathfrak{F}}(R^{1\otimes d_1}) \times \mathcal{F}^2_{\mathfrak{F}}(R^1)$, and it satisfies (1).

From Definition 1.1 it is seen that for discussing the solution of (1) we always need to assume that b satisfies the following assumption

(A)$_1$

$$b : [0,T] \times R^1 \times R^{1\otimes d_1} \times L^2_{\pi(\cdot)}(R^1) \times \Omega \to R^d$$

is jointly measurable, $\mathfrak{F}_t-$adapted, where

$$L^2_{\pi(\cdot)}(R^1) = \left\{ \begin{array}{c} f(z) : f(z) \text{ is } R^1 - \text{valued, and} \\ \|f\|^2 = \int_Z |f(z)|^2 \, \pi(dz) < \infty \end{array} \right\}.$$

One also sees that different from the solution of the usual stochastic differential equation with a given initial condition here we need three stochastic processes (x_t, q_t, p_t), where (x_t, q_t) are $\mathfrak{F}_t-$ adapted, and p_t is $\mathfrak{F}_t-$predictable, to satisfy one BSDE due to the termianal condition is given: $X \in \mathfrak{F}_T$. Roughly speaking, the financial meaning of a BSDE and its solution is as follows: If we explain x_t as a wealth process of a small investor in a financial market with some continuous and jump perturbation, and explain X as a future wealth target of the small investor at a future time T, then his wealth process in the financial market will satisfy some BSDE with jumps as (1). The solution (x_t, q_t, p_t) of this BSDE is just the right investment and portfolio of the small investor at time t, where x_t is the right total investment which should be made, and (q_t, p_t) is the right portfolio in this investment such that, as time t evolves, this right investment can help the investor to arrive at his target X at the future time T.[5,7,8]

2. A Comparison Theorem and An Approximation Lemma

Firstly, we will give a comparison theorem on solutions of BSDEs with jumps. Actually, it is very useful not only in here.

Suppose that $(x_t^i, q_t^i, p_t^i), i = 1, 2$, are solutions of the following BSDEs for $t \in [0, T]; i = 1, 2$

$$x_t^i = X^i + \int_t^T b^i(s, x_s^i, q_s^i, p_s^i, \omega) ds - \int_t^T q_s^i dw_s - \int_t^T \int_Z p_s^i(z) \tilde{N}_k(ds, dz). \tag{2}$$

Theorem 2.1. *(Comparison Theorem). Assume that*

1°

$$b^1(t, x_t^2, q_t^2, p_t^2, \omega) \geqslant b^2(t, x_t^2, q_t^2, p_t^2, \omega),$$

where (x_t^2, q_t^2, p_t^2) *is the solution of (2) with* $i = 2$;

2°

$$|b^1(t, x_1, q_1, p, \omega) - b^1(t, x_2, q_2, p, \omega)| \leqslant c_1(t) |x_1 - x_2| + c_2(t) |q_1 - q_2|,$$

$$|b^1(t, x, q, p_1, \omega) - b^1(t, x, q, p_2, \omega)| \leqslant \int_Z |C_t(z, \omega)| \, |(p_1(z) - p_2(z))| \, \pi(dz),$$

where $c_1(t)$ *and* $c_2(t)$ *are non-negative and non-random such that*

$$\int_0^T c_1(t)dt + \int_0^T c_2(t)^2 dt < \infty;$$

and $C_t(z,\omega)$ *satisfies the condition*
(B)

$$|C_t(z,\omega)| \leqslant 1, C_t(z) \in \mathcal{F}^2_{\mathfrak{F}}(R^1);$$

3°

$$X^i \in \mathfrak{F}_T, E\left|X^i\right|^2 < \infty, i = 1, 2,$$
$$X^1 \geqslant X^2.$$

Then there exists a probability measure $\widetilde{P}$, *which is equivalent to* P *such that* $\widetilde{P} - a.s.$

$$x_t^1 - x_t^2 \geq E^{\widetilde{P}}[X^1 - X^2|\mathfrak{F}_t]e^{-\int_t^T c_1(s)ds} \geq 0, \forall t \in [0,T],$$

where $E^{\widetilde{P}}[\cdot|\mathfrak{F}_t]$ *is the conditional expectation under the probability* $\widetilde{P}$.

Theorem 2.1 can be proved by Girsanov's theorem and taking limit as in Ref. 4. It has an immediate application as follows. If we take

$$b^1(t,x,q,p,\omega) = b(t,x,q,p,\omega) + c^1(t),$$
$$b^2(t,x,q,p,\omega) = b(t,x,q,p,\omega) + c^2(t),$$

where b satisfies the condition 2° in Theorem 2.1, then we will have the following result: For $X^i \in \mathfrak{F}_T, E\left|X^i\right|^2 < \infty, \int_0^T \left|c^i(t)\right| dt < \infty, i = 1, 2$

$$X^1 \geqslant X^2, \text{ and } c^1(t) \geq c^2(t), t \in [0,T]$$

$$\Longrightarrow x_t^1 \geq x_t^2, \forall t \in [0,T].$$

Now suppose that $c^2(t) \geq 0$, and we explain $c^i(t)$ as the consumption process of the small investor. Then an immediate financial meaning of the above result is as follows:[5,7,8]

For a small investor in the financial market, if he wants to consume more all the time or arrive at a higher level target at the future time, then he must invest more now.

Secondly, let us introduce an approximation lemma which is useful later in deriving the existence of solutions to BSDE with non-linear coefficients.

Lemma 2.1. *Assume that $b(t,x,q,p,\omega)$ satisfies conditions*
1° $b(t,x,q,p,\omega)$ is jointly continuous in $(x,q,p) \in R^1 \times R^{1\otimes d_1} \times L^2_{\pi(.)}(R^1) \backslash G$, where $G \subset R^1 \times R^{1\otimes d_1} \times L^2_{\pi(.)}(R^1)$ is a Borel measurable set such that $(x,q,p) \in G \Longrightarrow |q| > 0$, and $m_1 G_1 = 0$, where $G_1 = \{x : (x,q,p) \in G\}$, and m_1 is the Lebesgue measure in R^1; moreover, $b(t,x,q,p,\omega)$ is a separable process with respect to (x,q), (i.e. there exists a countable set $\{(x_i,q_i)\}_{i=1}^{\infty}$ such that for any Borel set $A \subset R^1$ and any open rectangle $B \subset R^1 \times R^{1\otimes d_1}$ the $\omega-$sets

$$\{\omega : b(t,x,q,p,\omega) \in A, (x,q) \in B\},$$
$$\{\omega : b(t,x,q,p,\omega) \in A, (x_i,q_i) \in B, \forall i\}$$

only differs a zero-probability $\omega-$set;
2°

$$|b(t,x,q,p,\omega)| \leqslant c_1(t)(1+|x|) + c_2(t)(|q| + \|p\|),$$
$$|b(t,x,q,p_1,\omega) - b(t,x,q,p_2,\omega)| \leqslant \int_Z |C_t(z,\omega)(p_1(z) - p_2(z))| \, \pi(dz),$$

where $c_1(t), c_2(t) \geqslant 0$ are non-random such that $\int_0^{T_0}(c_1(t) + c_2(t)^2)dt < \infty$, and $C_t(z,\omega)$ satisfies that
$C_t(z,\omega) \in \mathcal{F}^2_{\mathfrak{F}}(R^1)$, $|C_t(z,\omega)| \leqslant 1$.
Let

$$b_n(t,x,q,p,\omega) = \inf_{y,\widetilde{q}}(b(t,y,\widetilde{q},p,\omega) + (n \vee c_1(t))\,|y-x| + (n \vee c_2(t))\,|q-\widetilde{q}|), n \geqslant 1.$$

Then $\forall n$, b_n is still $\mathfrak{F}_t-$ adapted, and
1)

$$b_n(t,x,q,p,\omega) \uparrow \text{ , as } n \uparrow, \forall t \geqslant 0, (x,q,p) \notin G;$$
$$|b_n(t,x,q,p,\omega)| \leqslant c_1(t)(1+|x|) + c_2(t)(|q| + \|p\|);$$

2)

$$|b_n(t,x,q,p,\omega) - b_n(t,y,q',p,\omega)| \leqslant (n \vee c_1(t))(|y-x| + (n \vee c_2(t))\,|q-q'|),$$

3) $b_n(t,x,q,p,\omega)$ still satisfies the second inequality in condition 2° above, for all $n \geqslant 1$;
4)

$$(x_n,q_n,p_n) \to (x,q,p) \text{ in } R^1 \times R^{1\otimes d_1} \times L^1_{\pi(.)}(R^{1\otimes d_2})$$
$$\Rightarrow b_n(t,x_n,q_n,p_n,\omega) \to b(t,x,q,p,\omega). \forall t \geqslant 0, (x,q,p) \notin G.$$

This Lemma can be proved by using Theorem 2.1 similarly as the proof of Lemma 51 in Ref. 4. Roughly speaking, this lemma tells us that a jointly continuous coefficient (may have some discontinuous points) under some mild appropriate conditions can be monotonely approximated by a sequence of Lipschitzian coefficients.

3. Existence of Solution for BSDE with Quadratic Growth in *q*

In this section we will give the idea: how to derive solutions for BSDE with quadratic growth coefficients in q.

Assume it is satisfied that assumption (A):

$\widetilde{1}^\circ$

$$|b(t,x,q,p,\omega)| \leqslant c_1(t)(1+|x|) + \widetilde{c}_2(t)(1+|q|+\|p\|),$$

where $c_1(t) \geqslant 0$ and $\widetilde{c}_2(t) \geqslant 0$ are non-random such that

$$\int_0^T c_1(t)dt + \int_0^T \widetilde{c}_2(t)^2 dt < \infty;$$

$\widetilde{2}^\circ$

$$\begin{aligned}|b(t,x_1,q_1,p,\omega) - b(t,x_2,q_2,p,\omega)| &\leqslant c_1(t)\,|x_1-x_2| \\ &\quad + \widetilde{c}_2(t)[|q_1-q_2| + \|p_1-p_2\|],\end{aligned}$$

where $c_1(t)$ and $\widetilde{c}_2(t)$ satisfy the same conditions in $\widetilde{1}^\circ$, furthermore,

$$|b(t,x,q,p_1,\omega) - (b(t,x,q,p_2,\omega)| \leqslant \int_Z C_t(z)\,|(p_1(z)-p_2(z))|\,\pi(dz)$$

and $0 \leq C_t(z) \in \mathcal{F}^2_{\mathfrak{F}}(R^1), C_t(z) \leq 1$;

$\widetilde{3}^\circ$

$$b(s,0,0,0,\omega) \geq 0,$$

$\widetilde{4}^\circ$

$$X \in \mathfrak{F}_T, E\,|X|^2 < \infty, \text{ and } X \geqslant \widetilde{r}_0 > 0.$$

Then there exists a unique solution (x_t, q_t, p_t) solving the following BSDE[4]

$$x_t = X + \int_t^T b(s,x_s,q_s,p_s,\omega)ds - \int_t^T q_s dw_s - \int_t^T \int_Z p_s(z)\widetilde{N}_k(ds,dz), \quad (3)$$

and by the comparison theorem (Theorem 2.1) there exists a constant $r_0 = \widetilde{r}_0 e^{-\int_0^T c_1(t)dt} > 0$ such that

$$x_t \geqslant r_0, \quad \forall t \in [0,T].$$

Now let

$$y_t = 1/x_t.$$

Then $0 < y_t \leqslant 1/r_0$, and by Ito's formula

$$dy_t = -\frac{1}{x_{t-}^2} dx_t + \frac{1}{2} \cdot 2 \frac{1}{x_t^3} \left| q_t^2 \right| dt + \int_Z \left(\frac{1}{x_{t-} + p_t(z)} - \frac{1}{x_{t-}} + \frac{1}{x_{t-}^2} p_t(z) \right) N_k(dt, dz).$$

Let us discuss the last term in this formula. Notice that

$$x_t = x_{t-} + \int_Z p_t(z) \widetilde{N}_k(\{t\}, dz) \geqslant r_0 > 0, \forall t \in [0,T],$$

$$\frac{1}{x_{t-} + p_t(z)} - \frac{1}{x_{t-}} + \frac{1}{x_{t-}^2} p_t(z) = \frac{p_t(z)^2}{x_{t-}^2 (x_{t-} + p_t(z))},$$

and

$$\int_0^t \int_Z \frac{p_t(z)^2}{x_{t-}^2 (x_{t-} + p_t(z))} N_k(dt, dz) = \sum_{0 < \sigma_m \leqslant t} \frac{p_{\sigma_m}(k(\sigma_m))^2}{x_{\sigma_m -}^3 (1 + \frac{p_{\sigma_m}(k(\sigma_m))}{x_{\sigma_m -}})},$$

where the domain of point process $k(.)$ is

$$D_k = \{0 < \sigma_1 < \sigma_2 < \cdots < \sigma_m < \cdots\}.$$

and $1 + \frac{p_{\sigma_m}(k(\sigma_m))}{x_{\sigma_m -}} > 0$, $\forall m$; since

$$0 < r_0 \leqslant x_{\sigma_m} = x_{\sigma_m -} \left(1 + \frac{p_{\sigma_m}(k(\sigma_m))}{x_{\sigma_m -}}\right). \tag{4}$$

So,

$$E \int_0^t \int_Z \left| \frac{p_t(z)^2}{x_{t-}^2 (x_{t-} + p_t(z))} \right| N_k(dt, dz) = E \int_0^t \int_Z \frac{p_t(z)^2}{x_{t-}^2 (x_{t-} + p_t(z))} N_k(dt, dz)$$

$$\leq \frac{1}{r_0^3} E \int_0^t \int_Z p_t(z)^2 N_k(dt, dz) = \frac{1}{r_0^3} E \int_0^t \int_Z p_t(z)^2 \pi(dz) dt < \infty.$$

Hence, the last term in the above Ito's formula is finite and we can rewrite the expression as

$$\begin{aligned}
dy_t =&[y_t^2 b(t,1/y_t,q_t,p_t,\omega)+y_t^3\left|q_t\right|^2]dt-y_t^2 q_t dw_t\\
&-\int_Z y_{t-}^2 p_t(z)\widetilde{N}_k(dt,dz)+\int_Z \frac{y_{t-}^3 p_t(z)^2}{1+y_{t-}p_t(z)}N_k(dt,dz)\\
=&[y_t^2 b(t,1/y_t,q_t,p_t,\omega)+y_t^3\left|q_t\right|^2+\int_Z \frac{y_{t-}^3 p_t(z)^2}{1+y_{t-}p_t(z)}\pi(dz)]dt\\
&-y_t^2 q_t dw_t-\int_Z \frac{y_{t-}^2 p_t(z)}{(1+y_{t-}p_t(z))}\widetilde{N}_k(dt,dz),\\
y_T =&1/X.
\end{aligned}$$

Denote

$$\widetilde{q}_t=-y_t^2 q_t,\widetilde{p}_t(z)=-\frac{y_{t-}^2 p_t(z)}{(1+y_{t-}p_t(z))}. \tag{5}$$

By this one sees that

$$p_t(z)(\widetilde{p}_t(z)+y_{t-})y_{t-}=-\widetilde{p}_t(z).$$

Notice that $\widetilde{p}_t(z)+y_{t-}=0$ is impossible. Otherwise, we have $\widetilde{p}_t(z)=0$. Hence $y_{t-}=0$. It is a contradiction. So we have that

$$p_t(z)=-\frac{\widetilde{p}_t(z)}{(\widetilde{p}_t(z)+y_{t-})y_{t-}}.$$

By (5) one also sees that

$$q_t=-\widetilde{q}_t/y_t^2.$$

Furthermore,

$$\begin{aligned}
&0<y_t\leqslant 1/r_0,\\
&E\int_0^T\left|\widetilde{q}_t\right|^2 dt=E\int_0^T y_t^4\left|q_t\right|^2 dt\leq\frac{1}{r_0^4}E\int_0^T\left|q_t\right|^2<\infty,\\
&E\int_0^T\int_Z\left|\widetilde{p}_t(z)\right|^2\pi(dz)dt=E\int_0^T\int_Z\left|\widetilde{p}_t(z)\right|^2 N_k(dt,dz)\\
=&E\sum_{0<\sigma_m\leqslant t}\left|\widetilde{p}_{\sigma_m}(k(\sigma_m))\right|^2
\end{aligned}$$

$$
\begin{aligned}
&= E \sum_{0<\sigma_m \leqslant t} y^2_{\sigma_m -} \left| \frac{p_{\sigma_m}(k(\sigma_m))}{(x_{\sigma_m -} + p_{\sigma_m}(k(\sigma_m)))} \right|^2 \\
&\leq r_0^{-4} E \sum_{0<\sigma_m \leqslant t} |p_{\sigma_m}(k(\sigma_m))|^2 \\
&= r_0^{-4} E \int_0^T \int_Z |p_t(z)|^2 \, \pi(dz)dt < \infty.
\end{aligned}
$$

Then we arrive at the following Theorem.

Theorem 3.1. *Under $\widetilde{1}^\circ - \widetilde{4}^\circ$ of assumption (A) for b and X above then there exists a solution $(y_t, \widetilde{q}_t, \widetilde{p}_t)$ with $0 < y_t \leqslant 1/r_0$ and $\widetilde{p}_t(z) + y_{t-} \neq 0$ satisfying the following BSDE with jumps:*

$$
dy_t = -\widetilde{b}(t, y_t, \widetilde{q}_t, \widetilde{p}_t, \omega)dt + \widetilde{q}_t dw_t + \int_Z \widetilde{p}_t(z)\widetilde{N}_k(dt, dz), \; y_T = 1/X, \tag{6}
$$

where $r_0 = \widetilde{r}_0 e^{-\int_0^T c_1(t)dt} > 0$ and

$$
\begin{aligned}
-\widetilde{b}(t, y, \widetilde{q}, \widetilde{p}, \omega) =& y^2 b(t, I_{y\neq 0} 1/y, -I_{y\neq 0}\widetilde{q}/y^2, -I_{y\neq 0} I_{\widetilde{p}(\cdot)+y\neq 0} \frac{\widetilde{p}(\cdot)}{(\,\widetilde{p}(\cdot) + y)y}, \omega) \\
&+ y^{-1} |\widetilde{q}|^2 I_{y\neq 0} + \int_Z \frac{\widetilde{p}(z)^2}{\widetilde{p}(z) + y} I_{\widetilde{p}(z)+y\neq 0} \pi(dz),
\end{aligned}
$$

has a quadratic growth in $\widetilde{q}$, and it is unbounded in y, as y closes to 0, and it is also non-linear in p, and $I_{y\neq 0} I_{\widetilde{p}(\cdot)+y\neq 0} \frac{\widetilde{p}(\cdot)}{(\,\widetilde{p}(\cdot)+y)y}$ is a function defined by

$$
I_{y\neq 0} I_{\widetilde{p}(z)\ +y\neq 0} \frac{\widetilde{p}(z)}{(\,\widetilde{p}(z)+y)y} = \begin{cases} \frac{\widetilde{p}(z)}{(\,\widetilde{p}(z)+y)y}, & \text{as } y \neq 0 \text{ and } \widetilde{p}(z) + y \neq 0, \\ 0, & \text{as } y = 0 \text{ or } \widetilde{p}(z) + y = 0. \end{cases}
$$

Remark 3.1. In case that the coefficient $b(t, x, q, \omega)$ does not depend on p, all conditions on p can be erased, e.g. the condoition $\widetilde{2}^\circ$ in assumption (A) can be simplified as

$\widetilde{2}^{\circ\prime}$

$$
|b(t, x_1, q_1, \omega) - b(t, x_2, q_2, \omega)| \leqslant c_1(t) |x_1 - x_2| + \widetilde{c}_2(t) |q_1 - q_2| ;
$$

etc.

In the following Theorems the similar remarks can also be made, and we will omit them. Now let us give some examples.

Example 3.1. If

$$
b(t, x, q, p) = c_1(t)x + \widetilde{c}_2(t) |q| ,
$$

where $\int_0^T |c_1(t)|\,dt + \int_0^T \widetilde{c}_2(t)^2 dt < \infty$, then assumption (A) for b is satisfied. In this case

$$\begin{aligned}-\widetilde{b}(t,y,\widetilde{q},\widetilde{p}) =& b(t,x,q,p)y^2 + I_{y\neq 0}\left|\widetilde{q}\right|^2/y + \int_Z \frac{\widetilde{p}(z)^2}{\widetilde{p}(z)+y} I_{\widetilde{p}(z)+y\neq 0}\pi(dz) \\ =& c_1(t)y + \widetilde{c}_2(t)\left|\widetilde{q}\right| + I_{y\neq 0}\left|\widetilde{q}\right|^2/y + \int_Z \frac{\widetilde{p}(z)^2}{\widetilde{p}(z)+y} I_{\widetilde{p}(z)+y\neq 0}\pi(dz)\end{aligned}$$

has a quadratic growth in $\widetilde{q}$, and it is unbounded in y, as y closes to 0, and it is also non-linear in $\widetilde{p}$. Now if $0 < \widetilde{r}_0 \leqslant X \in \mathfrak{F}_T, E\left|X\right|^2 < \infty$,, then BSDE (6) has a solution $(y_t, \widetilde{q}_t, \widetilde{p}_t)$ with $0 < y_t \leqslant 1/r_0$ and $\widetilde{p}_t(z) + y_{t-} \neq 0$, where $r_0 = \widetilde{r}_0 e^{-\int_0^T c_1(t)dt} > 0$.

Example 3.2. Let

$$b(t,x,q,p) = c_1(t)x + \widetilde{c}_2(t)\left|q\right| - \int_Z p(z)\pi(dz),$$

where $c_1(t)$ and $\widetilde{c}_2(t)$ have the same properties as those in Example 3.1, and assume that $\pi(Z) < \infty$. Then assumption (A) for b is still satisfied. In this case one easily sees that

$$-\widetilde{b}(t,y,\widetilde{q},\widetilde{p}) = c_1(t)y + \widetilde{c}_2(t)\left|\widetilde{q}\right| + I_{y\neq 0}\left|\widetilde{q}\right|^2/y + \int_Z \widetilde{p}(z)\pi(dz).$$

4. Uniqueness of Solution to BSDE with Quadratic Growth in q

For the uniqueness of solution to (6) under the conditions of Theorem 3.1 if we require that the solutions satisfy more conditions, then they should be unique.

Theorem 4.1. *Under all assumptions in Theorem 3.1 the solution* $(y_t, \widetilde{q}_t, \widetilde{p}_t)$ *of (6), which satisfies the following two condtions:*
1) y_t *is greater than a positive constant;*
2) $y_{t-} + \widetilde{p}_t(z) \neq 0$;
is unique.

Proof. In fact, suppose that $(y_t, \widetilde{q}_t, \widetilde{p}_t)$ is a solution of (6) satisfying the above two conditions with $0 < \delta_0 \leq y_t$, where δ_0 is a constant. Let $x_t = \frac{1}{y_t}$. Then by Ito's formula one easily derives that

$$dx_t = -b(t,x_t,q_t,p_t,\omega)dt + q_t dw_t + \int_Z p_t(z)\widetilde{N}_k(dt,dz), x_T = X, \qquad (7)$$

where $q_t = -\widetilde{q}_t/y_t^2$,

$$p_t(z) = -\frac{\widetilde{p}_t(z)}{(\widetilde{p}_t(z) + y_{t-})y_{t-}}.$$

Moreover, $0 < x_t \leq \frac{1}{\delta_0}$ is also positive, and

$$\begin{aligned}
&E\int_0^T |q_t|^2 \, dt \leq \frac{1}{\delta_0^4} E\int_0^T |\widetilde{q}_t|^2 \, dt < \infty, \\
&E\int_0^T \int_Z |p_t(z)|^2 \, \pi(dz)dt = E\int_0^T \int_Z |p_t(z)|^2 \, N_k(dt, dz) \\
&= E\sum_{0<\sigma_m\leq T} |p_{\sigma_m}(k(\sigma_m))|^2 \leq \frac{1}{\delta_0^4} E\sum_{0<\sigma_m\leq T} |\widetilde{p}_{\sigma_m}(k(\sigma_m))|^2 \\
&= \frac{1}{\delta_0^4} E\int_0^T \int_Z |\widetilde{p}_t(z)|^2 \, N_k(dt, dz) = \frac{1}{\delta_0^4} E\int_0^T \int_Z |\widetilde{p}_t(z)|^2 \, \pi(dz)dt < \infty,
\end{aligned}$$

where we have applied that

$$0 < \delta_0 \leq y_t, \forall t \in [0, T] \Longrightarrow y_{\sigma_m -} + \widetilde{p}_{\sigma_m}(k(\sigma_m)) \geq \delta_0, \forall m.$$

Thus, (x_t, q_t, p_t) is a solution of (7). However, under the assumptions of Theorem 3.1 the solution of (7) is unique.[4,8] Hence the solution $(y_t, \widetilde{q}_t, \widetilde{p}_t)$ of (6) must be unique. □

Now we can have the following existence and uniqueness theorem for the BSDE with jumps and with quadratic growth in q.

Theorem 4.2. *Under all conditions of Theorem 3.1 if in addition, it also satisfies that*

$$\widetilde{5}^\circ \quad X \leq \widetilde{k}_0,$$

where $\widetilde{k}_0 \geq r_0 > 0$ is a constant, then the BSDE (6) has a unique solution $(y_t, \widetilde{q}_t, \widetilde{p}_t)$ with the properties 1) and 2) in Theorem 4.1.

Proof. The uniqueness is derived by Theorem 4.1. Let us show the existence. In fact, by Theorem 3.1 it is already known that there exists a unique solution (x_t, q_t, p_t) of (3) satisfying that

$$x_t \geq r_0 = \widetilde{r}_0 e^{-\int_0^T c_1(t)dt} > 0,$$

and there also exists a solution $(y_t, \widetilde{q}_t, \widetilde{p}_t)$ with $0 < y_t \leqslant 1/r_0$ and $\widetilde{p}_t(z) + y_{t-} \neq 0$ satisfying (6). Moreover, $y_t = \frac{1}{x_t}$. However, applying Gronwall's inequality under the given assumption one easily sees that Refs. 4,8

$$x_t = |x_t| \leq (\widetilde{k}_0^2 + 2\int_0^T c_1(t)dt)e^{\int_0^T (3c_1(t)+4\widetilde{c}_2(t)^2)dt} = k_0 < \infty.$$

Hence $y_t = \frac{1}{x_t} \geq \frac{1}{k_0} > 0$. That is, y_t and $\widetilde{p}_t(z)$ satisfies the properties 1) and 2) in Theorem 4.1. The proof is complete. □

This Theorem can be applied to the above Examples 3.1 and 3.2. In fact, if we also assume that the terminal random value $X \leq \widetilde{k}_0$, then the solutions with the properties 1) and 2) mentioned in Theorem 4.1 exist and they are also unique in Examples 3.1 and 3.2, respectively.

5. Existence of Solution for BSDE with Quadratic Growth in q and y

A more interesting thing is that we can also get results on the existence of solution $(y_t, \widetilde{q}_t, \widetilde{p}_t)$ for BSDE with the drift coefficient

$$\begin{aligned}\widetilde{b}(t, y, \widetilde{q}, \widetilde{p}, \omega) = &- y^2 b((t, I_{y\neq 0}1/y, -I_{y\neq 0}\widetilde{q}/y^2, \\ &- I_{y\neq 0} I_{\widetilde{p}(\cdot)+y\neq 0}\frac{\widetilde{p}(\cdot)}{(\widetilde{p}(\cdot)+y)y}, \omega) \\ &- y^{-1}|\widetilde{q}|^2 I_{y\neq 0} - \int_Z \frac{\widetilde{p}(z)^2}{\widetilde{p}(z)+y} I_{\widetilde{p}_t(z)+y\neq 0}\pi(dz)),\end{aligned}$$

which can have a greater than linear growth in y. For this we need to work a little bit more. Now let us make the following Assumption (B):

1° Conditions $\widetilde{1}°, \widetilde{3}°$, and $\widetilde{4}°$ in Assumption (A) holds;

2° $b(t, x, q, p, \omega)$ is jointly continuous in $(x, q, p) \in R^1 \times R^{1\otimes d_1} \times L^2_{\pi(.)}(R^1) \backslash G$, where $G \subset R^1 \times R^{1\otimes d_1} \times L^2_{\pi(.)}(R^1)$ is a Borel measurable set such that $(x, q, p) \in G \Longrightarrow |q| > 0$, and $m_1 G_1 = 0$, where $G_1 = \{x : (x, q, p) \in G\}$, and m_1 is the Lebesgue measure in R^1; moreover, $b(t, x, q, p, \omega)$ is a separable process with respect to (x, q), (i.e. there exists a countable set $\{(x_i, q_i)\}_{i=1}^{\infty}$ such that for any Borel set $A \subset R^1$ and any open rectangle $B \subset R^1 \times R^{1\otimes d_1}$ the ω−sets

$$\{\omega : b(t, x, q, p, \omega) \in A, (x, q) \in B\},$$
$$\{\omega : b(t, x, q, p, \omega) \in A, (x_i, q_i) \in B, \forall i\}$$

only differs a zero-probability ω−set;

3° b satisfise Lipschitzian condition only for p. i.e.

$$|b(t,x,q,p_1,\omega) - b(t,x,q,p_2,\omega)| \leqslant \widetilde{c}_2(t) \|p_1 - p_2\|,$$

where $\widetilde{c}_2(t) \geqslant 0$ is non-random such that $\int_0^T \widetilde{c}_2(t)^2 dt < \infty$; moreover,

$$\begin{aligned} &|b(t,x,q,p_1,\omega) - (b(t,x,q,p_2,\omega)| \\ &\leqslant \int_Z C_t(z,\omega) |(p_1(z) - p_2(z))| \pi(dz), \end{aligned}$$

where $C_t(z) \in \mathcal{F}^2_{\mathfrak{F}}(R^1)$ such that $0 \leq C_t(z) \leqslant 1$.

Theorem 5.1. *Assume that*

$$b(t,x,q,p,\omega) = b^1(t,x,q,p,\omega) + b^2(t,x,q,p,\omega),$$

where b^1 *satisfies conditions* $1° - 3°$ *in Assumption (B), and* b^2 *satisfies assumptions* $\widetilde{1}° - \widetilde{3}°$ *in Assumption (A); and assume that*

$$b^1(t,x,q,p,\omega) \geqslant 0, b^2(t,0,0,0,\omega) \geq 0.$$

and $0 < r_0 \leq X \in \mathfrak{F}_T, E|X|^2 < \infty$, *then BSDE* (6) *has a solution* $(y_t, \widetilde{q}_t, \widetilde{p}_t) \in S^2_{\mathfrak{F}}(R^1) \times L^2_{\mathfrak{F}}(R^{1\otimes d_1}) \times \mathcal{F}^2_{\mathfrak{F}}(R^1)$ *with* $0 < y_t \leq \frac{1}{r_0}$ *and* $\widetilde{p}_t(z) + y_{t-} \neq 0$, *where*

$$r_0 = \widetilde{r}_0 e^{-\int_0^T 2c_1(t))dt} > 0.$$

Theorem 5.1 can be proved by applying Theorem 3.1 and Lemma 2.1. Now let us give some examples.

Example 5.1. Let

$$\begin{aligned} b(s,x,q,p) = \widetilde{c}_0(s)[1 + |x| - I_{x\neq 0}x/|x|^{\beta_0}] + c_0 I_{s\neq 0} s^{-\alpha_1} |x|^{1-\beta} \\ + c_0' I_{s\neq 0} s^{-\alpha_2} |q|^{1-\beta_1} + \widetilde{c}_1(s)x - \widetilde{c}_2(s)q \end{aligned}$$

where $\alpha_1 < 1; \alpha_2 < 1/2; 0 < \beta, \beta_1 \leq 1, 0 < \beta_0 < 1$; and $\widetilde{c}_0(s) \geq 0$, $\int_0^T (\widetilde{c}_0(s) + |\widetilde{c}_1(s)| + \widetilde{c}_2(s)^2) ds < \infty$; denote $c_1(t) = \widetilde{c}_0(t) + \widetilde{c}_1(t) + c_0 I_{t\neq 0} t^{-\alpha_1}$, $c_2(t) = c_0' I_{t\neq 0} t^{-\alpha_2} - \widetilde{c}_2(t)$; and assume that $c_0, c_0' \geq 0$ are constants, and $0 < \widetilde{r}_0 \leqslant X \in \mathfrak{F}_T, EX^2 < \infty$.

Then Theorem 5.1 applies. So BSDE (6) has a solution $(y_t, \widetilde{q}_t, \widetilde{p}_t) \in S^2_{\mathfrak{F}}(R^1) \times L^2_{\mathfrak{F}}(R^{1\otimes d_1}) \times \mathcal{F}^2_{\mathfrak{F}}(R^1)$ with $0 < y_t \leq \frac{1}{r_0}$ and $\widetilde{p}_t(z) + y_{t-} \neq 0$. However, the drift coefficient in (6) has a greater than linear growth in y (if let $\beta = 1$, then it has a quadratic growth in y) and is unbounded in y belonging to any finite interval $(-\varepsilon, \varepsilon)$, for $\varepsilon > 0$, and it also has a quadratic growth in q as follows:

$$\begin{aligned}
&-\widetilde{b}(t,y,\widetilde{q},\widetilde{p})=b(t,I_{y\neq 0}1/y,-I_{y\neq 0}\widetilde{q}/y^2,\\
&-I_{y\neq 0}I_{\widetilde{p}(\cdot)+y\neq 0}\frac{\widetilde{p}(\cdot)}{(\widetilde{p}(\cdot)+y)y},\omega)y^2\\
&+I_{y\neq 0}\left|\widetilde{q}\right|^2/y+\int_Z\frac{\widetilde{p}(z)^2}{\widetilde{p}(z)+y}I_{\widetilde{p}(z)+y\neq 0}\pi(dz)\\
=&\widetilde{c}_0(s)[y^2+|y|-y\,|y|^{\beta_0}]+c_0I_{s\neq 0}s^{-\alpha_1}\,|y|^{1+\beta}\\
&+c_0'I_{s\neq 0}s^{-\alpha_2}\,|\widetilde{q}|^{1-\beta_1}\,|y|^{2\beta_1}+I_{y\neq 0}\,|\widetilde{q}|^2/y\\
&+\widetilde{c}_1(s)y+\widetilde{c}_2(s)\widetilde{q}+\int_Z\frac{\widetilde{p}(z)^2}{\widetilde{p}(z)+y}I_{\widetilde{p}(z)+y\neq 0}\pi(dz).
\end{aligned}$$

As Example 3.2 we also can give a similar example as follows:

Example 5.2. Let

$$\begin{aligned}
b(s,x,q,p)=&\widetilde{c}_0(s)[1+|x|-x/\,|x|^{\beta_0}]+c_0I_{s\neq 0}s^{-\alpha_1}\,|x|^{1-\beta}\\
&+c_0'I_{s\neq 0}s^{-\alpha_2}\,|q|^{1-\beta_1}+\widetilde{c}_1(s)x-\widetilde{c}_2(s)q\\
&-\int_Z p(z)\pi(dz),
\end{aligned}$$

where $c_0,c_0',\alpha_1,\alpha_2,\beta_0,\beta,\beta_1$ and $\widetilde{c}_0(s),\widetilde{c}_1(s),\widetilde{c}_2(s)$ have the same properties as those in Example 5.1, and assume that $0<\widetilde{r}_0\leqslant X\in\mathfrak{F}_T, EX^2<\infty$, and $\pi(Z)<\infty$. Then one easily sees that

$$\begin{aligned}
-\widetilde{b}(s,y,\widetilde{q},\widetilde{p})=&\widetilde{c}_0(s)[y^2+|y|-y\,|y|^{\beta_0}]+c_0I_{s\neq 0}s^{-\alpha_1}\,|y|^{1+\beta}\\
&+c_0'I_{s\neq 0}s^{-\alpha_2}\,|\widetilde{q}|^{1-\beta_1}\,|y|^{2\beta_1}+I_{y\neq 0}\,|\widetilde{q}|^2/y\\
&+\widetilde{c}_1(s)y+\widetilde{c}_2(s)\widetilde{q}+\int_Z\widetilde{p}(z)\pi(dz),
\end{aligned}$$

and Theorem 5.1 still applies.

6. Uniqueness of Solutions for BSDE with Jumps and with Quadratic Growth in q and y

For the uniqueness of solution to (6) the assumption of Theorem 5.1 is not enough. We need more conditions. Actually, we have the following theorem.

Theorem 6.1. *Under all assumptions in Theorem 5.1 and assumption that* $\forall(x_1,q_1,p_1),(x_2,q_2,p_2)\in R^1_+\times R^{1\otimes d_1}\times L^2_{\pi(\cdot)}(R^1)$,

$$\begin{aligned}
&(x_1-x_2)\cdot(b(t,x_1,q_1,p_1\omega)-b(t,x_2,q_2,p_2,\omega))\\
&\leqslant c_1(t)\rho(|x_1-x_2|^2)+c_2(t)\,|x_1-x_2|\,(|q_1-q_2|+\|p_1-p_2\|)],
\end{aligned}\tag{8}$$

where $R_+^1 = (0, \infty)$, *and* $c_i(t) \geq 0, i = 1, 2$, *are non-random such that*

$$\int_0^T [c_1(t) + c_2(t)^2]dt < \infty$$

and $\rho(u) \geqslant 0$, *as* $u \geqslant 0$, *is non-random, increasing, continuous and concave such that*

$$\int_{0+} du/\rho(u) = \infty;$$

the solution $(y_t, \widetilde{q}_t, \widetilde{p}_t)$ *of (6), which satisfies the following two condtions:*
1) y_t *is greater than a positive constant;*
2) $y_{t-} + \widetilde{p}_t(z) \neq 0$;
is unique.

Notice by the proof of Theorem 5.1 it is already known that BSDE (7) has a solution (x_t, q_t, p_t). However, from the condition (8) the solution (x_t, q_t, p_t) must be unique.[4] So Theorem 6.1 can be proved similarly as the proof of Theorem 4.1. Now we can also have the following existence and uniqueness theorem for BSDE with jumps and with quadratic growth in q and y.

Theorem 6.2. *Under all assumptions in Theorem 6.1 if in addition, it also satisfies that*

$$\widetilde{5}^\circ \quad X \leq \widetilde{k}_0,$$

where $\widetilde{k}_0 \geq r_0 > 0$ *is a constant, then the BSDE (6) has a unique solution* $(y_t, \widetilde{q}_t, \widetilde{p}_t)$ *with the properties 1) and 2) in Theorem 6.1.*

The proof of Theorem 6.2 is completely the same as Theorem 4.2. In the above Examples 5.1 and 5.2 one sees that if $c_0, \beta_1 = 0$, then the condition (8) is satisfied. So we have that in Example 5.1 if

$$\begin{aligned} -\widetilde{b}(s, y, \widetilde{q}, \widetilde{p}) =& \widetilde{c}_0(s)[y^2 + |y| - y\,|y|^{\beta_0}] + c_0' I_{s\neq 0} s^{-\alpha_2} |\widetilde{q}| \\ &+ I_{y\neq 0} |\widetilde{q}|^2 / y + \widetilde{c}_1(s) y + \widetilde{c}_2(s)\widetilde{q} + \int_Z \frac{\widetilde{p}(z)^2}{\widetilde{p}(z) + y} I_{\widetilde{p}(z)+y\neq 0}\pi(dz), \end{aligned}$$

and assume that $X \in \mathfrak{F}_T, E(X^2) < \infty$, such that $0 < \widetilde{r}_0 \leq X \leq \widetilde{k}_0$, where $\widetilde{r}_0$ and $\widetilde{k}_0$ are constants, and $c_0', \alpha_2, \beta_0, \widetilde{c}_0(s), \widetilde{c}_1(s), \widetilde{c}_2(s)$ satisfy all conditions explained in Example 5.1, then BSDE (6) with this $\widetilde{b}$ has a unique solution $(y_t, \widetilde{q}_t, \widetilde{p}_t)$ with the properties 1) and 2) stated in Theorem

6.1. Notice that in this result $0 < \beta_0 < 1$, so $b(s,x,q,p)$ is non-Lipschitzian in x, where

$$\begin{aligned} b(s,x,q,p) =& \widetilde{c}_0(s)[1+|x| - x/|x|^{\beta_0}] \\ &+ c_0' I_{s\neq 0} s^{-\alpha_2} |q| + \widetilde{c}_1(s)x - \widetilde{c}_2(s)q. \end{aligned}$$

Similar result holds for Example 5.2.

7. Optimal Consumption

Consider the following SDE system with jumps:

$$\left\{ \begin{array}{c} dy_t = \{\widetilde{c}_0(t)[y_t^2 + |y_t| - y_t |y_t|^{\beta_0}] + c_0' I_{t\neq 0} t^{-\alpha_2} |q_t| \\ \widetilde{c}_1(t) y_t + \widetilde{c}_2(t) q_t + I_{y_t \neq 0} |q_t|^2 / y_t \\ + \int_Z \frac{p_t(z)^2}{p_t(z)+y_{t-}} I_{p_t(z)+y_{t-}\neq 0} \pi(dz) - B_t u_t(y_t)\} dt + q_t dw_t \\ + \int_Z p_t(z) \widetilde{N}_k(dt,dz), \quad y_T = Y, t \in [0,T], \end{array} \right. \tag{9}$$

where we assume that for simplicity all processes and functions are $1-$dimensional, $\widetilde{c}_0(t), \widetilde{c}_1(t), \widetilde{c}_2(t)$ and B_t are non-random such that $B_t > 0$ is bounded, $\widetilde{c}_0(t) \geq 0$, and $\int_0^T (\widetilde{c}_0(s) + |\widetilde{c}_1(s)| + \widetilde{c}_2(s)^2) ds < \infty$, and $Y \in \mathfrak{F}_T^{w,\widetilde{N}_k}, 0 < \overline{r}_0 \leq Y \leq \overline{k}_0$, where $\overline{r}_0, \overline{k}_0, c_0'$ and α_2 are all constants such that $\alpha_2 < 1/2, 0 < \beta_0 < 1$, and $u_t(y_t)$ is any linear feedback control of the solution of (9), i.e. $u_t(y_t) \in \mathcal{U}$, where

$$\mathcal{U} = \left\{ \begin{array}{c} u_t(y_t) = a_t y_t : (y_t, \widetilde{q}_t, \widetilde{p}_t) \text{ is the unique solution of (9) for this } u(\cdot) \\ \text{with properties that 1) } y_t \text{ is greater than a positive constant,} \\ \text{2) } y_{t-} + \widetilde{p}_t(z) \neq 0; \\ \text{and } a_t \text{ is non-random such that } |a_t| \leq 1 \end{array} \right\}. \tag{10}$$

Obviously, the controlled system (9) for any $u(\cdot) \in \mathcal{U}$ has a unique solution $(y_t^u, \widetilde{q}_t^u, \widetilde{p}_t^u)$ with the properties 1) and 2) by Theorem 6.2 and Example 5.2, and the system is a very non-linear system. Moreover, when $c_0', \widetilde{c}_0(t) = 0, \pi(Z) = 0$ and $Y \in \mathfrak{F}_T^w$, it may reduce to some continuous wealth process in some continuous financial market.[7]

Let us denote

$$U = \{u_t = a_t y : a_t \text{ is non-random such that } |a_t| \leq 1, \text{ and } y \in R^1\}.$$

We have the following

Lemma 7.1. *Let*

$$V(t,y) = \frac{1}{2}|y|^2 + T - t,$$

for $y \in R^1$, $t \in [0,T]$, *then* $V(t,y)$ *satisfies*

$$\begin{cases} (\partial/\partial t)V - \inf_{u_t \in U}(B_t u_t \cdot \frac{\partial}{\partial y}V) + 1 - B_t |y|^2 = 0, \ 0 \leqslant t \leqslant T, \\ \qquad V(T,y) = \frac{1}{2}|y|^2. \end{cases}$$

Proof. The conclusion can be checked directly. □

Now let

$$u_t^0(y) = -y. \tag{11}$$

Then by Theorem 6.2 (9) has a unique solution with properties 1) and 2) for such control u^0. Denote it by (y_t^0, q_t^0, p_t^0). Then by Ito's formula

$$\begin{aligned} V(T,y_T^0) - V(t.y_t^0) &= \int_t^T \{(\frac{\partial}{\partial s}V - B_s u_s^0(y_s^0) \cdot \frac{\partial}{\partial y}V + 1 - B_s |y_s^0|^2) \\ &+\widetilde{c}_0(s)[(y_s^0)^3 + y_s^0 |y_s^0| - (y_s^0)^2 |y_s^0|^{\beta_0}] + c_0' I_{s \neq 0} s^{-\alpha_2} y_s^0 |q_s^0| \\ &+\widetilde{c}_1(s) |y_s^0|^2 + \widetilde{c}_2(s) y_s^0 q_s^0 + y_s^0 \int_Z \frac{p_s^0(z)^2}{p_s^0(z) + y_s^0} I_{p_s^0(z) + y_s^0 \neq 0} \pi(dz)\} ds \\ &+\text{martingale} + \int_t^T (\frac{3}{2}|q_s^0|^2 + \frac{1}{2}\int_Z |p_s^0|^2 \pi(dz)) ds - \int_t^T (1 - B_s |y_s^0|^2) ds \\ &= \text{martingale} + \int_t^T \{\widetilde{c}_0(s)[(y_s^0)^3 + y_s^0 |y_s^0| - (y_s^0)^2 |y_s^0|^{\beta_0}] + c_0' I_{s \neq 0} s^{-\alpha_2} y_s^0 |q_s^0| \\ &+\widetilde{c}_1(s) |y_s^0|^2 + \widetilde{c}_2(s) y_s^0 q_s^0 + y_s^0 \int_Z \frac{p_s^0(z)^2}{p_s^0(z) + y_s^0} I_{p_s^0(z) + y_s^0 \neq 0} \pi(dz)\} ds \\ &+ \int_t^T (\frac{3}{2}|q_s^0|^2 + \frac{1}{2}\int_Z |p_s^0|^2 \pi(dz)) ds - \int_t^T (1 - B_s |y_s^0|^2) ds. \end{aligned}$$

Similarly, for any $u(\cdot) \in \mathcal{U}$

$$\begin{aligned} &V(T,y_T^u) - V(t.y_t^u) \\ &\leqslant \text{martingale} + \int_t^T \{\widetilde{c}_0(s)[(y_s^u)^3 + y_s^u |y_s^u| - (y_s^u)^2 |y_s^u|^{\beta_0}] + c_0' I_{s \neq 0} s^{-\alpha_2} y_s^u |q_s^u| \\ &+\widetilde{c}_1(s) |y_s^u|^2 + \widetilde{c}_2(s) y_s^u q_s^u + y_s^u \int_Z \frac{p_s^u(z)^2}{p_s^u(z) + y_s^u} I_{p_s^u(z) + y_s^u \neq 0} \pi(dz)\} ds \\ &+ \int_t^T (\frac{3}{2}|q_s^u|^2 + \frac{1}{2}\int_Z |p_s^u|^2 \pi(dz)) ds - \int_t^T (1 - B_s |y_s^u|^2) ds. \end{aligned}$$

Now for any $u \in \mathcal{U}$, where $\mathcal{U}$ is defined by (10), let

$$J(u) = E(\frac{1}{2}|y_0^u|^2 + \int_0^T (\frac{3}{2}|q_s^u|^2 + \frac{1}{2}\int_Z |p_s^u|^2 \pi(dz))ds + \int_0^T B_s |y_s^u|^2 ds)$$
$$+ \int_0^T \{\widetilde{c}_0(s)[(y_s^u)^3 + y_s^u |y_s^u| - (y_s^u)^2 |y_s^u|^{\beta_0}] + c_0' I_{s\neq 0} s^{-\alpha_2} y_s^u |q_s^u|$$
$$+\widetilde{c}_1(s)|y_s^u|^2 + \widetilde{c}_2(s) y_s^u q_s^u + y_s^u \int_Z \frac{p_s^u(z)^2}{p_s^u(z)+y_s^u} I_{p_s^u(z)+y_s^u\neq 0}\pi(dz)\} ds. \quad (12)$$

Then, for all $u \in \mathcal{U}$

$$J(u) \geqslant \frac{E|Y|^2}{2} + T = J(u^0)$$

Thus we have proved the following Theorem 7.1, which shows that a feedback optimal stochastic control exists.

Theorem 7.1. *Define $u_t^0(y)$ by (11), and for any $u \in \mathcal{U}$, where $\mathcal{U}$ is defined by (10), denote $J(u)$ by (12), where (y_t^u, q_t^u, p_t^u) is the unique solution of (9) with the properties 1) and 2) for $u \in \mathcal{U}$.*

Then

1) $u_t^0 \in \mathcal{U}$,

2) $J(u) \geqslant J(u^0)$, for all $u \in \mathcal{U}$.

The above target functional $J(u)$ can be reduced to an energy functional in some continuous financial market, if we set
$c_0' = 0; \widetilde{c}_0(s), \widetilde{c}_2(s) = 0,\ \widetilde{c}_1(s) \geq 0$, and $\pi(Z) = 0, Y \in \mathfrak{F}_T^w$;
and we explain the term $C_t(y_t, q_t) - \int_0^t B_s u_s(y_s)ds$ as a consumption process, where $dC_t(y_t, q_t) = I_{y_t\neq 0}|q_t|^2/y_t$ and $B_t > 0$. Then $C_t(y_t^0, q_t^0) - \int_0^t B_s u_s^0(y_s^0)ds$ is an optimal consumption among all $u(\cdot) \in \mathcal{U}$, which can minimize this energy functional in this continuous financial market.

References

1. N. El Karoui, S. Peng, and M.C. Quenez,*Math. Finance*, **7**, 1 (1997).
2. M. Kobylanski, *The Annals of Probability*, **28**, 558 (2000).
3. R. Rouge and N. El Karoui, *Math. Finance*, **10**, 259 (2000).
4. Situ Rong, *Backward Stochastic Differential Equations with Jumps and Applications* (Guangdong Science & Technology Press, 2000).
5. Situ Rong, *Vietnam J. Math.*, **30**, 103 (2002).
6. Situ Rong, Statist. & Probab. Letters, **60**,279 (2002).
7. Situ Rong, *Abstract and Applied Analysis - Proceedings of the International Conference,* eds., N.M. Chuong, L. Nirenberg and W. Tutschke (World Scientific, 2004) pp. 515–532.

8. Situ Rong, *Theory of Stochastic Differential Equations with Jumps and Applications* (Springer, 2005).
9. Situ Rong and Huang Wei, *Acta Scient. Natur. Univ. Sunyatseni,* **43**, 41 (2004); **44**, 1 (2005).

Harmonic, Wavelet and p-Adic Analysis
Eds. N. M. Chuong *et al.* (pp. 363–381)

§17 INSIDER PROBLEMS FOR MARKETS DRIVEN BY LÉVY PROCESSES

ARTURO KOHATSU-HIGA
Osaka University, Japan

MAKOTO YAMAZATO
University of the Ryukyus, Japan
mail: yamazato@math.u-ryukyus.ac.jp

We study a semilinear elliptic boundary value problem with critical exponents both in the equation and in the boundary condition. We don't suppose that the energy functional is always positive and prove the existence of two positive solutions.

1. Introduction

The problem of asymmetric markets in continuous time mathematical finance has been considered since Karazas and Pikovsky Ref. 2. They regarded the insider's information as an enlargement of filtration. Corcuera et al. Ref. 1 considered insiders whose knowledge of asset price at maturity period is perturbed by an additional noise which vanishes at maturity period T. Without such a noise, insider's optimal logarithmic utility up to maturity period is infinite. They showed that under a condition on the strength of the noise, the optimal logarithmic utility up to maturity period become finite.

The markets considered above are driven by Brownian motion. In this paper, we try to extend Corcuera et al.'s results to markets driven by Lévy processes. First, we consider a progressive enlargement by final asset price disturbed by an additive process which vanishes at maturity period T. We decompose a given Lévy process as a sum of martingale part and bounded variation part with respect to enlarged filtrations. In a forthcoming paper,[3] we give semimartingale decomposition for general semimartingales with respect to filtrations enlarged by various random times and additional noises. Next, we consider utility optimization problem for logarithmic utility function. Formally, the result is parallel to non-insider case (Kunita Ref. 6). However, there arize some difficulties. Main reason of the difficulties is that the compensator with respect to enlarged filtration (insider's filtration) is a random measure. In the final section, we consider simple Lévy processes which have at most one type positive jumps and at most one type of negative jumps as both asset price processes and additional noise processes. For such markets, we completely determine whether the optimal logarithmic utility is finite or not. Calculation for general Lévy processes is complicated. It will be done in Ref. 4.

Kunita[6] showed that optimal logarithmic utility for non-insider is represented as a minimum of relative entropies of base probability measure w.r.t. equivalent martingale measures. Parallel result is obtained for insider (Ref. 5). Since, for insider model, compensators are not deterministic, integrability of the optimal portfolio is not clear and the class of equivalent martingale measures is not determined. This means that the martingale representation theorem is not known in contrast to non-insider case (see Kunita Ref. 7).

2. Enlargement of Filtration with Respect to Semimartingales

Let $Z = \{Z_t, 0 \leq t \leq T\}$ be a d-dimensional semimartingale defined on a complete probability space $(\Omega, \mathcal{F}, P)$. Here, $(\mathcal{F}_t)_{t\in[0,T]} \equiv \left(\mathcal{F}_t^Z\right)_{t\in[0,T]}$ is the filtration generated by the process Z.

Assume that the additional information of an insider until time t is given by a family of d-dimensional random variables $\{I_s\ , s \leq t\}$. Suppose that these random variables have the following structure:

$$I_t = G(X, Y_t),$$

where $G : \mathbb{R}^{2d} \to \mathbb{R}^d$ is a given measurable function, X is an $\mathcal{F}_T^Z$-measurable random variable on $\mathbb{R}^d$ and the process $Y = \{Y_t, 0 \leq t \leq T\}$ is a stochastic process on $\mathbb{R}^d$ adapted to a filtration $\mathcal{H} \supseteq \mathcal{F}$ independent of Z. We define $\mathcal{G}_t$ as the smallest filtration, satisfying the usual conditions that contains the filtration $\mathcal{F}_t^Z \vee \sigma(I_s, s \leq t)$ (see Ref. 8, Sec. II.67).

Semimartingale decomposition in this general setting will be discussed in Ref. 3. In this paper, we treat simpler case: We assume that $\{Z_t; t \in [0,T]\}$ is an $\mathbb{R}^d$-valued Lévy process with characteristic function $E(e^{i\langle\theta,Z_t\rangle}) = e^{t\psi(\theta)}$ where

$$\psi(\theta) = i\langle b, \theta\rangle - \langle c\theta, c\theta\rangle/2 + \int (e^{i\langle\theta,x\rangle} - 1 - i\langle\theta, x\rangle 1_{\{|x|\leq 1\}}(x))\nu(dx).$$

Here, $b \in \mathbb{R}^d$, c is a nonnegative definite $d \times d$ matrix and ν is a measure on $\mathbb{R}^d \backslash \{0\}$ satisfying $\int |x| \wedge |x|^2 \nu(dx) < \infty$. Note that b by this assumption, $E(|Z_t|) < \infty$ for $0 \leq t \leq T$. Also, we assume that $Y_t = Z'(T-t)$ where Z' is an additive process (that is, process with independent increments, but not necessarily have stationary increments), $X = Z_T$ and $G(x,y) = x + y$, so that $I_t = Z_T + Z'(T-t)$.

Theorem 2.1. *$Z_t - \int_0^t E[\frac{Z_T - Z_u}{T-u}|\mathcal{G}_u]du$ is a $\mathcal{G}$-martingale in $[0,T]$.*

Proof. Let $0 \leq s_1 < \cdots < s_n \leq s$ and let $X_j = Z_T$ and $Y_{s_j} = Z'(T - s_j)$ for $j = 1, \ldots, n$. For $x_j \in \mathbb{R}^d$, let $\phi(x_1, \ldots, x_n) = \prod_{j=1}^n e^{i\langle \theta_j x_j \rangle}$ where $\theta_j \in \mathbb{R}^d$, $j = 1, 2, \ldots, n$. Let $X = (X_j)$ and $Y = (Y_{s_j})$. We have for $s \leq u < t \leq T$ and bounded $\mathcal{F}_s$-measurable function h_s,

$$
\begin{aligned}
&E[\phi(X+Y)h_s(Z_t - Z_u)] \\
&= E\Big[h_s \prod_{j=1}^n \exp(i\langle \theta_j, Z_u + Z_T - Z_t + Z'(T-s_j)\rangle) \\
&\quad \times E[(Z_t - Z_u) \prod_{j=1}^n \exp\{i\langle \theta_j, Z_t - Z_u\rangle\}\Big] \\
&= E\Big[h_s \prod_{j=1}^n \exp[i\langle \theta_j, Z_u + Z_T - Z_t + Z'(T-s_j)\rangle\Big] \\
&\quad \times \frac{1}{i}\psi'(\sum_{j=1}^n \theta_j) \exp\{(t-u)\psi(\sum_{j=1}^n \theta_j)\} \\
&= \frac{1}{i}\psi'(\sum_{j=1}^n \theta_j) E[\phi(X+Y)h_s\Big],
\end{aligned}
$$

where $\psi'(\tau) = \left(\frac{\partial \psi(\tau)}{\partial \tau_k}\right)_{k=1}^d \in \mathbb{R}^d$ for $\tau = (\tau_1, \ldots, \tau_d) \in \mathbb{R}^d$. By letting $t = T$, we have

$$
\begin{aligned}
\int_s^t E[\phi(X+Y)h_s \frac{Z_T - Z_u}{T-u}]du &= \frac{1}{i}(t-s)\psi'(\sum_{j=1}^n \theta_j) E[\phi(X+Y)h_s] \\
&= E[\phi(X+Y)h_s(Z_t - Z_s)].
\end{aligned}
$$

Note that $E(\int_0^T |\frac{Z_T - Z_u}{T-u}|du) < \infty$. Hence, we have $Z_t - \int_0^t E[\frac{Z_T - Z_u}{T-u}|\mathcal{G}_u]du$ is a $\mathcal{G}$-martingale in $[0,T]$. □

Remark 2.1. By the above proof, we get that $Z_t - \int_0^t \frac{Z_T - Z_u}{T-u}du$ is a $\mathcal{F} \vee \sigma(Z_T)$- martingale in $[0,T]$.

We denote $\alpha_t(x) = \frac{x-Z_t}{T-t}$. Let Q_t be the law of $Z'(t)$, $P_t(\omega, dx)$ be the regular condisional law of X given $\mathcal{F}_t$ and $\mathcal{P}$ be the $\mathcal{F}_t$-progressive σ-field. Define measures by

$$\mu_1(B,t,\omega) = \int_{\mathbb{R}^d}\int_{\mathbb{R}^d} 1_B(x+y)Q_{T-t}(dy)P_t(\omega,dx)$$

and

$$\mu_2(B,t,\omega) = \int_{\mathbb{R}^d}\int_{\mathbb{R}^d} 1_B(x+y)\alpha_t(x)Q_{T-t}(dy)P_t(\omega,dx)$$

for $B \in \mathcal{B}(\mathbb{R}^d)$ and $t \in [0,T)$. The random measures μ_1 and μ_2 are $\mathcal{P}$-measurable for fixed B and μ_2 is absolutely continuous with respect to μ_1 for almost all $(t,\omega) \in [0,T) \times \Omega$. We define $\varphi_t(\omega, x)$ as a $\mathcal{P} \otimes \mathcal{B}(\mathbb{R}^d)$ -measurable version of the Radon-Nikodym derivative $\frac{d\mu_2}{d\mu_1}(x,t,\omega)$. Then it satisfies

$$\begin{aligned}&\int_{\mathbb{R}^d}\int_{\mathbb{R}^d} 1_B(x+y)\varphi_t(x+y)Q_{T-t}(dy)P_t(dx)\\ &= \int_{\mathbb{R}^d}\int_{\mathbb{R}^d} 1_B(x+y)\alpha_t(x)Q_{T-t}(dy)P_t(dx),\end{aligned} \tag{1}$$

for any $B \in \mathcal{B}(\mathbb{R}^d)$.

Theorem 2.2. *Let $\beta_t = \varphi_t(\omega, I_t)$. Then, $Z_t - \int_0^t \beta_u du$ is a $\mathcal{G}$-martingale in $[0,T)$.*

Proof. For $t \in [0,T)$ we may write, using the independent increments property of Z' and the independence of $\mathcal{F}_T$ and Y,

$$E(\alpha_t(X)|\mathcal{F}_t \vee \sigma(I_s : s \le t)) = E(\alpha_t(X)|\mathcal{F}_t \vee \sigma(I_t)). \tag{2}$$

We have

$$\begin{aligned}E(\alpha_t(X)1_B(I_t)|\mathcal{F}_t) &= \int_{\mathbb{R}^d}\int_{\mathbb{R}^d} 1_B(x+y)\alpha_t(x)Q_{T-t}(dy)P_t(dx)\\ &= \int_{\mathbb{R}^d}\int_{\mathbb{R}^d} 1_B(x+y)\varphi_t(x+y)Q_{T-t}(dy)P_t(dx)\\ &= E(\varphi_t(I_t)1_B(I_t)|\mathcal{F}_t)\end{aligned}$$

by (1) and the independence of $\mathcal{F}_t$ and Y_t. As φ is an $\mathcal{F}$-adapted process, we have

$$E(\alpha_t(X)|\mathcal{F}_t \vee \sigma(I_t)) = \varphi_t(I_t), \ \ a.e.$$

Hence $\varphi_t(I_t) = E(\alpha_t(X)|\mathcal{G}_t)$ a.e. by (2). □

We compute β as explicitly as possible using Theorem 2.2. Let $t \in [0, T)$. Let $L_t = Z_T - Z_t + Z'(T-t)$. U_{T-t} will denote the law of L_t. Let R_{T-t} be the law of$Z_T - Z_t$ and let $\widetilde{R}_{T-t}(dx) = xR_{T-t}(dx)$.

Theorem 2.3. *The signed measure $\widetilde{R}_{T-t}$ is a finite measure, $Q_{T-t} * \widetilde{R}_{T-t}$ is absolutely continuous with respect to U_{T-t}. Furthermore, $Z_t - \int_0^t \beta_u du$ is a $\mathcal{G}$-martingale in $[0, T]$, where*

$$\beta_t = \frac{1}{T-t}\frac{d(Q_{T-t} * \widetilde{R}_{T-t})}{dU_{T-t}}(L_t).$$

Proof. In order to compute φ in Theorem 2.2, note that

$$\begin{aligned}\mu_2(B,t) &= \int \alpha_t(x)Q_{T-t}(B-x)P_t(dx) \\ &= \int \alpha_t(z+Z_t)Q_{T-t}(B-z-Z_t)R_{T-t}(dz) \\ &= \frac{1}{T-t}\int Q_{T-t}(B-z-Z_t)zR_{T-t}(dz) \\ &= \frac{1}{T-t}Q_{T-t} * \widetilde{R}(B-Z_t) \qquad (3)\end{aligned}$$

and

$$\begin{aligned}\mu_1(B,t) &= \int Q_{T-t}(B-x)P_t(dx) \\ &= \int Q_{T-t}(B-z-Z_t)R_{T-t}(dz) \\ &= U_{T-t}(B-Z_t).\end{aligned}$$

We have

$$\frac{d\mu_2}{d\mu_1}(x) = \frac{1}{T-t}\frac{dQ * \widetilde{R}}{dU}(x - Z_t).$$

Hence $\varphi(I_t) = \frac{1}{T-t}\frac{dQ*\widetilde{R}}{dU}(L_t)$, which is obviously $\mathcal{G}_t$-measurable. By Theorem 2.1, $Z_t - \int_0^t \beta_t dt$ is a $\mathcal{G}$-martingale for $t \in [0,T]$. □

Proposition 2.1. *Without the assumption $\int_{|x|>1} |x|\nu(dx) < \infty$, Z_t is a $\mathcal{G}$-semimartingale in $[0.T]$.*

Proof. Define $Z_t^2 = \sum_{s \le t} \Delta Z_s 1(|\Delta Z_s| > 1)$ and $Z_t^1 = Z_t - Z_t^2$. Note that Z^1 and Z^2 are also independent Lévy processes with $E|Z^1| < \infty$. Regard Z_t^1 as Z_t and $Z_T^2 + Z'(T-t)$ as $Z'(T-t)$ in the previous arguments of Theorem 2.3 and the proof of 1. This gives that Z_t^1 is a semimartingale in the filtration $\mathcal{G}$. As Z_t^2 is adapted to the filtration $\mathcal{G}$ and it is a process of bounded variation then $Z_t = Z_t^1 + Z_t^2$ is an $\mathcal{G}$-semimartingale. □

Theorem 2.4. *Let g be a continuous incresing function on $[0,T]$ with $g(0)=0$. If the law of Z' is identical with the law of Z and $Y_t = Z'(g(T-t))$, then $Z_t - \int_0^t \beta_u du$ is a $\mathcal{G}$-martingale in $[0,T)$, where*

$$\beta_t = \frac{L_t}{T-t+g(T-t)}.$$

Proof. We calculate the Fourier transform of the measure μ_2 defined in (3).

$$\begin{aligned}\int e^{i\langle\theta,u\rangle}\mu_2(du,t) &= \int e^{i\langle\theta,u\rangle}\frac{1}{T-t}\int Q_{T-t}(du-z-Z_t)zR_{T-t}(dz)\\ &= \frac{1}{i}\psi'(\theta)e^{i\langle\theta,Z_t\rangle}E(e^{i\langle\theta,L_t\rangle}) \qquad (4)\\ &= \frac{1}{T-t+g(T-t)}\int e^{i\langle\theta,u\rangle}(u-Z_t)U_{T-t}(du-Z_t)\\ &= \int e^{i\langle\theta,u\rangle}\frac{u-Z_t}{T-t+g(T-t)}\int Q_{T-t}(du-x)P_t(dx).\end{aligned}$$

Hence $\varphi_t(\omega,x) = \frac{x-Z_t(\omega)}{T-t+g(T-t)}$. Therefore by Theorem 2.2 , we have

$$\beta_t = \frac{L_t}{T-t+g(T-t)}.$$

□

Note that in Theorem 2.4, the process L_t is not necessarily additive in time as the function g is not necessarily linear. Also, note that in general, L is not a Lévy process.

We remark that $E\int_0^T \left(\frac{Z_T-Z_u}{T-u}\right)^2 du = \infty$. This quantity is of importance when considering applications in insider problems. In particular, if the previous integral is finite it implies that the bounded variation part of the semimartingale decomposition of Z in the enlarged filtration is square integrable. We have

$$E\int_0^T \beta_s^2 ds = \int_0^T \left((b+\int_{|x|>1} x\nu(dx))^2 + \frac{c+\int x^2\nu(dx)}{T-t+g(T-t)}\right)dt < \infty$$

if $\int_0^T \frac{1}{g(T-t)}dt < \infty$ and $\int_0^T x^2\nu(dx) < \infty$. Therefore adding the Lévy process Z' is justified if we want to obtain that the bounded variation part of the semimartingale decomposition of Z is square integrable.

Let $(b, cW_t, N(dt, dx))$ be the Lévy-Itô decomposition of Z, where $b \in \mathbb{R}^d$, c is a nonnegative definite $d \times d$ matrix, W is a $\mathcal{F}$-adapted d-dimensional Brownian motion and N is a $\mathcal{F}$-adapted Poisson random measure on $[0, T] \times \mathbb{R}^d$ with compensator $dt\nu(dx)$ independent of W. That is,

$$Z_t = bt + cW_t + \int_0^t \int_{|x| \le 1} x\widetilde{N}(ds, dx) + \int_0^t \int_{|x|>1} xN(ds, dx)$$

where $\widetilde{N} = N(ds, dx) - ds\nu(dx)$ is a martingale part of N. We assume that Z' is identical in law as above theorem. Hence it has the Lévy-Itô decomposition $(b, cW'_t, N'(dt, dx))$, where W' is a d-dimensional Brownian motion and N' is a Poisson random measure on $[0, T] \times \mathbb{R}^d$ with compensator $dt\nu(dx)$. Note that W, N, W' and N' are mutually independent. We consider at first, for fixed $B \in \mathcal{B}(\mathbb{R}^d)$ satisfying $d(0, B) > 0$, a filtration

$$\mathcal{H}_t^B = \mathcal{F}_t \vee \sigma(W_T + N((0, T], B) + W'(g(T-u)) + N'((0, g(T-u)], B); \\ 0 \le u \le t).$$

Here $d(0, B)$ denotes the distance between 0 and B. Let

$$B_t = W_t - \int_0^t \beta(s) ds$$

and

$$M((0, t], B) = N((0, t], B) - \int_0^t F_s(B) ds$$

where

$$\beta(s) = \frac{W(T) - W(s) + W'(g(T-s))}{T - s + g(T-s)}$$

and

$$F_s(B) = \frac{N((s, T], B) + N'(0, g(T-s), B)}{T - s + g(T-s)}.$$

Then both B_t and $M((0, t], B)$ are $\mathcal{H}^B$-martingale as above theorem. Now, we consider wider filtration so that $M(ds, dx)$ become adapted. Let

$$\mathcal{H}_t = \mathcal{F}_t \vee \sigma(W_T + N((0, T], B) + W'(g(T-u)) + N'((0, g(T-u)], B); \\ 0 \le u \le t, B \in \mathcal{B}(\mathbb{R}^d)).$$

Theorem 2.5. *Both* B_t *and* $M((0, t]), B)$ *are* $\mathcal{H}$*-martingales for every* $B \in \mathcal{B}(\mathbb{R}^d)$ *satisfying* $d(0, B) > 0$.

Proof. Let $0 \le s_1 < \cdots < s_n \le s$ and let $X_j = N(A_j,(0,T])$ and $Y_{s_j} = N'(A_j,(0,g(T-s_j)])$ for $j=1,\dots,n$ with $A_i \cap A_j = \emptyset$ for $i \ne j$ and $d(A_j,0) > 0$. Let $\phi(x_1,\dots,x_n) = \prod_{j=1}^n e^{i\theta_j x_j}$ and let $X=(X_j)$, $Y=(Y_{s_j})$, $A = \cup_{j=1}^n A_j$, $A_{n+1} = \mathbb{R}^d \backslash A$. We have, for $s \le u < t \le T$, bounded $\mathcal{F}_s$-measurable function h_s and bounded $\mathcal{B}(\mathbb{R}^d))$-measurable function f,

$$\begin{aligned} &E[\phi(X+Y)h_s \int f(x)N(dx,(u,t])] \\ &= E\Big[h_s \prod_{j=1}^n \exp\Big(i\theta_j\{N(A_j,(0,u]\cup(t,T]) + Y_{s_j}\}\Big)\Big] \\ &\qquad \times E\Big[\int f(x)N(dx,(u,t]) \prod_{j=1}^n \exp\{i\theta_j N(A_j,(u,t])\}\Big]. \end{aligned}$$

We have

$$\begin{aligned} &E\Big[\{\sum_{k=1}^n \int_{A_k} f(x)N(dx,(u,t])\} \prod_{j=1}^n \exp\{i\theta_j N(A_j,(u,t])\}\Big] \\ &= E\Big[\{\sum_{k=1}^n \int_{A_k} f(x)N(dx,(u,t])\} \exp\{i\theta_k N(A_k,(u,t])\}\Big] E\Big[\prod_{j\ne k} \exp\{i\theta_j N(A_j,(u,t \\ &= (t-u)\Big\{\sum_{k=1}^n e^{i\theta_k} \int_{A_k} f(x)\nu(dx)\Big\} E\Big[\exp\{i\theta_k N(A_k,(u,t])\}\Big] \\ &= (t-u)\int_A f(x)\exp\{i\sum_{k=1}^n \theta_k 1_{A_k}(x)\}\nu(dx) E\Big[\exp\{i\theta_k N(A_k,(u,t])\}\Big]. \end{aligned}$$

Hence

$$\begin{aligned} &E\Big[\phi(X+Y)h_s \int f(x)N(dx,(u,t])\Big] \\ &= (t-u)\int f(x)\exp\{i\sum_{k=1}^n \theta_k 1_{A_k}(x)\}\nu(dx) E[\phi(X+Y)h_s]. \end{aligned} \tag{5}$$

By letting $t=T$, we have

$$\begin{aligned} &\int_s^t E\Big[\phi(X+Y)h_s \int f(x)\frac{N(dx,(u,T])}{T-u}\Big] du \\ &= (t-s)\int f(x)\exp\{i\sum_{k=1}^n \theta_k 1_{A_k}(x)\}\nu(dx) E[\phi(X+Y)h_s]. \end{aligned} \tag{6}$$

By an argument similar to (5), we have

$$E\Big[\phi(X+Y)h_s\int f(x)N'(dx,(0,g(T-u)])\Big]$$
$$= g(T-u)\int f(x)\exp\{i\sum_{k=1}^{n}\theta_k 1_{A_k}(x)\}\nu(dx)E[\phi(X+Y)h_s]$$

Hence we have

$$E\Big[\phi(X+Y)h_s\int f(x)(N(dx,(u,T])+N'(dx,(0,g(T-u)])\Big]$$
$$= (T-u+g(T-u))\int f(x)\exp\{i\sum_{k=1}^{n}\theta_k 1_{A_k}(x)\}\nu(dx)E[\phi(X+Y)h_s].$$

Therefore, we have

$$E\Big[\phi(X+Y)h_s\int f(x)\frac{N(dx,(u,T])+N'(dx,(0,g(T-u)])}{T-u+g(T-u)}\Big]$$
$$= \Big(\int f(x)\exp\{i\sum_{k=1}^{n}\theta_k 1_{A_k}(x)\}\nu(dx)\Big)E[\phi(X+Y)h_s]$$
$$= E\Big[\phi(X+Y)h_s\int f(x)\frac{N(dx,(u,T])}{T-u}\Big]$$

by (5). Integrating the both sides of the above equality w.r.t. u in $[s,t]$, we have that

$$N(B,(0,t]) - \int_0^t\int_B\Big\{\frac{N(dx,(u,T])+N'(dx,(0,g(T-u)])}{T-u+g(T-u)}\Big\}du$$

is an $\mathcal{H}$-martingale by (6). In the above argument, we did not consider W and W' because they are independent of N and N'. The proof for B is easier. □

3. Optimal Portfolios for Insiders

In this section, we summarize a part of results in Ref. 4. We use the notations W, N, $\widetilde{N}$, W', N', β, F, B and M after Theorem 5 in the previous section with $d=1$. Here we consider a Lévy process

$$Z_t = cW_t + \int_0^t\int_{|x|\le 1} x\widetilde{N}(dx,ds) + \int_0^t\int_{|x|>1} xN(dx,ds).$$

We define a stock price S by

$$S_t = S_0\exp\left(\left(b-\frac{c^2}{2}\right)t + Z_t\right).$$

We denote $\widehat{S}_t = e^{-rt} S_t$ the discounted stock price. Let

$$Z'_t = cW'_t + \int_0^t \int_{|x|\le 1} x\widetilde{N}'(dx, ds) + \int_0^t \int_{|x|>1} xN'(dx, ds)$$

be another Lévy process generated by W' and N'. Here, $\widetilde{N}'$ is the martingale part of $\widetilde{N}$. The process Z' is considered as an additional noise added to the information of an insider. Note that Z' and Z are identical in law. General case will be discussed in Ref. 4. Let $\mathcal{F}_t = \sigma\{Z_s : s \le t\}$ and let

$$\mathcal{G}_t = \mathcal{F}_t \vee \sigma\{Z_T + Z'(g(T-s)) : s \le t\}.$$

Let π_s be an insider's portfolio, i.e. proportion of stock assets to total assets, which is $(\mathcal{G}_s)$-predictable process. Discounted wealth process $\widehat{V}_t$ satisfies

$$\widehat{V}_t = V_0 + \int_0^t \frac{\pi_{s-}\widehat{V}_{s-}}{\widehat{S}_{s-}} d\widehat{S}_s.$$

By Itô's formula for semimartingale, discounted stock price satisfies the following equation :

$$\begin{aligned}
\widehat{S}_t &= S_0 + \int_0^t (b - r + c\beta(s))\widehat{S}_{s-} ds + c\int_0^t \widehat{S}_{s-} dB(s) \\
&\quad + \int_0^t \int_{|x|\le 1} (e^x - 1)\,\widehat{S}_{s-} M(dx, ds) + \int_0^t \int_{|x|>1} (e^x - 1)\,\widehat{S}_{s-} N(dx, ds) \\
&\quad + \int_0^t \int_{|x|\le 1} (e^x - 1 - x)\,\widehat{S}_{s-} F_s(dx) ds \\
&\quad + \int_0^t \int_{|x|\le 1} x\widehat{S}_{s-}\,(F_s(dx) - \nu(dx))\, ds.
\end{aligned}$$

Hence the wealth equation is

$$\begin{aligned}
\widehat{V}_t &= V_0 + \int_0^t (b - r + c\beta(s))\pi_{s-}\widehat{V}_{s-} ds + c\int_0^t \pi_{s-}\widehat{V}_{s-} dB(s) \\
&\quad + \int_0^t \int_{|x|\le 1} (e^x - 1)\,\pi_{s-}\widehat{V}_{s-} M(dx, ds) \\
&\quad + \int_0^t \int_{|x|>1} (e^x - 1)\,\pi_{s-}\widehat{V}_{s-} N(dx, ds) \\
&\quad + \int_0^t \int_{|x|\le 1} (e^x - 1 - x)\,\pi_{s-}\widehat{V}_{s-} F_s(dx) ds \\
&\quad + \int_0^t \int_{|x|\le 1} x\pi_{s-}\widehat{V}_{s-}\,(F_s(dx) - \nu(dx))\, ds = V_0 + \int_0^t \widehat{V}_{s-} d\widehat{R}_t
\end{aligned}$$

where

$$
\begin{aligned}
&\widehat{R}_t \\
&= \int_0^t (b - r + c\beta(s))\pi_{s-} ds + c\int_0^t \pi_{s-} dB(s) \\
&\quad + \int_0^t \int_{|x|\le 1} (e^x - 1)\,\pi_{s-} M(dx, ds) + \int_0^t \int_{|x|>1} (e^x - 1)\,\pi_{s-} N(dx, ds) \\
&\quad + \int_0^t \int_{|x|\le 1} (e^x - 1 - x)\,\pi_{s-} F_s(dx)ds + \int_0^t \int_{|x|\le 1} x\pi_{s-}\,(F_s(dx) - \nu(dx))\,ds
\end{aligned}
$$

Using Itô's formula wealth process can be written as : $\widehat{V}_t = V_0 \exp(R_t)$, where

$$
\begin{aligned}
R_t = &\int_0^t \left((b - r + c\beta(s))\pi_{s-} - \frac{c^2}{2}\pi_{s-}^2\right) ds + c\int_0^t \pi_{s-} dB(s) \\
&+ \int_0^t \int_{|x|\le 1} x\pi_{s-}\,(F_s(dx) - \nu(dx))\,ds \\
&+ \int_0^t \int_{|x|\le 1} \log\left(1 + (e^x - 1)\,\pi_{s-}\right) M(dx, ds) \\
&+ \int_0^t \int_{|x|>1} \log\left(1 + (e^x - 1)\,\pi_{s-}\right) N(dx, ds) \\
&+ \int_0^t \int_{|x|\le 1} \Big(\log(1 + (e^x - 1)\,\pi_{s-}) - x\pi_{s-}\Big) F_s(dx)ds.
\end{aligned}
$$

We set $V_0 = 1$. We say that a portfolio π is admissible ($\pi \in \mathcal{A}$) if π is self financing, $\mathcal{G}$-predictable, $V_t^\pi > 0$, and logarithmic utility $u(t, \pi) = E(\log(V_t)) = E(R_t)$ is finite for $t < T$. A self financing and $\mathcal{G}$-predictable π is admissible if $V_t^\pi > 0$,

$$
\begin{aligned}
&E\Big(\int_0^t |\pi_s|^2 ds\Big) < \infty, \\
&E\left(\int_0^t \int \{\log(1 + (e^x - 1)\pi_s)\}^2 \mathbf{1}(-1 < (e^x - 1)\pi_s < -\frac{1}{2}) F_s(dx)ds\right) < \infty, \\
&E\left(\int_0^t \int_{|x|>1} |\log(1 + (e^x - 1)\pi_s)| F_s(dx)ds\right) < \infty
\end{aligned}
$$

for all $t < T$. Here, it is implicitly assumed that

$$(e^x - 1)\pi_s(\omega) > -1 \text{ for } F_s(dx, \omega)dsP(d\omega)\text{-a.e. } (x, s, \omega).$$

We want to maximize the logarithmic utility

$$
\begin{aligned}
u(t,\pi) = \int_0^t \Big[& E\left((b-r+c\beta(s))\pi_{s-} - \frac{c^2}{2}\pi_{s-}^2\right) \\
& + E\left(\int_{|x|\le 1} x\pi_{s-}(F_s(dx)-\nu(dx))\right) \\
& + E\Big(\int_{\mathbb{R}} \{\log(1+(e^x-1)\pi_{s-}) - x1_{|x|\le 1}(x)\pi_{s-}\} F_s(dx)\Big)\Big] ds.
\end{aligned}
$$

Since π_s is $\mathcal{G}$-predictable, we consider

$$
\begin{aligned}
f(y) = {} & (b-r+c\beta(s))y - \frac{c^2}{2}y^2 \\
& + y\int_{|x|\le 1} xE(F_s(\omega,dx)-\nu(dx)|\mathcal{G}_{s-}) \\
& + \int_{\mathbb{R}} \{\log(1+(e^x-1)\,y) - yx1_{|x|\le 1}(x)\} E(F_s(dx)|\mathcal{G}_{s-}).
\end{aligned}
$$

Then

$$
f''(y) = -c^2 - \int \frac{(e^x-1)^2}{(1+(e^x-1)y)^2} E(F_s(dx)|\mathcal{G}_{s-}) \le 0.
$$

Hence $f(y)$ is concave. The maximal point of $f(y)$ satisfies

$$
\begin{aligned}
(b-r+c\beta(s)) - c^2 y + \int_{\mathbb{R}} \{ & \frac{e^x-1}{1+(e^x-1)y} - x1_{|x|\le 1}\} E(F_s(dx)|\mathcal{G}_{s-}) \\
& + \int_{|x|\le 1} xE(F_s(dx)-\nu(dx)|\mathcal{G}_{s-}) = 0 \qquad (7)
\end{aligned}
$$

This equation for noninsider is

$$
(b-r) - c^2 y + \int_{\mathbb{R}} \{\frac{e^x-1}{1+(e^x-1)\,y} - x1_{\{|x|\le 1\}}(x)\}\nu(dx) = 0.
$$

If $c \neq 0$ or, supp $\nu \cap (-\infty,0) \neq \emptyset$ and supp $\nu \cap (0,\infty) \neq \emptyset$, then the solution is unique. Obviously, we have

$$
\max_{\pi \in \mathcal{G}} E(R(t)) \ge \max_{\pi \in \mathcal{F}} E(R(t)).
$$

Proposition 3.1. *Let $\pi \in \mathcal{A}$ be an admissible portfolio such that there exists a positive constant M with $|\pi(s)| \le M$ for almost all $(s,\omega) \in [0,T] \times \Omega$. Then $-\infty \le u(T,\pi) < \infty$.*

Proof. We have

$$\int_0^T E(\int_{|x|\leq 1} x^2 F_s(dx))ds = T\int_{|x|\leq 1} x^2\nu(dx) < \infty, \tag{8}$$

$$\begin{aligned}&\int_0^T E\left(\int_{x<-1} F_s(dx) + \int_{x>1} xF_s(dx)\right)ds\\ &= T\Big(\nu((-\infty,-1)) + \int_1^\infty x\nu(dx)\Big) < \infty,\end{aligned} \tag{9}$$

$$\int_0^T E\,|\beta(s)|\,ds = \int_0^T \Big(\frac{2}{\pi(T-s+g(T-s))}\Big)^{1/2} ds < \infty. \tag{10}$$

By (8), we have that $\int_0^T\int_{|x|\leq 1} x(N-\nu)(dx,ds)$ and $\int_0^T\int_{|x|\leq 1} x(N-F_s)(dx,ds)$ are square integrable martingales. Hence $\int_0^T\int_{|x|\leq 1} x(\nu-F_s)(dx,ds)$ is square integrable and hence, $\int_0^T E\left|\int_{|x|\leq 1} x(F_s(dx)-\nu(dx))\right| ds < \infty$. As $\log(1+(e^x-1)y) \leq x(1+y)$ for $x \geq 0$ and $y > \frac{-1}{e^x-1}$, we have

$$\log(1+(e^x-1)\pi_s) \leq x(1+\pi_s). \tag{11}$$

Using the boundedness of the portfolio and (8) – (12), we bound the utility as follows,

$$\begin{aligned}u(t,\pi) \leq \int_0^t \Big[&E\left((b-r+c\beta(s))\pi_s - \frac{c^2}{2}\pi_s^2\right) + E\left(\int_{|x|\leq 1} x\pi_s\,(F_s(dx)-F(dx))\right)\\ &+E\left(\int_{[-1,1]} (e^x-1-x)\,F_s(dx)\pi_s\right) + E\left(\int_{x>1} xF_s(dx)(1+\pi_s)\right)\\ &+E\left(\int_{x<-1}(1-e^x)F_s(dx)|\pi_s|\right)\Big]ds\\ \leq M\int_0^t \Big[&E\left((b+r+c|\beta(s)|)\right) + E\left(\Big|\int_{|x|\leq 1} x(F_s(dx)-F(dx))\Big|\right)\\ &+E\left(\int_{[-1,1]} x^2F_s(dx) + \int_{x>1} xF_s(dx) + \int_{x<-1} F_s(dx)\right)\Big]ds < \infty.\end{aligned}$$

□

4. Examples of Simple Compound Poisson Processes

Note that in the Wiener case ($Z = W$) it was proven in Corcuera et al. Ref. 1 that if

$$\mathcal{H}_t = \mathcal{G}_t = \mathcal{F}_t \vee \sigma(W(T) + W'((T-s)^\alpha); s \leq t)$$

then the optimal portfolio of the insider is

$$\pi^*(t) = \frac{b - r + c\beta(t)}{c^2} = \frac{b-r}{c^2} + \frac{W_T - W_t + W'((T-t)^\alpha)}{c(T - t + (T-t)^\alpha)}.$$

Optimal utility is

$$\begin{aligned} u(t, \pi^*) &= \int_0^t E\Big(\frac{(b - r + c\beta(s))^2}{2c^2}\Big) ds \\ &= \frac{(b-r)^2}{2c^2} t + \int_0^t \frac{c^2\{E(W_T - W_s)^2 + E(W'((T-s)^\alpha)^2)\}}{2c^2(T - s + (T-s)^\alpha)^2} ds \\ &= \frac{(b-r)^2}{2c^2} t + \int_0^t \frac{1}{2(T - s + (T-s)^\alpha)} ds. \end{aligned}$$

The optimal utility $u(T, \pi^*)$ is finite if $\alpha < 1$ and infinite if $\alpha \geq 1$. This financial market does not allow for arbitrage if $\alpha < 1$ in the interval $[0, T]$ and still is realistic enough. Nevertheless one has the undesirable characteristic that $\limsup_{t \to T} \pi^*(t) = +\infty$ and $\liminf_{t \to T} \pi^*(t) = -\infty$.

To illustrate how these results may change with the introduction of jumps in the model we give simple examples.

In this section we will consider first a pure jump case ($c = 0$) in order to simplify calculations. Let us suppose that we are given two independent compound Poisson processes Z and Z' which have only two types of jumps. One of size $a_1 = a \in (0, \log 2)$ and the other of size $a_2 = \log(2 - e^a) < 0$. That is,

$$\begin{aligned} N((0,T], \mathbb{R}) &= N(\{a_1\}, (0,T]) + N(\{a_2\}, (0,T]), \\ N'((0, g(T)], \mathbb{R}) &= N'(\{a_1\}, (0, g(T)]) + N'(\{a_2\}, (0, g(T)]), \\ Z_t &= a_1 N((0,t], \{a_1\}) + a_2 N((0,t], \{a_2\}), \\ Z'(t) &= a_1 N'((0,t], \{a_1\}) + a_2 N'((0,t], \{a_2\}). \end{aligned}$$

Then, $S(t) = S_0 \exp(bt + N_t)$. This particular choice simplifies the calculations. Furthermore suppose that the rates of jumps for each type are

$$\lambda_+ = E(N((0,1], \{a_1\}) = E(N((0,1], \{a_1\}) > 0$$

and

$$\lambda_- = E(N((0,1], \{a_2\}) = E(N((0,1], \{a_2\}) > 0,$$

respectively. Then $S(t) = S_0 \exp(bt + Z_t)$ and there is an insider in the market who has information about the final value of the stock at time t in the form of $Z_T + Z'(g(T-t))$.

In fact the goal of this section is to show that if the insider has information about the number of jumps left to happen in the future of the stock price then he can create an arbitrage in the market. This depends strongly on the algebraic structure of the value of the jumps a and $\log(2 - e^a)$.

The insider has an additional information flow of the form $\mathcal{G}_t = \mathcal{F}_t \vee \sigma(I(s); s \le t)$ where $I(s) = Z_T + Z'(g(T-s))$. In this case,

$$\begin{aligned}\mathcal{H}_t = \mathcal{F}_t &\vee \sigma(N((0,T],\{a_1\}) + N'((0,g(T-s)],\{a_1\}); s \le t) \\ &\vee \sigma(N((0,T],\{a_2\}) + N'((0,g(T-s)],\{a_2\}); s \le t).\end{aligned}$$

Let

$$F_s(\{x\}) = \frac{N((s,T],\{x\}) + N'((0,g(T-s)],\{x\})}{T - s + g(T-s)}.$$

Then, by Theorem 6, we have that

$$N((0,t],\{x\}) - \int_0^t F_s(\{x\})ds$$

is an $\mathcal{H}$-martingale and

$$N((0,t],\{x\}) - \int_0^t E(F_s(\{x\})|\mathcal{G}_s)ds$$

is a $\mathcal{G}$-martingale. Set

$$B_+ \equiv B_+(s) = E\left(F_s(\{a_1\})|\mathcal{G}_s\right), \;\; B_- \equiv B_-(s) = E\left(F_s(\{a_2\})|\mathcal{G}_s\right).$$

In this model, logarithmic utility of the insider is

$$u(t,\pi) = \int_0^t \left[E\Big((b-r)\pi_{s-} + \int_{\mathbb{R}} \log(1 + (e^x - 1)\pi_{s-})F_s(dx)\Big)\right]ds.$$

First we start considering the solution to the portfolio optimization for the non-insider.

Proposition 4.1. *The non-insider has as optimal portfolio*

$$\pi^* = \begin{cases} y_+ & \text{if } \mu > 0, \\ \frac{\lambda_+ - \lambda_-}{\lambda_+ + \lambda_-}(e^a - 1)^{-1} & \text{if } \mu = 0, \\ y_- & \text{if } \mu < 0 \end{cases}$$

where

$$y_\pm = -\frac{\lambda_+ + \lambda_-}{2\mu} \pm \sqrt{\left(\frac{\lambda_+ + \lambda_-}{2\mu}\right)^2 + \frac{\lambda_+ - \lambda_-}{\mu(e^a - 1)} + \frac{1}{(e^a - 1)^2}}. \tag{12}$$

The optimal logarithmic utility is finite and given by

$$T\Big(\mu\pi^* + \lambda_+ \log(1 + (e^a - 1)\pi^*) + \lambda_- \log(1 + (1 - e^a)\pi^*)\Big).$$

Proof. We maximize the following concave function:

$$\begin{aligned} f_s(\pi) &= \mu\pi + \int_{\mathbb{R}} \log\left(1 + (e^x - 1)\,\pi\right)\nu(dx) \\ &= \mu\pi + \lambda_+ \log(1 + (e^a - 1)\pi) + \lambda_- \log(1 + (1 - e^a)\pi) \end{aligned}$$

where $\mu = b - r$ and $\nu(dx) = \lambda_+\delta_{\{a_1\}}(dx) + \lambda_-\delta_{\{a_2\}}(dx)$. As $\lim_{\pi\to\pm(e^a-1)^{-1}\mp} f_s(\pi) = -\infty$ therefore the optimal portfolio value is a solution of the equation $f_s'(\pi) = 0$. This gives

$$\mu + \lambda_+ \frac{(e^a - 1)}{1 + (e^a - 1)y} + \lambda_- \frac{(1 - e^a)}{1 + (1 - e^a)y} = 0.$$

This equation is a quadratic equation with two solutions $y_\pm$ in (12) if $\mu \neq 0$. The restriction, $\frac{1}{e^a-1} > y > -\frac{1}{e^a-1}$, determines the optimal portfolio given in the statement of the theorem. The calculation of the optimal utility is straightforward. □

Remark 4.1. Note that this result is valid as long as $\lambda_+ > 0$ and $\lambda_- > 0$. Otherwise, if $\lambda_- = 0$ and $\mu \geq 0$, then the optimal utility is infinite since $\lim_{\pi\to\infty} f_s(\pi) = \infty$. This will be useful in the insider case that follows.

We have the following result.

Proposition 4.2. *Assume that there exists k_1, $k_2 \in \mathbb{N}$ such that $k_1 a + k_2 \log(2 - e^a) = 0$. Then the insider with information given by $(\mathcal{G}_t)_{t\in[0,T]}$ has as optimal portfolio*

$$\pi^*(s) = \begin{cases} y_+(s) & \text{if } \mu > 0, \\ \frac{B_+ - B_-}{B_+ + B_-}(s)(e^a - 1)^{-1} & \text{if } \mu = 0, \\ y_-(s) & \text{if } \mu < 0 \end{cases}$$

where

$$y_\pm = -\frac{B_+ + B_-}{2\mu} \pm \sqrt{\left(\frac{B_+ + B_-}{2\mu}\right)^2 + \frac{B_+ - B_-}{\mu(e^a - 1)} + \frac{1}{(e^a - 1)^2}}\,.$$

The maximal utility is finite and given by

$$\int_0^T E\Big(\mu\pi^* + B_+ \log(1 + (e^a - 1)\pi^*) + B_- \log(1 + (1 - e^a)\pi^*\Big)ds. \tag{13}$$

Proof. We maximize the following concave function:

$$\begin{aligned} f_s(\pi) &= \mu\pi + \int_{\mathbb{R}} \log\left(1 + (e^x - 1)\,\pi\right) E(F_s(dx)|\mathcal{G}_{s-}) \\ &= \mu\pi + B_+ \log(1 + (e^a - 1)\pi) + B_- \log(1 + (1 - e^a)\pi). \end{aligned}$$

Hence the optimal portfolio and maximal utility are obtained as for non-insider case. The maximal utility is given by (13). For any $x \in \mathbb{N}a + \mathbb{N}b$, with positive probability, there are both positive and negative jumps under the condition $Z_{T-s} + Z'(g(T-s) = x$. Hence the conditional expectations $B_+(s)$ and $B_-(s)$ are both positive a.s. under the above condition. That is, $P\{\omega;\ B_+(s) > 0$ and $B_-(s) > 0$ for all $s \in [0,T]\} = 1$. Therefore portfolios are bounded $(-\frac{1}{e^a-1} < \pi^*(s) < \frac{1}{e^a-1})$. By Proposition 3.1, maximal logarithmic utility is bounded from above. Since $u(T,0) = 0$, $u(T,\pi^*)$ is finite. The optimal portfolio is obtained by the same way as in the proof of Proposition 4.1. □

Note that the existence of a such that there exists $k_1,\ k_2 \in \mathbb{N}$ with $k_1 a + k_2 \log(2-e^a) = 0$ is assured by the continuity of the function $h(a) = -a^{-1}\log(2-e^a)$ for $a \in (0, \log 2)$, $\lim_{a\downarrow 0} h(a) = -\infty$ and $\lim_{a\uparrow \log 2} = \infty$.

If we consider for simplicity the case $\mu = 0$ in the above proposition, we see that for all the values of (B_+, B_-) such that $B_+ = cB_-$ for some constant $c \in [0,+\infty]$, the value of the portfolio remains constant. That is, the value of the optimal portfolio ratio is determined by the ratio between expected future positive and negative jumps. The portfolio value is an increasing function of c. Furthermore note that if $\mu = 0$ then $\lim_{B_\pm \to +\infty} \pi^* = \pm(e^a - 1)^{-1}$. That is, as the number of jumps of one type increase and the other remains constant the optimal portfolio tends to the opposite risk jump values. This is natural because that risk will tend to disappear when most of the jumps become only positive or negative. For other values of μ a similar reasoning holds.

The case where there is an algebraic structure on the jumps (that is, the insider can count the jumps in order to know when to use his advantage optimally) can be sometimes used to obtain infinite logarithmic utility and therefore generate arbitrage in the model.

Proposition 4.3. *There is no $k_1,\ k_2 \in \mathbb{N}$ such that $k_1 a + k_2 \log(2-e^a) = 0$ then the maximal logarithmic utility of the $\mathcal{G}$-investor is infinite.*

Proof. Assume that $\mu \geq 0$. We choose a portfolio

$$\pi_{s-} = \begin{cases} \frac{\exp((T-s)^{-1})-1}{T-s} & \text{if } N(s-) = a, \\ 0 & \text{otherwise,} \end{cases}$$

which is $\mathcal{G}$-predictable. Then we have

$$\begin{aligned} u(T,\pi) &\geq \int_0^T \log(1+(e^a-1)\pi_{s-})E(F_s(dx):N(s-)=a)ds \\ &\geq \int_0^T (T-s)^{-1}\lambda_+ \exp\{-(\lambda_+ + \lambda_-)(T-s+g(T-s)\}ds \\ &= \infty. \end{aligned}$$

In the case $\mu < 0$, by a similar calculation we have that the optimal utility is infinite. □

In both propositions whether $g \neq 0$ or not (i.e., extra noise exists or not) is not concerned to the conclusions, in contrast to the Wiener case considered in Corcuera et. al., where $g(t) = t^\alpha$ played an important role in order to obtain finite logarithmic utility (if $\alpha < 1$). This effect is obviously due to the algebraic structure of the support of the Lévy process. If we add Brownian motion, to our model, then the situation will be changed as follows.

Proposition 4.4. *If $c \neq 0$ and $\int_0^T \frac{1}{g(T-s)}ds < \infty$, then the optimal utiliy is finite.*

Proof. We have, by (11),

$$\begin{aligned} u(t,\pi) &= \int_0^t E\Big((b-r+c\beta(s))\pi_{s-} - \frac{c^2}{2}\pi_s^2 + \log(1+(e^a-1)\pi_{s-})F_s(\{a_1\}) \\ &\qquad + \log(1+(1-e^a)\pi_{s-})F_s(\{a_2\})\Big)ds \\ &\leq \int_0^T E\Big((b-r+c\beta(s))\pi_s - \frac{c^2}{2}\pi_{s-}^2 + a(1+\pi_{s-})F_s(\{a_1\}) \\ &\qquad +(1-e^a)\pi_{s-}F_s(\{a_2\})\Big)ds =: u^1(t,\pi). \end{aligned}$$

optimal portfolio for the right-hand side of the above inequality is given by

$$\pi_{s-}^o = \frac{1}{c^2}\Big(b-r+c\beta(s)+aF_s(\{a_1\})+(1-e^a)F_s(\{a_2\})\Big)$$

and the optimal utility is

$$\begin{aligned} u^1(T,\pi^o) &= \int_0^T E\Big(\frac{c^2}{2}(\pi_{s-}^o)^2 + aF_s(\{a_1\})\Big)ds \\ &= \int_0^T E\Big((b-r+\beta(s))^2 + MF_s(\mathbb{R})^2 + aF(\{a\})\Big) \\ &< M' \int_0^T \frac{1}{T-s+g(T-s)}ds < \infty. \end{aligned}$$

Here, M and M' are positive constants. □

The following two propositions are valid. These completes the classification of finiteness and infinitness of optimal logarithmic utility for geometric Lévy markets with at most one size positive jumps and at most one type negative jumps. We omit the proof of the following two propositions which will be given in Ref. 4 under more general setting.

Proposition 4.5. *If $c \neq 0$, $g \equiv 0$, $\lambda_+ > 0$ and $\lambda_- > 0$, then the optimal utility is finite.*

Proposition 4.6. *If $c \neq 0$, $g \equiv 0$, $\lambda_+ > 0$ and $\lambda_- = 0$, then the optimal utility is infinite.*

References

1. Corcuera J. M., Imkeller P., Kohatsu-Higa A., Nualart D., *Additional utility of insiders with imperfect dynamical information*, (Finance and Stochastics, 2004) (to appear).
2. I. Karatzas, I. Pikovsky, *Adv. Appl. Prob.* **28**, 1095 (1996).
3. A. Kohatsu-Higa, M. Yamazato, *Enlargement of filtrations with random times for processes with jumps* (preprint).
4. A. Kohatsu-Higa, M. Yamazato, *Insider modelling and finite logarithmic utility in markets with jumps* (preprint).
5. A. Kohatsu-Higa, M. Yamazatom, *Entropy representation of optimal logarithmic utility for insiders in Lévy markets*, (RIMS Kokyuroku) (to appear).
6. H. Kunita, *Mathematical finance for price processes with jumps,* Abstract in *Stochastic processes and applications to Mathematical finince* (2003).
7. H. Kunita, *Proc. Kyoto Conf. Stoch. Anal. Rel. topics*, (2002).
8. L.C.G. Rogers, D. Williams, *Diffusions, Markov Processes, and Martingales,* (John Wiley and Sons, 1994).